MANAGEMENT THROUGH SYSTEMS AND PROCEDURES

Management through Systems and Procedures

The Total Systems Concept

WILLIAM F. KELLY, M.B.A.

Management Consultant

WILEY-INTERSCIENCE

JOHN WILEY & SONS

NEW YORK · LONDON · SYDNEY · TORONTO

Library of Congress Catalog Card Number: 69-19925
SBN 471 46810 X

Printed in the United States of America

To Chris and Gerry

Preface

One of the most significant developments in business and public administration during the last quarter century has been the realization that a body of knowledge called systems and procedures is a dynamic force for improving management control and information, reducing costs and inefficiencies, and upgrading the performance of operational techniques and practices. The proper utilization of the disciplines in this field—such as systems analysis, work simplification, and automatic data processing—can result in a major improvement in performance for companies that take advantage of them.

The substantial use of systems-and-procedures techniques by business organizations is made manifest by the large increase in demand for people specializing in this field. There appears to be no end in sight to this expansion. Concurrent with this growth has been a parallel increase in the number of educational institutions that now offer courses in systems and procedures as an integral part of their business curriculum.

Although the study of systems and procedures is necessary for those who become systems analysts, it is also valuable for others who hold or aspire to positions in fields such as accounting, industrial management, and marketing. The transcendant thought has been the total systems concept—the viewing of a business as a single whole made up of interrelated parts.

Based on the assumption that a comprehensive approach will best serve to introduce the reader to this field of study, *Management Through Systems and Procedures* is a survey presentation of the full panorama of the systems-and-procedures field. In carrying out this broad plan it has been necessary occasionally to include subject matter found elsewhere, such as in books on management and accounting. The use of this material is justified on two counts: first, some persons may have no knowledge in these areas; second, the subject matter presented from the point of view of other fields may not be properly suited to systems and procedures.

This book, in more specific application, lays stress on the analytical and creative processes. Ideas are the fountainhead of systems-and-procedures work. Accordingly, the aim throughout is to provide the reader with basic concepts, guidance, skills, principles, and applications that can

be applied in numerous situations. The presentation slants not so much on a "how to" basis as on the realization of achievement through thought-provoking techniques.

The table of contents lists topics in the sequence that I feel to be most logical and conducive to the learning process.

Physically the book is divided into eight major segments. In each I deal with the principles, practices, and techniques that vitally affect the operations of an enterprise. Cause and effect relationships are stressed wherever possible.

Part 1 introduces the reader to systems and procedures.

Part 2 is concerned with the organization and appraisal of manpower and physical facilities.

Part 3 is basic to all systems studies, regardless of type or use of equipment, and introduces an elaboration of 10 well-defined steps on how to perform such studies effectively.

Part 4 contains a discussion of special techniques: motion study, kinds and designs of printed forms, forms creation, forms management, and the preparation of manuals and written procedures.

Parts 5, 6, and 7 present an explanation of automatic data processing as it affects, and is affected by, systems-and-procedures work.

Part 8 deals with the presentation and implementation of recommendations.

I wish to express my appreciation to the individuals and business organizations whose permission has been granted to use the illustrations in this book.

In a book of this kind there is always the possibility of mistakes going undetected, despite the effort made to eliminate them. In keeping with tradition I absolve all and accept full responsibility.

William F. Kelly

Union, New Jersey
May, 1969

Contents

MANAGEMENT THROUGH SYSTEMS AND PROCEDURES

part 1

Systems-and-Procedures Work:

Its Role, Nature, and People

1

The Field of Systems and Procedures

Systems-and-procedures administration is a broad discipline, ranging from the traditional study of manual techniques to newer automatic-data-processing methods. The discipline itself is not new. It had its origin in very early history when man's activities became increasingly complex and the limitations of the human mind became apparent; for example, simple filing systems were handed down from antiquity. Of more recent origin, but still remote, was the development of sophisticated procedures for keeping financial records: basic accounting systems have been in evidence for over 600 years.

A direct lineage to the modern systems-and-procedures field can be traced from the scientific management movement of the late 1800s, when scientific principles were being applied toward the management of plant operating systems. In the years that followed both accounting and office systems became more complex and involved. Ultimately, the application of scientific controlling principles spread from the plant to the office, where their successful utilization led to widespread adoption throughout business and industry.

Since World War II the growth and expansion of the systems-and-procedures field has been at an accelerated pace. This is attributable to the following:

1. The size, rapid growth, and greater complexity of business enterprises.
2. Advancements within the framework of scientific management.
3. Economic factors.
4. Trends and technological developments in data processing.

WHAT ARE SYSTEMS AND PROCEDURES?

A *system* is an integrated set of related procedures designed to accomplish a specific phase of a total business operation. Following is a list of major systems:

Financial
- General accounting
- Budgetary control
- Property accountability
- Accounts payable and disbursements
- Accounts receivable
- Sales statistics
- Credit control
- Payroll
- Cost accounting

Marketing
- Sales order processing
- Customer service
- Special order handling
- Customer claims and complaints
- Market and economic trends

Production and Procurement
- Inventory control
- Procurement (purchasing, receiving)
- Labor distribution and timekeeping
- Work order processing
- Inspection
- Quality control
- Requisitioning—tools, fixtures, etc.

Warehousing
- Shipping
- Storing and handling

Manpower
- Personnel records
- Job evaluation

Whereas a *system* is a set of related procedures, a *procedure* is a combination of unified elementary processes that carry out some specific objective. Just as there are machine assemblies that include one or more subassemblies, there are systems that have one or more subsystems.

WHAT IS SYSTEMS-AND-PROCEDURES ADMINISTRATION?

The term "administration" literally means the management of business activities. Administration by systems and procedures is the actual planning and controlling of operational guides in a business enterprise. It is directly concerned with the following:

1. Acquiring optimum performance through systems and procedures.

2. Improving source data.
3. Data processing—manual and automatic.
4. Work measurement and effectiveness.
5. Improving management analysis and control techniques.
6. Conservation of resources and facilities—elimination of unnecessary documents and duplication of processing activities.
7. Preparation and use of organizational and operational manuals.
8. Forms design and control.
9. Layout planning.
10. Developing the most effective use of new systems machines and equipment.

Each of these functions has a direct bearing on the overall efficiency and profitability of a business. The scope of their application depends on the size and pursuit of an enterprise.

Administration through systems and procedures is but a part of the broader field of business management. Business management controls such factors as key personnel, basic policies, line of products, wages and salaries, and capital expenditures. The structural organization of business management provides for both line and staff activities. The line organization is the core, and the staff organization serves it. Systems-and-procedures organizations function as staff agencies, providing counsel and services for other departments.

OBJECTIVES OF SYSTEMS-AND-PROCEDURES WORK

The major objective of systems-and-procedures work is to develop systems that provide optimum effectiveness in management. The achievement of this objective is related to the following specific goals:

1. To promote the economical use of all resources—money, manpower, and facilities. This is done through the application of expense controls, cost-reduction programs, work-simplification methods, systems-planning techniques, improved plant and office layouts, and sound organizational plans.
2. To provide data to management that is useful, accurate, and timely.
3. To simplify and standardize procedures.
4. To provide for the continuity of operations in the event of death, retirement, or discharge of key personnel. This requires written procedures to serve as a safeguard against the serious disruption of a business enterprise.

5. To promote the optimum use of data-processing equipment and communications facilities.
6. To study current developments in equipment and techniques so that full advantage of improvements may be obtained.
7. To promote the effectiveness of systems-and-procedures work through conferences, written instructions, and visual aids.
8. To assist in the guidance and orientation of management personnel by providing current operational guides; for example, written instructions in policy and procedures manuals.
9. To establish and keep updated procedural manuals in compliance with approved policies and practices.
10. To promote in management personnel the potential benefits derived from successful systems and procedures.

In practice, the objectives of systems-and-procedures work are seldom so conveniently separated. Usually a systems assignment will have any number of these goals, because of their interrelationship.

TYPES OF ORGANIZATIONS THAT PERFORM SYSTEMS-AND-PROCEDURES WORK

Among the primary organizations that do systems-and-procedures work are the following:

1. Central systems-and-procedures departments.
2. Decentralized departments.
3. Public accounting firms.
4. Management consulting firms.
5. Suppliers of forms.
6. Manufacturers and distributors of data-processing equipment.

Central Systems-and-Procedures Departments. Depending on the size and organizational structure of a firm, one individual or a large staff organization may constitute a central systems-and-procedures department. This commonly includes a manager, a staff of systems analysts, automatic-data-processing specialists, and clerical personnel.

The prime advantage of a central systems-and-procedures agency is its potential to synthesize entire operations without interfering with organizational structures or patterns. Part II, Chapter 5 presents a detailed discussion of the placement of systems-and-procedures departments within company organizations.

Decentralized Departments. In many companies major departments have a systems analyst or systems-and-procedures staff solely for their own requirements. This is particularly true for departments such as accounting, production control, administrative engineering, quality control, and marketing.

Decentralized staffs afford the advantages that accrue from being close to the source of local problems and requirements; they are fostered by such advantages as a high degree of work specialization, nearness to operating personnel, and the on-the-spot ease with which decisions can be made and corrective measures taken. There is, however, the other side of the coin. Here the considerations include the possible drawbacks that are associated with decentralization; namely, lack of coordination among the various units, duplication of activities, and a tendency to lose the overall view and to become strictly provincial in thought and deeds. Accordingly, there is no substitute for a central systems-and-procedures department that stresses the "total systems concept" without compromising the specialized requirements of the departments that need their own systems personnel.

Public Accounting Firms. These firms ordinarily perform a good deal of accounting systems-and-procedures work for their clients. Because of their contributions in providing financial data and interpretations they are in a unique position to advise management on a wide variety of business problems. Many of the larger accounting firms employ systems analysts, automatic-data-processing specialists, and other types of management consultants in order to service the needs of clients in practically every area of business activity.

Management Consulting Firms. Firms of this type commonly include the terms "business," "engineering," or "industrial" in their titles, but the broad and prevalent classification is "management consulting firm." Some specialize in limited areas, whereas others are extremely diversified and capable of virtually taking over the complete operation of a going concern. Whether one systems analyst or a team of experts is assigned to a client company depends on the number and kinds of problems involved. Some management consulting firms specialize in the rehabilitation of faltering or sick companies.

Certain of the services rendered by public accounting houses and those of management consulting firms are, in essence, similar. By their very nature public accounting houses are management consulting firms. However, unlike management consulting firms, they often have a prolonged and close association with their clients. Although this affords them numerous advantages in making worthwhile contributions, it also makes them vulnerable to the same ills that commonly befall management; for example, an attitude of complacency, lack of a detached point of view, and limitations on time and qualified personnel.

Suppliers of Forms. To increase their sales many printing houses and other suppliers of forms employ systems analysts to assist their customers. Usually these systems analysts function in an advisory capacity. Their purpose is not to act as a substitute for a firm's own systems-and-procedures personnel but rather to provide basic customer service—as, for example, by helping in the design of a form to be used in connection with automated equipment. Generally, their services are rendered without obligation.

Manufacturers and Distributors of Data-Processing Equipment. Most manufacturers and distributors of data-processing equipment lend technical assistance to their clients. In cases involving complex installations they provide advisory services in support of the systems and procedures affected. Representatives of some concerns will assist in designing entire systems as an inducement to purchase or rent their equipment.

NATURE AND CAUSES OF DEFICIENT SYSTEMS

As a prelude to the study of systems and procedures it will be of interest to delineate the salient reasons why businesses have inadequate or faulty systems. Part of the explanation lies in the very dynamic nature of business itself. In today's fast-moving world of business, a company must continue to fulfill efficiently the satisfaction of customer needs or desires in order to stay in the running. To remain successful or, indeed, just to hold its relative position in the industry, it must have people who are dynamic; but it must also have systems and procedures that are geared to its requirements.

Progress demands change. Yet not uncommonly the systems of a concern will remain static long after the company has grown and the character of the business operations has changed. Upon investigation it is frequently found that no real attempt had been made to keep abreast of new business conditions and the growing needs of management. Makeshift procedures resulting from expediency become the accepted practice, and the affected systems become complex, error prone, and expensive to operate.

Another type of situation that leads to outmoded systems is brought about by changes in management personnel. Some systems are instituted to serve the special talents of individuals in the organization and their specific ways of doing things. Although these systems may initially have been good ones, the systems lose much of their value when the persons involved leave the company or transfer to another position.

Sometimes poor systems are attributable to lack of qualified manpower. An enterprise may not have trained personnel who can effectively analyze and evaluate existing systems; or else, it may be that the management personnel are working to capacity on regular responsibilities and do not have the time to make thorough studies.

Still another important reason why businesses have poor systems is an attitude of complacency on the part of management. The underlying philosophy is expressed in statements such as, "The existing system works," or, "We are perfectly satisfied with our present way of doing things." Seldom, however, will the systems analyst find that any system is as efficient as it should be. Even the most up-to-date systems continually become obsolete because the characteristics of businesses are continually changing. Also there is the obsolescence that is brought about by advancements within the framework of scientific management and by the availability of new equipment.

One of the most outstanding attributes of good management is the ability to improve performance on a continuing basis. In a properly managed organization operations are subjected to periodic evaluations, and there is a deliberate pursuit of improvement. Without the introduction of improved operations an organization becomes rigid and unchanging, and optimum results cannot be achieved.

PEOPLE WHO DO SYSTEMS-AND-PROCEDURES WORK

Practically everyone within a firm does some form of systems-and-procedures work. If this were not the case, it would take a massive systems-and-procedures organization to handle the tasks. This part of systems-and-procedures work, however, is only a by-product of other particular job functions. Systems-and-procedures work—as a profession—must be performed on a regular and continuing basis, and as the primary function of the individual.

The titles presented below are examples of the occupations found in the systems-and-procedures field.

The *director of systems* governs and directs overall systems activities, which include the study, installation, modification, and operation of data-processing systems.

The *project leader*, who is a systems analyst, acts as the head of a team and is responsible for coordinating and carrying out the strategy of systems-and-procedures assignments.

The *systems analyst* reviews, evaluates, develops, and prepares procedures and controls for conducting the operations of a business.

The *programmer* develops and prepares the precise sequence of events that a computer must follow to process a given problem.

The *forms specialist* designs and controls the use of forms for the purpose of providing better information and improved service.

The above titles may not correspond exactly to the ones used in a particular enterprise. Systems analysts doing essentially the same work are known by a variety of titles, such as management planning and controls specialists, procedures planners, and methods analysts.

Methods analyst has been the most widely used alternative. However, a distinction arising in industry today designates technicians trained in the mechanics of production processes as *methods analysts*. This differentiates them from *systems analysts*, who perform managerial or administrative functions. The mark of the methods analyst is his direct and specialized concern for the technical operations of production. As this is extraneous to the systems and procedures field, the work of the methods analyst is outside the scope of this book except where the two disciplines coincide.

Functional Types of Systems Analysts. As in many professions, there are *general practitioners* and *specialists*. The general practitioner handles a wide range of subject matter, and his work takes him into areas such as accounting, administrative engineering, production control, and market research. Usually the experience and technical knowledge of the individual general practitioner will determine the scope of his assignments.

The work of the specialist, however, is concentrated in one area or a limited number of activities. One type of specialization is in systems and procedures for the operations of a particular department, such as accounting, production, engineering, or marketing. Another type is based on the separation of the tasks of a systems-and-procedures department. This type results in specialists concerned with data-processing equipment, the preparation of manuals, forms design and control, or organizational studies.

Typical Job Descriptions

Director of Systems. The director of systems is responsible for the supervision and performance of overall data-processing-systems activities. In particular the director is responsible for the following:

1. Administrative duties such as cost budgeting for systems work.
2. Preparation of work procedures and schedules.
3. Review of proposals for new systems and changes to existing systems.

4. Developing and overseeing management reporting programs.

5. Disseminating educational material to management and operating personnel; for example, by providing information on the latest data-processing techniques, both manual and automatic.

6. Exercising authority over the selection and purchase of automatic-data-processing equipment.

Systems Analysts. Systems analysts are directly concerned with the analytical and creative processes of systems work. Through research they obtain an understanding of the problem, perform a systematic review of the data-processing requirements; and when necessary either modify an existing systems design or create a new one.

This work involves a great deal of personal contact with managerial and administrative personnel to secure information, cooperation, and support in the planning and carrying out of projects. The analyst must work with considerable independence under the general direction of the director of systems and procedures. His duties may include the scheduling of data-processing programs and supervising the execution of projects.

The normal requirements for systems-analysis work is graduation from college with courses in business administration, such as accounting, marketing, and management. The individual should have originality and initiative, and ability to analyze, synthesize, and express ideas logically. A number of companies accept specific and meaningful experience as a substitute, in whole or in part, for educational requirements. To qualify as a *senior systems analyst* the individual should have several years of experience in the systems-and-procedures field.

Trainees. Trainees work under the guidance of professional systems analysts. They learn by doing; for example, they learn how to design forms by completing drafts that were designed and sketched by experienced systems analysts.

Training for a general practitioner's position will require more comprehensive coverage of work areas than for specialization.

Programmers. Programmers formulate machine instructions for the execution of operations by means of automatic-data-processing equipment. A set of such instructions is called a *program;* in machine language it directs the computer to perform each detail step of processing. The systems design, outlined by the systems analyst, is presented to the programmer in general flow charts or diagrams. Programming involves refining these into detail charts and diagrams that indicate the computations and sequence of machine steps necessary to accomplish an operation. Each step is translated into coded machine language.

Programmers verify the accuracy and adequacy of programs by testing them on the computer. They make program corrections and changes, called "debugging," until the program is operational. Preparation of instruction sheets is necessary to direct the console operator during the production run. Programmers also review and modify existing programs to improve computing procedures.

For this position it is desirable to have at least two years of college, but a number of employers accept pertinent experience in lieu of formal education. The knowledge required usually includes basic business administration, data elements and codes, fundamental numerical analysis, programming techniques, and equipment utilization.

PROFESSIONAL ASSOCIATIONS

Professional societies and organizations have contributed greatly to the advancement of the field of systems and procedures, both as a science and an art. With a number of these groups the contributions have been general and often interwoven with the more extensive achievements of the progressive management movement. With others the contributions have been of a more specific nature.

It seems certain that professional associations will continue to play a major role in the evolution and progress of systems-and-procedures work. They serve a much needed function by disseminating great quantities of information that helps the reader keep abreast of the developments within his field. This includes information on advance management methods, new operational techniques, and the latest in equipment.

Of special relevance has been the work of the Systems and Procedures Association (SPA). Figure 1-1 lists the objectives of this organization. The SPA code of ethics is set forth in Figure 1-2; it states a definite declaration of the professional standards of conduct of the association. Other professional organizations with a considerable interest in systems and procedures work include the following:

Administrative Management Society (AMS)
American Management Association (AMA)
Society for the Advancement of Management (SAM)
American Institute of Management (AIM)
American Institute of Industrial Engineers (AIIE)
The Institute of Management Sciences (IMS)
Controllers Institute of America (CIA)
National Association of Accountants (NAA)

National Machine Accountants Association (NMAA)
National Tabulating Management Society (NTMS)

OBJECTIVES

To assume responsibility for leadership and enlightenment of the business public in areas where the systems profession has demonstrated a special competency;

To promote a broader understanding and acceptance of systems function as a component of effective management;

To encourage, establish, and maintain high standards of professional education, competence, and performance;

To further the exchange of professional knowledge;

To engage in research;

To conduct educational seminars and conferences;

To disseminate, by all appropriate means, accurate knowledge and information with respect to systems operations;

To do any and all things that are lawful and appropriate in the furtherance of these purposes.

Figure 1-1. Stated objectives of the Systems and Procedures Association.

AS A MEMBER OF THIS ORGANIZATION I RECOGNIZE:

THAT I have an obligation to promote the advancement of systems and procedures throughout all management. I shall uphold the standards of our Association and cooperate with others in the dissemination of knowledge on developments in our field.

THAT I have an obligation to my employer and my fellow member whose trust I hold. I will endeavor to the best of my ability to warrant this confidence by advising them wisely and honestly.

THAT I have an obligation not to use my membership in this Association as an advertisement on my stationery to solicit business with clients.

THAT I have an obligation not to accept commissions for the sale of equipment to my employer or my clients.

THAT I have an obligation not to use any knowledge gained of the internal conditions of the business of a fellow member's employer to further my own personal interests with that company or discredit the fellow member before his employer.

I ACCEPT MEMBERSHIP IN THIS ASSOCIATION WITH A FULL KNOWLEDGE OF THE RESPONSIBILITY IT IMPOSES UPON ME AND I SHALL DEDICATE MYSELF TO THE DISCHARGE OF THESE OBLIGATIONS.

Figure 1-2. Code of ethics of the Systems and Procedures Association.

part 2

The Organization of Manpower and Physical Facilities

2

The Structural Organization of Manpower

In many respects human progress can be gaged by the ability of people to organize for a common purpose. The magnificent structures of ancient Greece, the imposing aqueducts of the Roman Empire, and the intricate development of the symphony orchestra are all examples. The widespread material benefits of present-day society certainly would be impossible without the collective works of many people.

To organize means to systematize. Therefore organizational planning itself becomes an important part of systems-and-procedures work.

The most efficient performance of any commercial activity requires that both people and work be properly coordinated. This chapter deals with the organization of people. The organization of work is considered later on.

The systems analyst must be concerned with a company's organization because it is within this context that all operational systems and procedures must operate. If the analyst has a thorough understanding of the types and principles of organizational structures, he will be able to evaluate present structures to see if they provide a suitable base for efficient systems. Where they do not, his observations will enable him to make valuable recommendations for improvement. Experience shows that good systems have been severely hampered, or have even failed, because of faulty organizational structures that contain improper work relationships, vagueness in assignment of responsibilities, poor communications, or incompetence of personnel. It has been estimated that one-half of all management problems have organizational defects as their source. Although this may be an exaggeration, it nevertheless emphasizes the need for a continuing study of organizational arrangements and functions.

TYPES OF ORGANIZATIONAL PATTERN

Organizational arrangements are classified as follows:

1. Line-and-staff designations:
   (*a*) Line
   (*b*) Staff
   (*c*) Line and staff
   (*d*) Line type
   (*e*) Line-and-staff type
2. Functional-and-product designations:
   (*a*) Functional
   (*b*) Product
   (*c*) Product and functional
   (*d*) Mixed product and functional
3. Committees
4. Project teams

Line-and-Staff Designations

Line and *staff* are terms that are commonly used in discussing matters of business organization. The structural concepts of these two words were derived from military usage. Armies represent the oldest type of large organizations. The pharaohs, the caesars, and the Bonapartes all had their military line-and-staff agencies. The magnitude of organizational problems is apparent when we consider that some major corporations today employ nearly as many people as Napoleon Bonaparte had in his *Grande Armée.*

In military terminology the word "line" is used to distinguish combatant forces from the staff corps and supply departments. In business, "line" and "staff" have analogous meanings in the sense that the staff type of organization is used in support of the line.

The word "line" refers to the vertical line of authority that starts at the highest level of management and extends down to the lowest level of employees. It is used to designate a type of structure, and the type of authority and responsibility associated with such a structure. "Staff" refers to a horizontal relationship to this direct line of authority.

A line organization is that part of a business structure which is directly concerned with the production or marketing of the end product. (See Figure 2-1.)

Staff organizations, in contrast, render specialized knowledge and ef-

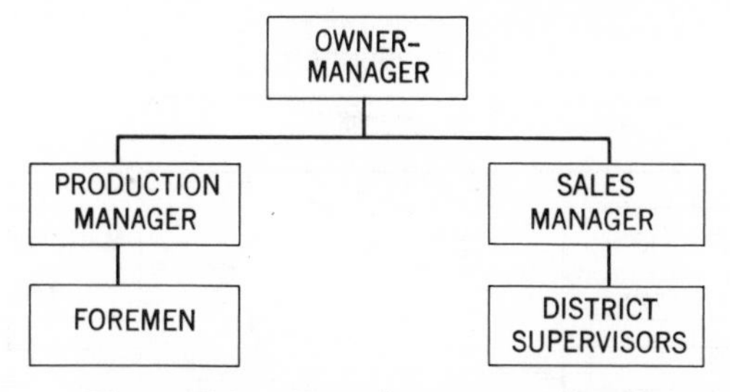

Figure 2-1. Basic line organizations.

fort in support of line activities. This support may be of either an indirect or a direct nature. Indirect support is the work performed by general staff agencies, such as accounting and personnel departments, which provide centralized services for an entire company. Direct support is the work performed by staff assistants, known also as administrators, who function to assist line managers in the performance of their work. Entire staff departments may be established to provide direct support; for example, marketing division managers may not only have sales organizations reporting to them, but also departments such as advertising, market research, customer service, and product promotion. Also, staff organizations responsible to the production manager may include research and development, engineering, production planning, and inspection. Two ways of portraying the line-and-staff organization are shown in Figure 2-2.

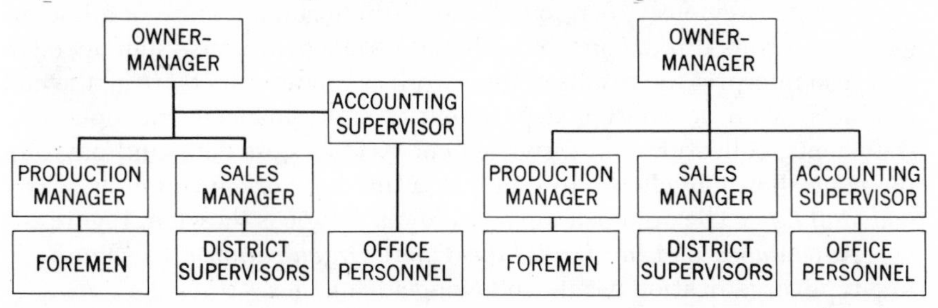

Figure 2-2. Two ways of portraying the same fundamental line-and-staff organization. In this example the chart on the left emphasizes the staff relationship of the accounting division. For reasons of simplicity and to reflect the importance of this division it is customary to show it as pictured in the chart on the right.

A staff agency that is in direct support of a line organization is an integral part of it. When identifying itself with the whole structure, it is classed with its organization. More precisely, it is a staff agency of a line-and-staff organization. (See Figure 2-3.)

In practice line organizations are seldom found in their pure form.

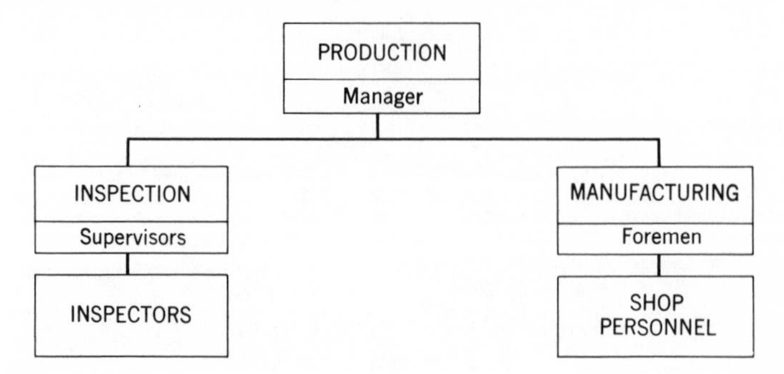

Figure 2-3. Inspection department as an example of a staff agency being an integral part of a line-and-staff organization.

Even companies of limited size have need of some staff services that must be performed by persons who are not directly concerned with either production or selling. Ordinarily, the first step is to segregate the accounting and office functions. As the company grows, its staff organizational pattern will be influenced by such distinctive factors as the nature of the business, its objectives, and the availability of qualified talent. The basic line organizations—production and sales—however, will always remain as the essential framework of the business. Figure 2-4 shows representative types of staff organizations.

Staff agencies may be modeled in accordance with the line-and-staff concept. As previously mentioned, the distinguishing feature of a line organization is its direct concern with end products. Because staff agencies may also interpret the results of their work as "end products" (e.g., the end products of an accounting department include financial and operating statements, collection and disbursement records, cost data, and payroll), the controller may choose to establish a *line-type organization* or a *line-and-staff type.* The organization chart for each type is shown in Figure 2-5.

Advantages of Line (and Line-Type) Organizations. The line (and line-type) organization has the following advantages:

1. It is simple and direct.
2. Authority and responsibility are fixed.
3. Quick decisions and actions are possible.
4. It is economical to operate.

Limitations of Line (and Line-Type) Organizations. The line (and line-type) organization has the following limitations:

1. Specialization is limited within the activities of the "line" division.
2. It is not suitable for complex operations.

Typical Staff Organization	Principal Function			
	Advising	Controlling	Coordinating	Servicing
Accounting, general		X		
Auditing		X		
Budgetary control		X		
Cost accounting		X		
Credit		X		
Economics	X			
Engineering, design				X
Engineering, manufacturing				X
Industrial engineering		X		
Insurance				X
Legal	X			
Market research			X	
Personnel and public relations	X			
Production planning			X	
Purchasing				X
Real estate				X
Research and development				X
Statistics				X
Systems and procedures		X		
Tax				X
Traffic				X

Figure 2-4. Representative types of staff organizations. The check marks indicate a general classification of primary functions. Certain of these organizations perform two or more of these functions.

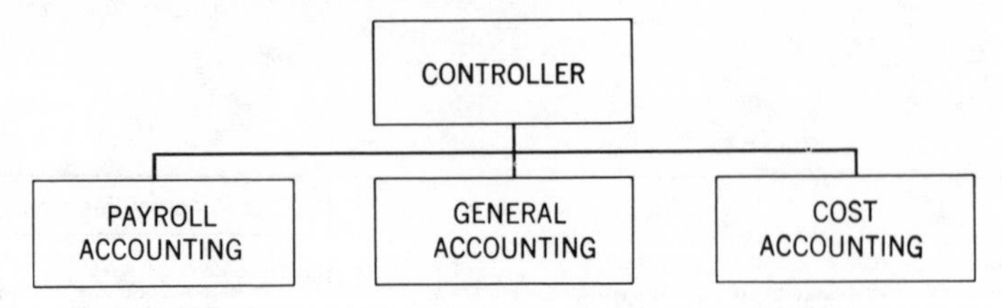

Figure 2-5*a*. Staff organization of the line type.

3. One man is apt to be burdened with excessive responsibilities.
4. Superior managerial abilities are necessary.
5. Operations are more likely to be impaired by the loss of a single manager.

Advantages of Line-and-Staff (and Line-and-Staff Type) Organizations. The line-and-staff (and line-and-staff type) organization has the following advantages:

1. It is possible to retain all the attributes of line and line-type organizations without suffering their limitations (listed below).
2. Line and line-type managers are relieved of work that is not directly related to their functions.
3. Both planning and control functions are divided into manageable parts.
4. The work load is distributed.
5. Specialization is enhanced.
6. Experts are developed.
7. Coordination is improved.

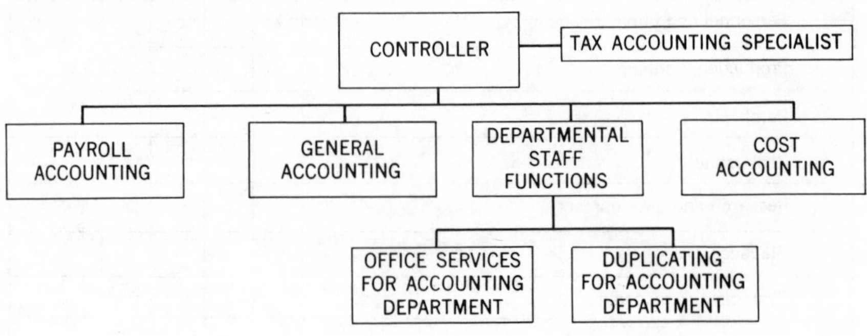

Figure 2-5*b*. Staff organization of the line-and-staff type.

Limitations of Line-and-Staff (and Line-and-Staff Type) Organizations. The line-and-staff (and line-and-staff type) organization has the following limitations:

1. It is more expensive to establish. (In operation, however, it can be far more economical.)
2. It causes conflicts of interest between line and staff personnel.
3. It impedes fast action.
4. It affords means to evade responsibilities.

Functional-and-Product Designations

Organizational classifications can be made on the basis of functional responsibilities and product associations.

Functional agencies are established under specialized activities such as manufacturing, marketing, sales, and engineering. Figure 2-6 shows two major divisions as classical examples of functional agencies.

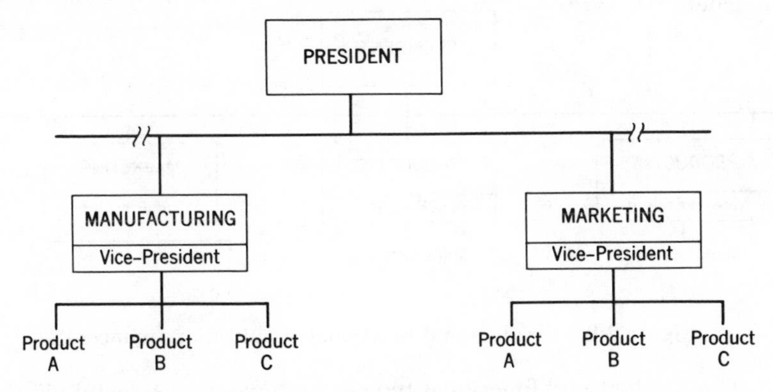

Figure 2-6. Manufacturing and marketing divisions as functional organizational agencies.

In turnabout fashion, product agencies are organized to allocate functional activities under product categories. An arrangement of this type makes divisional managers responsible for activities that are related to a product or a group of related products. Figure 2-7 illustrates an organizational breakdown on the basis of products. Accounting, engineering, and purchasing are examples of other functional activities that may fall within product classifications.

From a manufacturing point of view the products of a company may form logical groupings in the organizational structure. Such an arrangement, however, is apt not to be desirable for other organizational units (e.g., the marketing organization). Unless the products require salesmen with specialized skills or each product is sold to different types of customers, the best organizational plan for a marketing division is the functional arrangement.

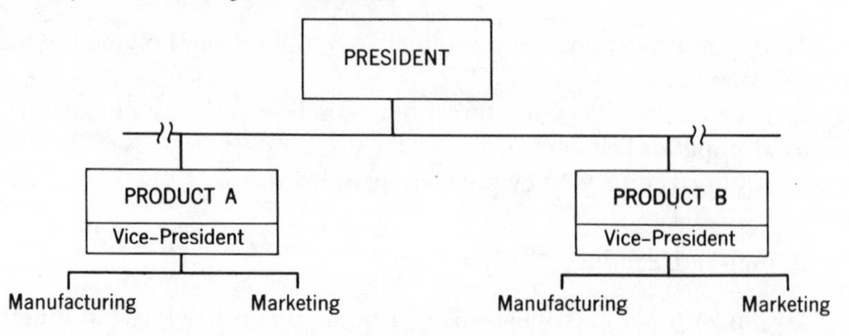

Figure 2-7. Product organizational agencies.

It is not uncommon for companies to combine product and functional designations. Figure 2-8 shows a combination product-and-functional arrangement.

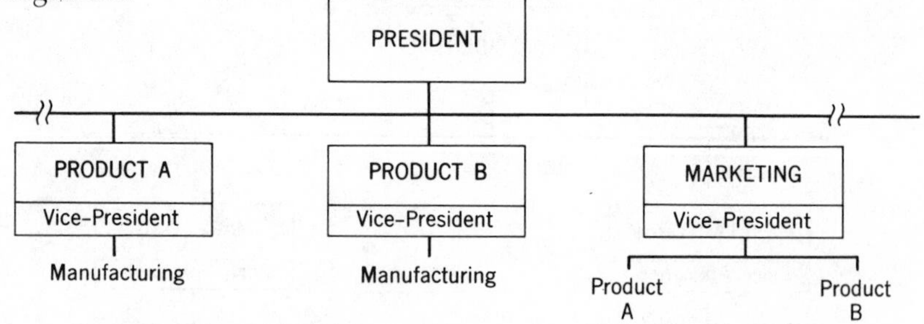

Figure 2-8. Product-and-functional organization (unmixed).

The product-and-functional type of organization is flexible in that it permits both *mixed* and *unmixed* arrangements. Figure 2-8 shows an arrangement that is typical of the *unmixed* type. The areas of responsibility are clear-cut, and there is no overlapping of responsibility and authority.

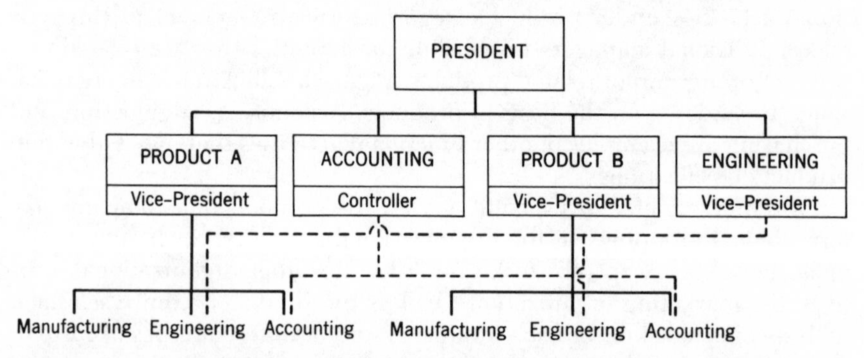

Figure 2-9. Mixed functional-and-product organization. The solid lines indicate *product responsibility;* and the dotted lines, *functional responsibility.*

Mixed product-and-functional organizations, on the other hand, provide certain executives with functional authority to cut across departmental lines. Figure 2-9 shows two examples of this type. Observe that the vice-president of engineering is given complete authority over engineering function for all departments. The same is true for the controller in matters relating to accounting and finance.

It should be noted that functional, product, and combined functional-and-product organizations are not restricted to the top structures of business. These basic formats are applicable at various levels of organization, including relatively small departments. Whether an organization should be operated on a functional, product, or some combined or mixed form depends on the situation.

Advantages of Functional Organizations. The functional agency has the following advantages:

1. It permits broad specialization within conventional areas such as accounting, engineering, and marketing.
2. It develops managerial talent in wide areas of operation.
3. It conforms to the prevailing categorizations used throughout industries.
4. Responsibilities are centralized according to well-defined functions.
5. Policies, systems, procedures, and operations are easier to standardize and coordinate.
6. It is especially suited to companies that manufacture one product or a group of related products in small quantities.

Limitations of Functional Organizations. The functional agency has the following limitations:

1. Responsibilities may be too great for any one man.
2. Exceptional leadership is required. In many situations this is not easily attainable.
3. A large number of subordinates are necessary.
4. It is not suitable for the decentralization of activities, particularly on a geographical basis.
5. The loss of a manager may profoundly disrupt the business.

Advantages of Product Organizations. The product agency has the following advantages:

1. It readily lends itself to decentralization.
2. Each product division may operate as a separate entity, with its own financial statements.

3. Authority is easily delegated, and responsibility is readily affixed.
4. It is suitable for companies with a diverse line of products.

Limitations of Product Organizations. The product agency has the following limitations:

1. It lacks flexibility in manpower utilization.
2. It is adaptable only by companies whose products fit into logical classifications.
3. It can be expensive to operate, because some duplication of effort is inevitable. (An example of duplication on a grand scale was found in the case of one well-known corporation with three subsidiaries, each manufacturing different but related products. At the time each subsidiary had not only its own sales and market-research departments but engaged the services of different advertising agencies. Losses appearing with monotonous regularity on the company profit-and-loss statements made a change mandatory. The marketing operations were consolidated; one advertising agency was selected, and the company subsequently prospered.)

Committee Organization

Committees are essentially a form of staff agency. Their hybrid and distinctive characteristics, however, entitle them to special treatment as a unique type of organization. They do not perform regular work in the sense ascribed to other organizational bodies. Instead, committees function to evaluate facts and ideas, and to render considered judgments.

Distinction Between Committees and Groups. During the course of a normal business day groups of people assemble to jointly discuss problems and business matters. Staff personnel call on department managers for information, line supervisors discuss operations with their subordinates, and divisional managers meet with general staff personnel in an attempt to resolve perplexing problems. These activities are those of *groups,* and not of *committees.* Despite their importance on the business scene, groups are not organizational units; they are informal gatherings of individuals.

Committees differ from groups in that they are formally recognized organizational bodies with assigned functions and responsibilities. A committee is a definite organizational entity with status.

Nature of Committees. General opinion of the value and importance of committee organizations varies considerably. Company managers who have experienced successful results from committee activities naturally feel that they are useful, even vital, administrative agencies. Conversely, others take a dim point of view and feel that committees are largely a waste of time and effort. In fact, some top-ranking company managers are so opposed to the committee form that they outlaw it altogether, contending that a manager has the authority and responsibility to make whatever decisions are normally rendered by a committee. The anticommittee policies of some concerns are so extensive that they discourage anything resembling committee activity, including group meetings and conferences.

Actually, committees can be of real service to a business enterprise. They promote teamwork and cooperation; provide an outlet for the airing of common problems; furnish a medium for the pooling of ideas, experiences, and plans; help in cutting across departmental lines; often prevent costly mistakes; serve to broaden the perspective of members; and develop managerial talents.

Because of the great benefits that can be derived from the committee type of organization, many managers promote its extensive use throughout all levels of the organizational structure. It should be kept in mind, however, that a good thing can be overdone. Some firms become overenthusiastic about committees and establish them for just about every activity of any importance, going so far as to form a committee just for the scheduling of committee meetings. An excessively large number of committees in a company is expensive and also a sign of weak management.

Another caution is to avoid the misuse of the committee form. As a rule committees should not be assigned regular work. They are notoriously poor agencies for gathering facts. The violation of this rule has brought the committee-type organization into much condemnation. The criticism is unjustified; it is simply not the nature of committees to perform in the manner of other organizations.

Committees function primarily to appraise facts, not to establish them. A committee should provide for the exchange of ideas, the airing of viewpoints, and the consideration of alternative actions. Its coordinated best judgment attempts to formulate the best course of action. Failing this, it should offer dissenting opinions that still prove useful to management in forming a solution.

In general unsound composition, improper use, and poor administration are the underlying reasons for unsatisfactory results of committee work. Committees that are correctly organized, properly used, and efficiently managed are assets to their companies.

Types of Committees. There are two principal types of committees: ad hoc (special) and regular (permanent).

Ad hoc committees are special committees established to handle situations that are not likely to be repeated. Upon completion of its particular assignment an ad hoc committee is dissolved; for example, an ad hoc committee comprised of members of such departments as design engineering, manufacturing, quality control, and sales might be formed to resolve a particular problem relating to the malfunctioning of a recently designed product. After corrective action is instituted the committee is abolished. In contrast, regular committees have an indefinite life and are a permanent part of the organizational structure of a company.

Advantages of Committee Organization. The committee organization has the following advantages:

1. Opinions are extensively considered prior to action.
2. The processes of coordination are centralized. A staff agency performing the same functions would have to coordinate activities and viewpoints by contacting each department separately.
3. Committee members can enlist the specialized services of their respective departments when the need arises to develop or acquire facts for committee evaluation and action.
4. Having been born of cooperative effort, plans will receive wide support.
5. The committee serves as an excellent medium for eliciting ideas. Also, proposals that lack individual merit may be combined into meaningful plans.
6. Composition is essentially democratic rather than autocratic. Committee rooms furnish an excellent environment for the liberal expression of ideas.
7. Committees provide good managerial training grounds.

Limitations of Committee Organizations. The committee organization has the following limitations:

1. It is susceptible to misuse. Committees serve as a means for shifting responsibilities. When faced with problems they cannot handle or wish to avoid some managers refer their responsibilities to committees for action.
2. It may lack skillful leadership. The chairman must know how to conduct meetings, formulate agendas, and meet objectives.
3. It may become too large for effectiveness. Membership must be limited in size.
4. Members may consider the activities to be secondary to their regular work.
5. Compromises may result in inferior plans.
6. It is slow in performing functions.
7. It is expensive to operate.

Basics for Establishing and Conducting Committee Meetings. A prime reason for the ineffectiveness of committee work is the lack of preparation on the part of the discussion leader. Every project that involves the work of a number of people requires groundwork. Without plans, difficulties arise and there is the strong possibility of being hurried by events into finding immediate answers to perplexing problems.

Committees, whether special or permanent, need leadership and at least rudimentary rules of procedure. It is the job of the chairman to make plans for committee meetings and to provide leadership and direction at the sessions.

Some ground rules for holding committee meetings are as follows:

1. Committee members should be selected so as to constitute a "central core of skills" to deal with the particular problem.
2. Joint effort must not be thwarted by excessive membership. Eight to twelve persons is generally satisfactory.
3. Individuals should be selected on the basis of their involvement with committee purposes and objectives. Provision may be made to draw on anyone in the company—on a temporary basis—if it is felt that his background and experience will make a worthwhile contribution.
4. Members should be provided with an agenda sufficiently in advance of meetings so that they will come prepared to participate in the topic for discussion.
5. The work of the committee should be planned. Figure 2-10*a* pre-

How to Get Ready to Lead a Discussion Meeting

1. DETERMINE PURPOSE OF MEETING:
   To develop support for required action.
   To consider unsolved problems.
   To settle disagreements and get group agreement.
2. EXPLORE THE SUBJECT:
   Get facts and information on the subject.
   Consider probable differences in viewpoints.
   Outline points that need discussion.
   Prepare presentation and materials.
3. OUTLINE THE DISCUSSION:
   Set the end objective.
   Set intermediate objectives.
   Make a timetable for the meeting.
   Plan opening, specifically.
   Plan close, specifically.
4. HAVE EVERYTHING READY:
   Issue, and check, announcements in advance of meeting.
   Arrange for room, table, chairs, blackboard; check light, ventilation.
   Prepare charts, outline blackboard sketches, or any other needed aids.

The Mutual Benefit Life Insurance Company

(*a*)

DISCUSSION LEADING

How to Lead a Discussion Meeting

1. OUTLINE SUBJECT CLEARLY.
   Start meeting promptly.
   State problem or situation clearly.
   Introduce topic for discussion.
2. DIRECT THE DISCUSSION.
   Draw out opinions, viewpoints, and experiences.
   Make sure all participate.
   Keep discussion on the subject.
   Avoid personal conflicts and arguments.
3. CRYSTALLIZE THE DISCUSSION.
   Present points of agreement and disagreement.
   Determine degrees of feeling—watch for shifts of opinion.
   State intermediate conclusions as reached.
   Make sure of understanding and acceptance.
4. GET ACCEPTANCE FOR ACTION.
   Summarize previous agreements and state conclusions clearly.
   Get agreement on action.
   Check to be sure all understand.
   Get group support, based on conviction.

Arrange additional meetings if required.

(*b*)

Figure 2-10. Planning a committee meeting: (*a*) how to prepare to lead a discussion meeting; (*b*) how to actually lead one.

sents an outline of steps on how to get ready to lead a discussion meeting. The accompanying illustration, Figure 2-10*b*, is on discussion leading.

6. Long, drawn-out meetings should be avoided. To obtain practical results the frequency and length of conferences should be kept within bounds so that they will not cause undue infringement on the time and effort of members.

Project Teams

The "project" form of organization is widely used throughout industry. Most people readily associate it with companies that are research conscious or employ a high degree of engineering in their product development. The form itself, however, is adaptable along other lines as well and may be used in whatever situations warrant the marshaling of diversified talents to bear on a problem.

A project team may be defined as a number of persons associated together to perform a definite task in such a way as to accomplish the established objective more effectively than would be the case with individuals working alone. Its membership may be strictly departmental or companywide, depending on the scope of the problem. The project leader may be an executive, a supervisor, or simply the "first" among equals. In all events, it is he who is given the task of coordinating and directing the interrelated aims of the work. As with the other members of the team, he should possess a high degree of competence in the assigned work; a talent for analysis and synthesis; and an abiding appreciation of truth, reality, and value. In addition, he should possess recognized skills in at least one particular kind of knowledge or discipline, because a project leader so qualified will command the respect of his colleagues.

Project teams have been formed in virtually every kind of business and for a variety of purposes. They can be brought into existence at any time, be of long or moderately short duration, and deal with various subjects. They have been very successful in aiding management in solving problems in such areas as research and development, facilities planning, marketing and distribution studies, selection of production equipment, and business-information-processing systems.

CONCLUSION

Our main interest in this chapter has been an explanation of the structural organization of manpower. The next chapter focuses on standards for the evaluation of organizational structures.

3

Standards for the Evaluation of Organizational Structures

In much the same manner as the architect describes the conceptual designs of different types of dwellings to his client, pointing out desirable and undesirable features, the discussion in the preceding chapter serves to explain basic types of organizational structures, and the inherent advantages and limitations associated with each.

No one type of architectural design is suitable for every family. Similarly, no particular organizational arrangement is suitable for every business. Although some form of the line-and-staff organization is now generally employed by firms of fair size, it has many variations even among concerns within the same industry or trade. For best results the individual company must build its organization to accomplish its objectives. Just as the custom-made home is tailored to the needs of the individuals involved, so also can be the company organization.

A well-designed organization is suited to its environment of place, time, and occasion; it is suited to the nature of the business, the collective characteristics of the management, and the policies and functions of the enterprise. The manner in which it is designed should achieve coordination, control, and an effective execution of plans.

EXTENT OF INTEREST

Systems analysts, being organizational architects, are interested in the following three aspects of manpower arrangements:

1. The capability of a whole company organization and its component parts to fulfill objectives.

2. The place of systems-and-procedures agencies within the entire organizational framework of the business concern.
3. The organization of systems-and-procedures agencies.

We shall consider these aspects of interest in the same order as listed above. Item 1 is covered in this chapter and in the one immediately following, entitled "The Management Audit." Items 2 and 3 are the subject of Chapter 5, which deals with the placement and organization of systems-and-procedures departments in an enterprise.

STANDARDS FOR THE EVALUATION OF AN ORGANIZATIONAL STRUCTURE

In evaluating any particular organization it is helpful to have standards to measure the competency of the structure. For this purpose the criteria listed in Figure 3-1 are useful.

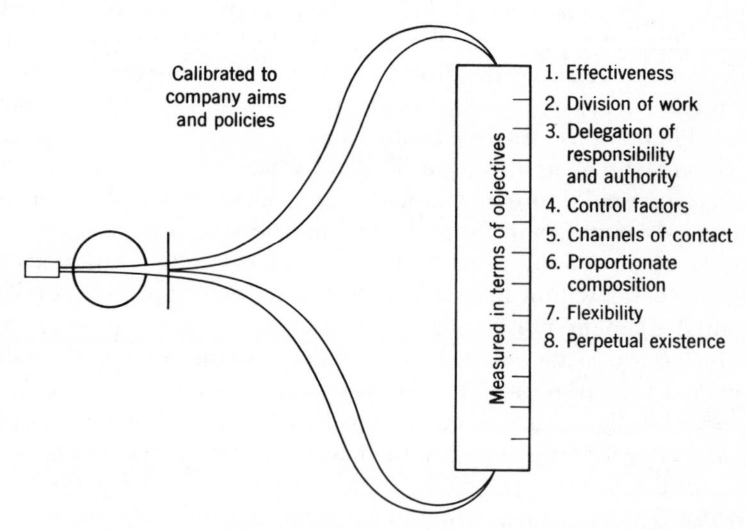

Figure 3-1. Criteria for evaluating organizational structure.

Although every organization problem has its own peculiarities, the standards listed in Figure 3-1 are applicable to all. Together they constitute a set of criteria for measuring the extent of organizational effectiveness. To be of value they must be calibrated in terms of company aims and policies.

Effectiveness

Factors of importance in evaluating the effectiveness of an entire enterprise include profits, sales, manufacturing costs, material costs, and inventory data. This constitutes a body of facts that management must constantly appraise in order to administer the business efficiently. Main features concerning each are discussed below.

Profits. Profit is usually held to be the object of a business enterprise. It functions as the common denominator for competitive industry, and the degree to which it is attained is a significant barometer of the managerial effectiveness of a particular company. A major objective of nearly all company managements is to keep the company operative, with the ability to compete. On this fundamental, vital part of business all else ultimately rests. Only when the business survives will production take place, jobs will be created, and money incomes will be made.

Profit-making implies more than mere economic survival. In the ordinary business sense *net profit* is the excess of business income over *all* business expenses calculated by acceptable accounting methods. Such profits—moderate or substantial—provide the incentive for establishing and operating any business, since it is the hope of an adequate return on their money that induces owners (stockholders in the case of corporations) to invest.

The making of money, though of profound concern to those who own private enterprises, is not the sole concern. Profits provide the necessary capital with which to expand operations and to produce more and better products at a fair price. Profits have a valid social objective because the material progress of a society as a whole is dependent on the bulk of its operations returning more than they consume.

Sales. The problems of merchandising are vital. A profitable sales department is a combination of successful salespeople and the application of scientific management principles. Scientific sales management includes, for example, the provision of visual aids to sales personnel in the form of records and graphs that show the quantity of sales per product in relative comparison to previous day, week, month, or yearly sales; also, the setting of a sales policy that determines selling prices, customer discounts, and merchandise returns and allowances.

Manufacturing Costs. A properly constructed cost-accounting system yields a wealth of dependable information, such as variations from standard costs, efficiency of operations and performance, profitability of departments and products, and bases for pricing.

With proper cost information, it is possible to relate the cost of manufacture for a given period to past periods and to standard costs, and to analyze variations from standard labor, material, and overhead costs.

Material Costs. Usually materials may be purchased from many sources. To continue to buy materials from the same vendors without considering alternative sources may mean higher cost, relatively poor quality material, and a decrease in sales. Effective procurement routines are needed to cover specification requirements, the processing of purchase requisitions, the selection of vendors, the issuance of purchase orders, the receiving and inspection of materials, and record-keeping activities.

Inventories. Inventory controls are extremely important as it is not unusual for inventories to comprise more than 25 percent of the total assets of a business. Unwarranted accumulation of manufactured stock reduces working capital. Other important reasons for inventory control are to supply the sales department with a record of the availability of stock, facilitate the scheduling of production, and prevent pilferage.

Division of Work

The effectiveness of business is often increased by specialization. The division of work and the division of labor go hand in hand. By breaking down large tasks into simple repetitive tasks individuals can learn to do them easily and rapidly. This basic principle was used by Henry Ford, whose assembly-line method of automobile manufacture serves as a classic example of mass production by specialized techniques.

The division of work on an organizational plane should follow the principles that are used in mass-production processes. The work should be arranged for the simultaneous and sequential performance of specialized functions leading to the completion of a product. The entire organizational structure should be determined with this basic idea in mind.

Some companies have people and units that are overspecialized, thus encumbering their operations. From the standpoint of efficiency a certain amount of diversification is desirable because it (*a*) reduces job fatigue; (*b*) yields more flexible use of manpower; (*c*) reduces the tools, machinery, and equipment needed; and (*d*) fosters quality of output.

Every organization is the sum of its peculiarities in respect to work activities, skills, experiences, and personalities. Under an optimum arrangement it should possess a smooth blend of specialization and diversification.

Delegation of Responsibility and Authority

Any manager has the alternative of doing a task himself or delegating it to his subordinates. This concept applies alike to presidents of large corporations, supervisors of small departments, and principals of owner-manager enterprises.

The most important thing a man in authority must do is to define the scope in which each organizational unit may operate. Every subordinate box on an organization chart is actually an indication of the delegation of duties and functions.

Prudent managers seek unique qualities in persons who will work for them, especially those who will be close to them. Assistants should possess attributes that differ from those of their employers and should be capable of doing work the leader is unable to do for himself. Organizations should be composed of persons whose abilities and activities supplement and complement one another.

Effective delegation of duties by management starts at the top of the organization structure; it requires a clear commissioning of official authority and responsibility throughout the chain of command. The term "authority" applies to the power or right of one person to require another to perform assigned duties. "Responsibility" refers to the transfer of accountability for the performance of duties, charges, or obligations.

If an individual is held responsible for carrying out work that involves the assistance of others, he should have whatever authority is appropriate for the purpose. Problems arise when executives and managers assign responsibility but fail to transmit the authority necessary to get the job done in an effective manner.

The leader's job is to get work done by other people. Having delegated work, he must have faith and confidence in his subordinates and strike a balance between the delegation of responsibility and the delegation of authority. As a rule he will retain a portion of the responsibility and delegate most of it to the person or persons assisting him, who in turn may pass it on to others.

Generally speaking both responsibility and authority should be delegated to as low a level in the organization structure as is practicable. The advantages of doing so are based on the following assumptions:

1. Persons making decisions should be those who are most familiar with the problems of a task because of their proximity to the work. Familiarity with work problems improves the quality of decisions made.
2. Economies of management time are realized. Each higher rung on the managerial ladder is permitted to concentrate its attention on successively more broad and important problems. At the top executives are free to create policies and to develop new ideas.
4. Decision-making is accelerated.
5. Employees are given the opportunity to learn and develop.

On the other hand, it should be noted that potential disadvantages are associated with delegating responsibility and authority at the lower levels of organization. These include the following:

1. Lack of uniformity in decisions due to individual approaches in accomplishing tasks.
2. Neglect to use the talents of available specialists.
3. Management may be inadequately informed. A curious and notable thing about a large business enterprise is the difference between what top management thinks is occurring at the lower levels of the structure, and what is actually occurring. Directives, budget controls, production goals, and other instructions may flow unhindered from the top to the bottom of the company organization. The upward flow of information is another matter. Each management member in the chain of command is prone to invoke censorship by relating only good tidings to his superior. As a result top executives tend to hear only good news or information about problems that have been divested of their significance.

Managers who exercise censorship controls do so to protect their positions. This is indirectly encouraged by their supervisors, who have a tendency to want to hear only good news. Many serious business situations can result from this. Although it is true that an extended period of compatibility is possible under such circumstances, in the long run the filtering of pertinent negative information can prove to be devastating.

Control Factors

In evaluating the competency of an organization it is necessary to consider the number of employees under the direct supervision of any one manager. This number is referred to as the "span of control." (See Figure 3-2.) Because of human limitations it is necessary that the span of control be restricted to a number of people that can be efficiently directed by a

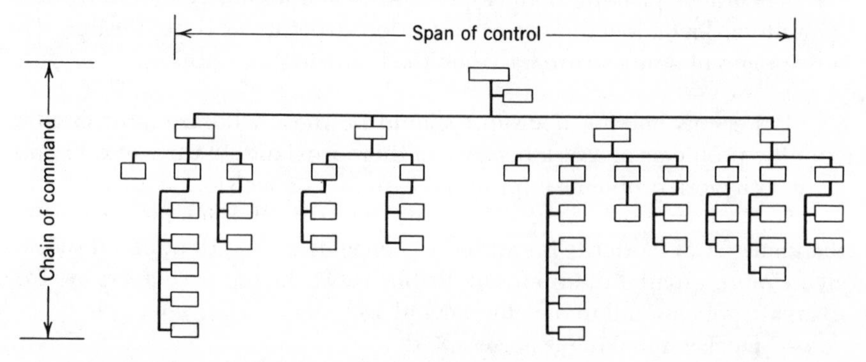

Figure 3-2. Organizational span of control and chain of command.

single individual. This number will differ for various reasons, among which are the following:

1. The complexity and nature of the work;
2. The knowledge, skill, and capabilities of the manager;
3. The knowledge, skill, and capabilities of his subordinates;
4. The degree to which the activities are of the same kind or nature. (With a high degree of similar activities the number of subordinates that can be adequately controlled is much greater than in a situation in which a manager coordinates activities of a diverse nature.)
5. The volume and frequency of problems and difficulties;
6. The amount of decision-making involved;
7. The ease with which communications flow.

For every level of organization there is an optimum span of control. Most authorities agree that at the top levels of management a span of control between four and seven individuals will permit an efficient operation. At lower levels, such as between a first-line supervisor and his workers, a span of control consisting of 10 to 20 individuals is considered not unreasonable. This is useful only as a guide and should not be considered to be inflexible. As a general rule, however, the greater the span of control, the greater the possibility of inefficiencies.

The span of control has to do with the width of an organizational structure. We now turn our attention to its length, called the chain of command.

The chain of command refers to the number of organizational levels under the direct supervision of an individual manager. It signifies the linkage of authority that extends down from the top position, link by link, to the lowest supervisory level. In some instances a chain of command may be lengthy and still effective. National governments, church organizations, and big business enterprises are known to function well with long chains of command.

Generally, however, efficient results are not obtained by lengthy chains of command. There are two good reasons for this: First, the problems of communication increase directly in relation to the length of the chain of command; second, swift action, vital to successful performance, is often hampered by the red tape that is peculiar to lengthy chains of command.

Channels of Contact

It would be impractical to expect that all communications follow the formality prescribed by the chain of command. This would stifle operations and cause delays. Common sense dictates the need for contacts, such as those illustrated in Figure 3-3, which are not confined or limited to the lines of authority and accountability of the formal structure. These types of contacts are necessary where duties interlock and must be coordinated.

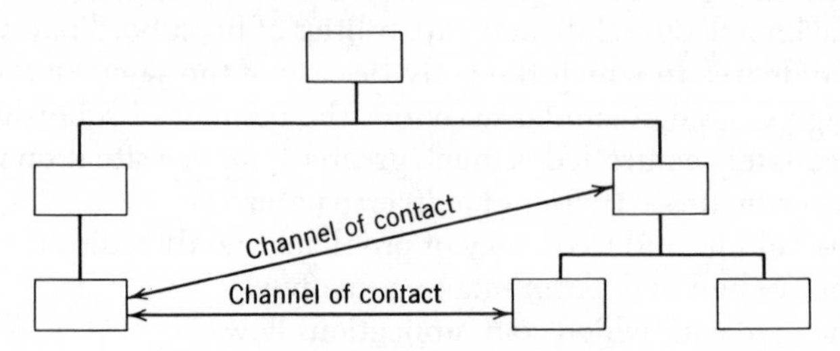

Figure 3-3. Channels of contact.

Establishment of lateral lines of contact, mutuality, and teamwork should not violate the formal organizational arrangement. There are three prerequisites for success in this sphere: First, the parties involved must have the sanction of their respective supervisors; second, all parties involved must be in agreement; and third, the parties must keep their respective supervisors informed on all matters of importance.

The relationships within a business are manifold. Coordination and cooperation are required wherever a position is involved with that of others. As stated, the relationships need not be limited to the lines of responsibility and authority, and may be crisscrossed wherever efficiency is increased, just as long as they are authorized and formal communications are maintained.

Proportionate Composition

Proportionate composition refers to the appropriate adaptation of parts to one another. This same meaning is retained in its application to organizational studies. It does not apply to any particular sort of organization, but to the special sort of way in which an organization has its units or departments arranged.

The proportionate organization is one that has the strength of its units in relative proportion to their real worth and contribution to the overall success of the enterprise. Each unit should be of sufficient size to fulfill its functions, and no larger.

Proportion implies a kind of balance. The most profound need of any organization is that the parts comprise a coordinate whole, with a balancing of qualities and abilities so as to constitute a well-rounded organization.

This concept does not have static attributes; they are subject to change with modifications in the organizational arrangement.

Progress will result in changes to an organization in different areas and to different extents. Product development through research, revised company aims, and current management interest are typical factors that influence or bring about such changes. The addition of a new item to the product line and an emphatic turning of managerial attention in that direction may result in the establishment of new departments. Quality control gradually assumes greater importance because of increased technological complexities of the product, and the organizational units concerned with this type of work are permitted to expand and exercise greater power.

Dynamic changes within a company can cause the entire organization to lose its proportionality. The addition or expansion of a unit may be required to make all the organizational units again compatible. Too often the change in a condition causes results that go beyond the point of proper proportion and brings about serious incompatibility among the units. Continual effort is needed to maintain an effective balance.

Proportionality can be difficult to discern, and great care should be exercised before alleging that it is lacking in a structure; for example, in manufacturing industries the production departments were long favored as organizational units demanding a concentration of effort. Back in 1935, by numerical standards, a firm with a ratio of one office employee to every 2.5 factory workers was likely to be classed as improperly proportioned; by today's standards this ratio would not necessarily mean a discordant structure, and it approximates the national average. In many instances the emphasis has shifted from the problems of manufacture to other vital areas. Companies manufacturing piece parts and components for space vehicles stress the quality and reliability of their products. Because of this they are apt to have quality-control departments and reliability units that at first glance appear to be out of proportion and in reality are not.

To make this concept clearer let us consider the manufacturing of transistors. These electronic devices are not difficult to manufacture; in fact, their construction is quite simple. Still, despite their basic simplicity, the quality and reliability of a transistor is important as a vital piece part to expensive equipment. Such piece parts are termed "critical," and seemingly out-of-proportion departments are needed to perform special testing and inspection functions. Wherever critical defects may occur maximum precaution needs to be taken. Thus proportional composition, although apparently lacking in an organization, may well be present. It must be viewed in relation to the objectives of these seemingly oversized departments and their contributions to the success of an entire company.

Still another situation could be a sales organization that appears to be unduly heavy. The nature of the product or the custom of the trade may dictate the need for a large sales force. In itself size is no criterion for measuring proportion within an organization.

The "empire builders" represent a very dynamic threat to well-formed company structures. These empire builders are departmental and organizational heads who promote the unjustifiable expansion of their units. Some are motivated by personal ambitions, others are honestly inclined to overrate the contribution of their own organizations. Whatever the cause, the threat is real and in need of tactful handling. The preservation of proportionate composition is paramount, but preventative measures need not deprive managers of their individual traits. A senior supervisor is in an excellent position to explain objectively the need for organizational coordination to empire builders. The very fact that the matter has been openly discussed acts as a deterrent.

Flexibility

A flexible organization is one in which temporary changes in work volume do not seriously impair its basic operating ability. As business conditions change the company organizational units expand or contract to meet current demands efficiently.

Business concerns should be prepared to take any number of different kinds of variations in stride: the gain or loss of contracts, changes in the economic climate, short-term fluctuations in available manpower, and seasonal variations in sales.

As fluctuations in the total work load are common to nearly all business concerns, a company should have built-in flexibility. The more flexible organizations are the more stable ones, as they basically remain the same and yet quickly adapt to changing conditions.

Several techniques are used to lessen the severity of seasonal and other short-run fluctuations in the total work volume. One method is to utilize personnel with diversified experience in different positions, as needed. Another method is to establish centralized service departments to perform the overflow of work within and among departments. The setting up of a typing pool is an example of this technique. Still another method is to hire temporary personnel to do the portion of the work load that can be readily learned and accomplished. A solution that sometimes surpasses any of these is simply improved scheduling of tasks.

Perpetual Existence

One of the advantages of the corporate form of business enterprise over that of a partnership or sole proprietorship is its legal capacity for perpetual existence. Its existence is similar to that of mankind: the human race remains, although the individuals within the race come and go. If this legal capacity for an indefinite existence is to become a reality, provision must be made by each organization to assure its own immortality. The state only grants the privilege and does not confer any obligation for a corporation to have an endless duration.

In this sense there is no such thing as an automatically self-perpetuating enterprise. This must come from proper organizational planning, which will ensure continuity of operation beyond the life span of any particular individual.

To achieve the perpetuation of an organization clear-cut policies and procedures must be reduced to writing. It is also the obligation of top management to have replacements in training for all key positions.

INVISIBLE ORGANIZATIONAL RELATIONSHIPS

As photographs picture people, the organization charts of companies reveal their formal structure. Both types of portrayals are informative and useful, and both have serious limitations. Neither one reveals the "personality" of the individuals with which they are concerned. Photographs do not reveal the invisible characteristics of an individual, nor do organization charts indicate the invisible or informal relationships within organizations.

These relationships are important because they may enhance or hinder the effective functioning of the formal organization. Relationships of this type manifest themselves in many different ways. Strong friendships between persons may bind them into an informal subgroup that exerts its influence independent of the formal company structure. There is also the possibility of having individuals with dominant personalities informally controlling small groups of people, or there may be unrecognized natural leaders whom individuals support.

Even in the case of official relationships there may be invisible forces at work. Charts of the type shown in Figure 3-4a are conventional and commonly depict a number of vice-presidents or top executives as having equal status in an organization. Although authoritative, such portrayals can be misleading. Organizational charts indicate only official power, and

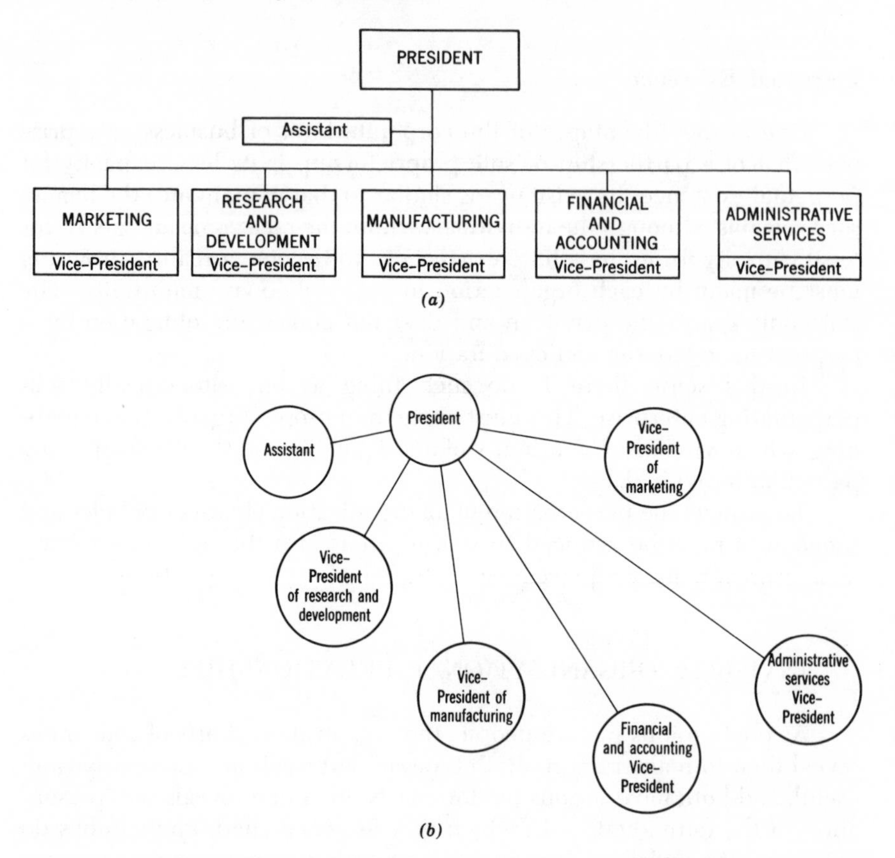

Figure 3-4. Conventional organization chart (*a*), and chart showing informal relationships (*b*).

not the power that emanates from additional influence and prestige. Hence, whereas a number of vice-presidents could be shown to have equal rank on a chart, one may have a great deal more authority than any of the others; for instance, in a company that is engineering oriented, the vice-president in charge of research and development may well have more power than any other executive on the same level. Such an arrangement is apparent in the Figure 3-4*b*.

Figure 3-4*a* shows a typical organization chart for a manufacturing concern. Figure 3-4*b* shows the informal relationships of this organization as determined from observations such as associations and number of personal contacts. It indicates decreasing grades of authority in terms of distance from the circle that designates the president.

The invisible relationships within organizations are not peculiar to business structures. They are common to all types of organizations and are discernible even in class rooms, fraternal societies, and other organized groups.

By their very nature, invisible organizations may be difficult to uncover. They are informal associations, and even the participants themselves may not be consciously aware that a particular relationship exists. This is true in the case of company personnel that lend unofficial support to natural leaders by seeking their counsel and following their examples. There is no exact method of unveiling these relationships. Yet, as intangible as the air we breathe, they are just as real. We do not see the wind, but we feel it and observe its influence on our surroundings. The systems analyst has to depend on inquiries and personal observations to ferret out the existence of these invisible organizations; then he must evaluate their influence on the organization.

In themselves, invisible organizational relationships are neither good nor bad. If these relationships support or complement the acknowledged structure, the arrangement can be beneficial. If, on the other hand, they generate conflicts or resistances that are not in accord with the policies and programs of the formal organization, adjustments are definitely in order.

CENTRALIZATION AND DECENTRALIZATION

Total centralization is present when organizational units are together in one place and under one control. Decentralization refers to the separation of physical facilities away from the main location, or to the delegation of managerial controls—responsibility and authority—from higher management to subordinates, or to both.

The possible organizational patterns are the following:

1. Centralization of both management controls and physical facilities;
2. Centralization of management controls and decentralization of physical facilities;
3. Decentralization of both management controls and physical facilities (*Note:* Some centralization of management controls is always necessary);
4. Decentralization of management controls and centralization of physical facilities.

Decentralization, both of management and physical facilities, is generally associated with large companies, but it does have application to many small companies and to many departments in large companies.

Four basic patterns have been mentioned. When considering any one of these patterns with respect to a particular organization it is also essential to consider the desirable degree of centralization and decentralization. The optimum extent varies from organization to organization and depends on many factors, such as the nature of the enterprise, the kind and amount of work involved, and the characteristics of the management.

For any specific organization it is useful to consider which of the advantages and disadvantages normally ascribed to centralization and decentralization, listed below, are applicable, and to evaluate the importance of each.

Advantages of Centralization. Potential advantages of centralization are as follows:

1. Administration is facilitated. Executives and other management personnel are "on the spot" and in a position to make sounder judgments.
2. Decision-making is expedited. There is no necessity to check with a home office that is located elsewhere, as is so often the case under decentralization.
3. There is greater interchange of managerial and technical information.
4. Plant economies—especially in service departments—are possible.
5. Labor specialization is enhanced. A sufficiently large operation makes it practical to develop highly trained and proficient employees.
6. Equipment economies where operations are similar or identical are possible.
7. There is increased justification for special-purpose machines where operations are similar or identical.
8. There is less duplication of functions.
9. More standardization of operations, equipment, and output is permitted.

Disadvantages of Centralization. Potential disadvantages of centralization are as follows:

1. The complexity of organizational patterns and operational functions is increased.
2. Executives, administrators, and supervisors are overburdened.
3. There is failure to develop executives through the delegation of managerial responsibility, authority, and accountability.
4. The flexibility of decentralized organizations is lacking.
5. Incentive for individual advancement in the organization through sufficient opportunities at succeeding levels is not provided.
6. Errors in the centralized control affect a wide area of operations.

Advantages of Decentralization. Potential advantages of decentralization are as follows:

1. Less precise coordination of systems and practices is required.
2. Decentralization lends itself to instituting plans on an experimental basis in one location (e.g., plant or department), so that they can be proven and "debugged" before wider adoption at similar locations.
3. Decentralization favors the delegation of authority, thus relieving the work load of top managers.
4. Prompt action is possible, and the quality of decisions is improved because they are based on a first-hand knowledge of local conditions and requirements.
5. Close cooperation among employees is encouraged.
6. More meaningful jobs at lower organizational levels are provided.
7. Where geographical dispersion of plants is the case, risks and possible losses involving personnel and company property are extended over a wider area.

Disadvantages of Decentralization. Potential disadvantages of decentralization are as follows:

1. Policies, practices, and decisions lack uniformity.
2. Facilities and functions are duplicated.
3. It is not conducive to developing management personnel with an overall company viewpoint.
4. There is a greater communications problem on consolidating efforts.
5. The skills, ideas, and advice of specialists are not fully employed.
6. There is inadequate supply of proper equipment.

4

The Management Audit

A management audit is a critical review of an organizational structure and its administration. Its purpose is making recommendations for adjustment and improvement. An audit may involve a whole company structure or be restricted to one of its parts, such as a division or department.

There are many reasons for initiating this type of study. Some organizations are improperly formed from the start; others become obsolete because of changes in their objectives, conditions, or personnel. Sometimes, as a stop-gap measure, functions are assigned to an organization on a temporary basis and become permanent, making the structure illogical, disproportionate, and difficult to manage.

The composition and effectiveness of an organization structure should be periodically examined to determine if changes are needed. Some companies assign this task to the systems-and-procedures staff, which may be expected to survey conditions periodically or as directed. On the basis of its findings and evaluations, it makes recommendations for improvement.

In the absence of a qualified internal staff companies will usually engage a management consulting firm for this job. The internal systems-and-procedures staff is expected to cooperate with the consultants, and to adjust and realign systems in accordance with management-approved organizational changes.

A management audit includes the following major considerations:

1. Appraisal of company policies;
2. Gathering, analyzing, and evaluating data about the existing organization structure;
3. Preparing recommendations for adjustments and improvements, coupled with a well-regulated conversion plan that will minimize disruptive influences; and
4. Follow-up.

APPRAISAL OF COMPANY POLICIES

Policies are general guides to action and delineate the functions, responsibilities, authority, and respective relationships required to achieve company objectives. To be effective they must be properly formulated, reduced to writing, and disseminated throughout the organization.

Before the actual study of an organization is undertaken the policies of the company should be reviewed to determine if they have been kept current and in accord with management objectives. Top management can be negligent in formulating original policies or policies may simply become obsolete. In any event a review of policies is necessary, since they are an expression of management planning and, in turn, affect organization.

When management is advised on the status of policies the facts and recommendations should be in written form. Depending on the facts revealed, the report may contain a recommendation that management institute wholly new policies or make extensive revisions in existing ones.

GATHERING, ANALYZING, AND EVALUATING ORGANIZATIONAL DATA

Special aids and techniques have been developed for conducting organizational studies. These include organization manuals and charts, questionnaires, function-allocation charts, work-distribution charts, and check lists. The bulk of this chapter is concerned with a description of these aids and techniques.

From the inception of an audit, when he first begins to acquire information, the systems analyst should seek and encourage the participation of affected management personnel. They should be persuaded to make suggestions, exchange ideas on the advantages and disadvantages of possible alternatives, and to participate in the development of any reorganization.

PREPARING RECOMMENDATIONS

The objective of the audit report is to provide information on which executive action may be taken. It tells what is right and what is wrong with the organization, and gives the systems analyst's opinions and recommendations to guide top management to the solution, should one be needed.

Where transition from an old to a new or modified organization structure is necessary, the systems analyst should propose a plan for an orderly changeover. The plan must take into account not only the effect the change will have on operations but also its impact on the morale of employees. In situations where the existing organization structure is grossly inefficient or costly to maintain the changeover may be made immediately. Such action is the exception rather than the rule, however. Most cases allow for the gradual shifting from the present to the planned organization in a series of phases or steps, which permits changes to be introduced over a period of time. During this transition the plan and its unfolding should be under constant surveillance and evaluation to insure success in its objective and to make whatever refinements and modifications are indicated. Carrying out the transition by phases also encourages greater acceptance by employees, who are given time to assimilate small changes. This prevents resistance to an innovation that they feel is too great or too sudden. Of course, it is possible to impose a plan on people; but if they do not like it, their resentment can lead to a lack of cooperation and even to ultimate failure.

THE FOLLOW-UP

The follow-up function is extremely important. Nothing should be taken for granted. A constant follow-up survey should be in effect to determine if the recommendations are being properly implemented, if desired results are being achieved, and if overall effects on morale are satisfactory. During the follow-up activity the systems analyst should also be alert in making additional specific improvements.

MECHANICS AND TECHNIQUES OF ORGANIZATIONAL STUDIES

Aids and techniques used in the performance of organizational studies will now be examined. As a brief introduction, these tools, along with their prime purposes, are presented below.

Tool	Purpose
Organization questionnaire	To obtain information from management on the nature and adequacy of an organization.
Organization manual	To provide an authoritative guide on the company's organization.
Organization chart	To show a view of the structure of an organization and its work relationships.
Function-allocation chart	To portray the allocations of functions to the various units of an organization.
Work-distribution chart	To assemble all of the tasks of the personnel of an organization or unit; it shows what is done, who does what, and how much time is spent doing it. The information is usually derived from task summary lists.
Time-task list	To itemize the different tasks performed by each employee during the course of one day. A number of *lists* are accumulated for the period of time under study, usually a week; the recorded daily information is summarized on a *task summary list*.
Task summary list	To summarize and total the information contained on a number of *time-task lists,* which is then transferred to the *work-distribution chart.*

QUESTIONNAIRES

Although the management audit or review may appropriately begin by consulting organization charts and manuals, they must not be accepted as reliable until their adequacy and correctness have been substantiated. Audits must be based on first-hand data, which requires that the systems analyst personally call on department and unit heads to obtain facts on basic organizational functions, division of work, lines of communication, and so on.

Organizational questionnaires are also used to obtain first-hand data. These contain significant questions regarding the composition and effectiveness of the organization. Replies to these questions are obtained either by personal interviews or through the mail.

Mail questionnaires have their greatest value when it is necessary to acquire a small amount of information from a large number of persons. Savings in time and effort result.

Such questionnaires commonly consist of five elements, which are discussed below.

Identification. Within this classification are spaces on the form that provide for the insertion of the title of the organization, organization number, name and title of the organization head, and the name and title of the supervisor to whom he is responsible.

Purpose of the Questionnaire. It is desirable to include a preface to the body of the questionnaire of a few brief statements regarding its purpose. The omission of this may result in undesirable interpretations on the part of organization heads.

Information. This category includes the explanation of terms, definition of words, and instructions needed to correctly complete the form. The information portions of a questionnaire may be of a general nature, applying to matters throughout the entire form or they may be specific in that they apply only to particular questions. If the information is of a general nature, it should be placed at the beginning of the material. Information that is applicable to a particular question or group of questions should, wherever possible, be placed near the questions to which it applies.

Body. This section of the form contains the questions which, if properly answered, will provide the information being sought. For best results queries should be made interesting and simple, and phrased to get the cooperation of the persons questioned. To prepare a good questionnaire we need to abide by certain "do's" and "don'ts." The rules for the preparation of questions are as follows:

1. Ask the organization head questions that are germane to his unit. He should not be expected to supply information on organizations not connected to his own.
2. Prepare questions so that their meanings cannot be misunderstood. A good means of finding out if a question is understandable is to ask it of a few persons who are representative of the whole group. Do not ask these persons their opinion about the question, just ask the question itself. If it produces fruitful replies, it may then be included on the form.
3. Avoid questions that are too general, since such questions yield answers that are too general. An example is a question such as "How effective is your organization in meeting its objectives?" The response can range from "very effective" to the relation of information that is more detailed than necessary.
4. Do not use leading questions. A leading question is one that suggests the answer. An example of a leading question would be, "What trouble do you have in delegating responsibility and authority?" This question implies that the organization head has such problems.
5. Make the questions as short and as simple as possible. Long and complex questions may confuse or mislead the reader. It is wiser to have a

series of short questions that will produce explicit answers than an involved query that produces ambiguous information.

6. Arrange the questions so as to facilitate the flow of thought. Questions on any series of items should be presented in a logical order—"logical order" to be interpreted from the viewpoint of the respondent.

7. Do not use questions that have implied questions within them. This type of question is exemplified by, "Why did you change your car to a Chevrolet?" Actually, there are two questions in one inquiry: "Why did you change your car?" and, "Why did you choose a Chevrolet?" This example illustrates the insidious nature of this type of question.

8. Eliminate questions whose answers are obvious, can be obtained in an easier way, or are not pertinent to the study. In many cases this rule is purposely broken in an attempt to fill up the page of a questionnaire to make it look impressive. This is a mistake. Not only do such inquiries turn out to be annoying to the respondent but they may well reduce his time in adequately replying to the significant questions. The possibility of receiving inadequate replies to truly important questions, plus the concealed cost of wasted time and effort on the part of all concerned, hardly warrants the construction of a form that is made superficially impressive by the inclusion of redundant inquiries.

9. Omit any questions that might cause friction between individuals or groups. If the answers to such questions are necessary to the study, it is better to obtain them in personal interviews.

Suggestion. The well-constructed questionnaire should give a section for the organization heads to make suggestions for improving their own organizations, and the relationships between their organizations and other units. By including this, not only recommendations will be received but often very informative facts may be obtained—facts that might not otherwise be given. In addition, a request for suggestions has a good human-relations aspect to it: it shows that the company values the opinions of its management personnel at all levels.

NOTE: A cover letter is sometimes used to present at least part of the *purpose* and *information* elements, in which case it forms a part of the questionnaire package. This letter should carry the signature of an official high enough to be given due attention.

ORGANIZATION MANUALS

Manuals specify the functions, responsibilities, authorities, and relationships of positions within a company. They enable executives and subordinates to comprehend the whole company structure and thereby to understanding better their individual places. Manuals also serve as a means of control by checking duplication of effort and illogical grouping of organizational structures. These manuals must be kept up to date to be useful.

Organization manuals may pertain to the entire company structure,

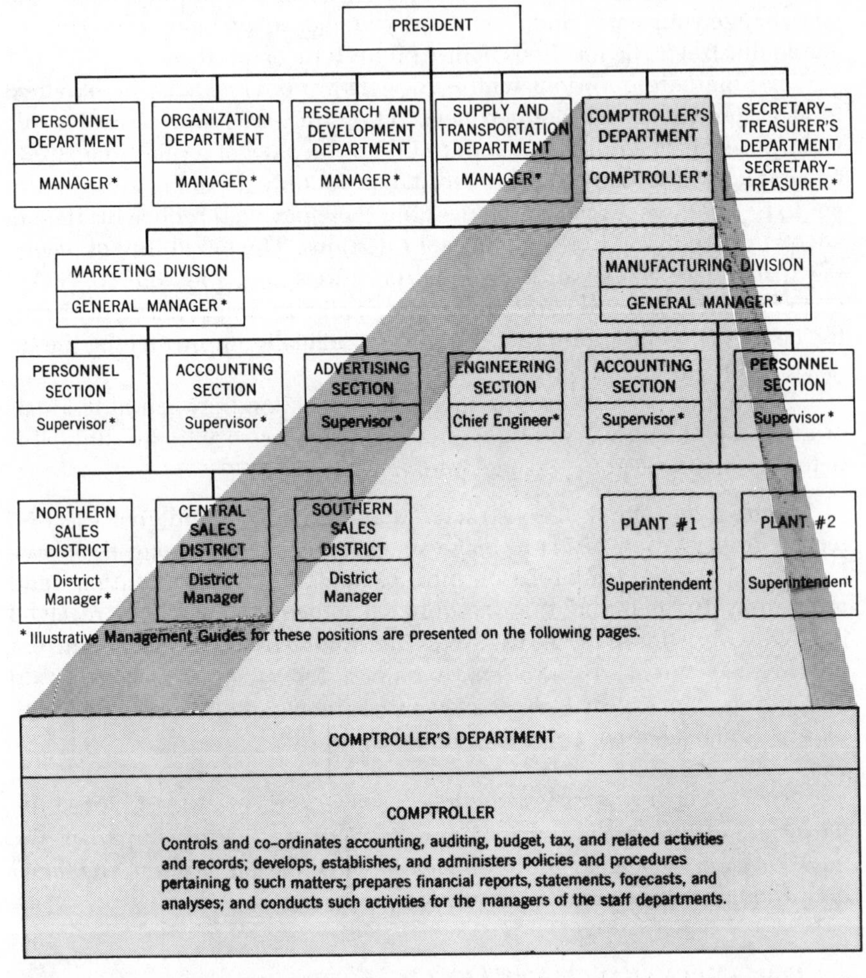

Figure 4-1. Organization chart featuring the comptroller's department. As noted on the chart itself, each block containing the symbol • is similarly featured on a separate page of the manual. Opposite each chart are listings giving the function, responsibilities and relationships of that unit.

MANAGEMENT GUIDE

COMPTROLLER

I. FUNCTION

Controls and co-ordinates accounting, auditing, budget, tax, and related activities and records; develops, establishes, and administers policies and procedures pertaining to such matters; prepares financial reports, statements, forecasts, and analyses; and conducts such activities for the Managers of the staff departments.

II. RESPONSIBILITIES AND AUTHORITY

The responsibilities and authority stated below are subject to established policies.

A. Activities

1. Formulates, or receives and recommends for approval, proposals for policies on accounting, auditing, the budget, the preparation and payment of payrolls, tax matters, the compilation of statistics, and office methods and procedures; and administers such policies when approved.
2. Establishes and administers procedures pertaining to accounting, auditing, the budget, the preparation and payment of payrolls, tax matters, the compilation of statistics, and office administration.
3. Provides services pertaining to accounting, the budget, the preparation and payment of payrolls, tax matters, and the compilation of statistics to the Managers of the staff departments.
4. Prepares and maintains the company cost accounting books and such other records as are necessary to the accomplishment of his function.
5. Prepares the principal financial statements of the company.
6. Consolidates the proposed annual budgets of the organizational components of the company into the proposed annual company budget, and prepares recommendations thereon.
7. Conducts the internal auditing program.

B. Organization of His Department

1. Recommends changes in the basic structure and complement.
2. Recommends placement of positions not subject to the provisions of the Fair Labor Standards Act in the salary structure.
3. Arranges for preparation of new and revised Management Guides and position and job descriptions.

C. Personnel of His Department

1. Having ascertained the availability of qualified talent from within the company, hires personnel for, or appoints employees to, positions other than in management within the limits of his approved basic organization.
2. Approves salary changes for personnel not subject to the provisions of the Fair Labor Standards Act who receive not over $____per month, and recommends salary changes for such personnel receiving in excess of that amount.

3. Approves wage changes for personnel subject to the provisions of the Fair Labor Standards Act.
4. Recommends promotion, demotion, and release of personnel not subject to the provisions of the Fair Labor Standards Act.
5. Approves promotion, demotion, and release of personnel subject to the provisions of the Fair Labor Standards Act.
6. Approves vacations and personal leaves, except his own.
7. Prepares necessary job and position descriptions.

D. Finances of His Department

1. Prepares the annual budget.
2. Administers funds allotted under the approved annual budget, or any approved extraordinary or capital expenditure program, or any appropriation.
3. Approves payment from allotted funds of operating expenses and capital expenditures not in excess of $, which are not covered by the approved budget, any approved expenditure program, or an appropriation.
4. Recommends extraordinary or capital expenditures.
5. Administers fiscal procedures.
6. Receives for review and recommendation the items of the annual budgets of other staff departments and the field divisions coming within his province.

III. RELATIONSHIPS

A. President

Reports to the President.

B. Other Department Managers and Division General Managers

Advises and assists other Department Managers and Division General Managers in the fulfillment of their respective functions in matters within his province, and co-ordinates his activities and co-operates with them in matters of mutual concern.

C. Federal and State Agencies

Conducts such relationships with representatives of federal and state agencies as are necessary to the accomplishment of his function.

D. Independent Auditors

Conducts such relationships with independent auditors as are necessary to the accomplishment of his function.

E. Others

Establishes and maintains those contacts necessary to the fulfillment of his function.

Figure 4-2. A description of the comptroller's function, responsibilities and authority, and organizational relationships, as given in *The Management Guide*, Second Edition (March 1956). Courtesy: Department on Organization, Standard Oil Company of California.

or just individual plants or divisions. Some contain only charts, whereas more comprehensive ones contain the following:

1. A brief history of the company and its organization, with summary information on its products.
2. Statements as to the nature and purpose of the manual.
3. A rather short, written discussion about the philosophy of management regarding such considerations as cooperation, services, and goodwill.
4. Organization charts. (See Figure 4-1.)
5. Descriptions of the objectives and functions of the organization units charted.
6. Clarification of names and titles. Nomenclature such as "division," "department," "bureau," "branch," "section," and "unit" need to be defined for particular applications. Also, titles that include the words "chief," "general," and "supervising" often need clarification.
7. Job descriptions. These delineate the responsibility, authority, and functions of the positions charted. They may also include information on the interrelationships of positions and written procedures for coordinating certain functions. (See Figure 4-2.)
8. The general objectives of management, as outlined in Figure 4-3.
9. A table of contents and an index.

Manuals are usually kept in loose-leaf binders to facilitate changes and deletions. Some companies have the front covers engraved with the name of the company and the title of the manual.

ORGANIZATION CHARTS

Charts are the substance of organization manuals. Essentially they are devices for showing "who's who" in the organization and how the responsibility is divided. The typical organization chart portrays a leader at the top, with his subordinates at the bottom. Grades of responsibility and authority are assigned by means of levels between the top and bottom.

Organizational charts make it possible to visualize and understand the manpower structure of an entire company. They serve not only as management guides but also as instruments for analysis and synthesis.

Sets of organization charts include (*a*) A master chart showing the overall breakdown into divisions, departments, and lesser organizational bodies. This chart customarily depicts the board of directors and the chairman of the board; the president, vice-presidents, and other company officers; staff and service managers; and heads of operating divisions, and (*b*) Detail or auxiliary charts that portray each major division and department, and their major subunits.

GENERAL OBJECTIVES OF MANAGEMENT

OBJECTIVES	REQUISITE CONDITIONS
1. ORGANIZATION—To develop and maintain a sound and clear-cut plan of organization through which Management can most easily and effectively direct and control the enterprise.	a. Organization structure (organizational components) designed best to facilitate management, prevent overlapping of functions, and duplication of effort. b. Function, responsibilities and authority, and relationships clearly defined for each management position. (See Management Guide.) c. Proper delegation of authority by management to permit decisions to be made at the lowest practicable level of management. d. Thorough understanding of the requirements and responsibilities of their positions on the part of personnel. e. Proper co-ordination of the entire organization plan.
2. PERSONNEL—To develop and administer a constructive personnel development and training program which will gradually ensure that all positions in the organization are filled by individuals fully qualified to meet the requirements of their respective positions.	a. Adequate control to ensure selection of best-qualified personnel available for the different types of work, first from within the organization, and then, if necessary, through outside hiring. b. Effective training by superiors, with the assistance of appropriate staff agencies, of all employees to meet the requirements of their jobs. c. Comprehensive annual rating of all employees in terms of job requirements. d. Positive action to correct deficiencies in qualifications and assignments as disclosed by the rating program. e. Carefully planned personnel utilization program to take best advantage of demonstrated abilities, develop each individual's full potentialities, ensure adequate potential material for responsible positions, and to ensure placement of the best-qualified individual in each job. f. Adequate control to ensure that all promotions and appointments are made from among the best-qualified candidates available. g. Full co-operation in effecting the most advantageous placement of personnel. h. Proper co-ordination of the entire personnel development and training program.
3. PLANNING—To formulate well-considered plans and objectives, covering all operations, activities, and expenditures for each year or longer ahead, as a basis for authorization, a guide to achievement, and a measure of performance.	a. Clear conception of essential needs and worthwhile objectives. b. Clear-cut plans for accomplishing these objectives. c. Sound analysis of requirements in terms of manpower, costs, facilities, and money. d. Good business judgment as to justification and extent of proposed undertakings. e. Effective participation of subordinates in formulation of their respective parts of the program. f. Proper co-ordination of each program. g. Appraisal of results compared with planning for these results.
4. ADMINISTRATION—To accomplish all functions and responsibilities fully, effectively, and harmoniously.	a. Guiding policies clearly stated, and well understood by all. b. Effective co-ordination and control of results. c. Prompt, well-considered management decisions. d. Close supervision affording first-hand familiarity and appraisal of operations, activities, and management problems on the ground without relieving subordinates of their proper responsibilities. e. Maximum use of best thought and capabilities of the entire organization in accomplishing the program. f. Assumption of proprietary responsibility for successful conduct of all activities under his control, relieving superiors of details, and presentation of matters of justifiable importance. g. Active co-operation in furthering the proper interest of other organizational components and of the enterprise as a whole. h. Maintenance of good public relations. i. Proper co-ordination of operations and activities.

5. COSTS—To keep all costs and manpower at an economic minimum, consistent with essential purposes.	a. Periodic analysis and appraisal of all functions and activities as to justification and required effort. b. Elimination of all unessential or ineffectual work, expense, and manpower as disclosed by such analysis. c. Establishment of most efficient methods for performing operations and activities. d. Establishment of suitable standards and measures as to what constitutes optimum performance and cost in regard to all operations, activities, and expenditures. e. An adequate control system through which actual results are currently evaluated against the optimum or planned expectations, and all deficiencies are brought to the attention of the proper person for corrective action. f. Proper co-ordination of cost and manpower control programs.
6. BETTERMENT—To plan, stimulate, and develop improvement in methods, products, facilities, and other fields as applicable, keeping abreast of the best thought and practice throughout the industry, and to ensure that out-moded procedures and uneconomical facilities are abandoned.	a. Clear-cut recognition of needs and limitations. b. Well-planned betterment program with clearly defined objectives. c. Solicitation of best thought and suggestions from the entire organization. d. Keeping abreast of best thought and practice throughout the industry. e. Effective action in putting desirable improvements into effect. f. Proper co-ordination of betterment program. g. Periodic appraisal of results.
7. EMPLOYEE RELATIONS—To make sure that all employees are accorded fair and equitable treatment, and that they are inspired to their best efforts.	a. Personnel policies and practices (including benefit plans, wage and salary schedules, and working hours and conditions) kept up to date and in favorable relation to competition, through well-considered changes as necessary. b. Enlightened supervision, ensuring that each employee is treated fairly and justly as an individual, with helpful consideration for his personal feelings, ambitions, and problems, within the scope of reasonably interpreted rules and policies. c. Adequate control to ensure that each employee is fairly and appropriately compensated in general conformity with the established rate structure and policies. d. Maintenance of close touch with personnel and their problems. e. Effective leadership and stimulation of morale. f. Confidence and respect on the part of superiors, subordinates, and associates. g. Proper co-ordination of entire employee relations program.

Figure 4-3. Chart showing the objectives of management and the requisite conditions for meeting these objectives. From *The Management Guide*, Second Edition (March 1956). Courtesy: Department on Organization, Standard Oil Company of California.

On all charts, master or detail, the following facts should appear:

1. Comprehensive identification so there is no doubt as to what the chart is intended to cover. This includes such information as the official title of the organization, the number of the unit (if numbers are also used for identification), and its locations in terms of plant or physical location.
2. An explanation of any symbolic markings, special kinds of lines, or other distinctive features that are used.
3. Date of preparation. If an organization chart is not dated, it may be difficult to determine its present validity.
4. Approvals. The approval of a detail organization chart should be designated by the signature of the top man of the particular structure and his immediate supervisor. Sometimes companies require also the approval of the president, a vice-president, or possibly a committee on organization.

Often charting variations are adopted to enhance a chart. One variation is to include in the individual boxes a brief description of the applicable responsibilities and functions. Another is to show the names, and even the photographs, of individuals within the boxes.

There are reasons for and against the inclusion of personal names and photographs on organization charts. The main disadvantage is that affected charts have to be redone and distributed when changes in personnel take place. For some companies this would not present any great handicap; for others, which have considerable turnover in personnel, it would be sufficient to preclude their use. In favor of the inclusion of names and photographs on organization charts is the fact that they become more valuable as reference sources.

Many top executives in large corporations have large-size organization charts hung on the wall, featuring both photographs and personal names. The photographs are often duplicates of the ones attached to personnel records.

Photo-charts can be particularly useful. As one manager put it, while pointing his finger at a large organization chart hung on the wall of an outer office, "This chart enables me to refresh my memory about a person's name and face before seeing him in person. In a way, it helps to humanize things. After all, we are a very large organization, and I feel that everyone in it is important."

Some large wall charts are constructed of magnetic steel boards framed with aluminum molding. Magnetic card holders are used to hold cards, photographs, and self-adhesive stickers. The cards may be outlined in the form of organizational blocks, and information may be typed or handwritten on them. Acetate tape is used to simulate lines that would appear on a regular chart.

Advantages and Limitations of Organization Manuals and Charts

Among the principal advantages of organization manuals and charts are the following:

1. Lines of responsibility and authority are clearly specified and acknowledged.
2. Relationships between organizational entities are described and made known to all concerned.
3. Training of new personnel is made easier. New employees find manuals and charts a useful source of information.
4. Mistakes made by uninformed personnel are prevented.
5. Excellent ready-reference guides for all management personnel are provided.
6. Possible problem areas such as gaps and overlaps in responsibility are uncovered.
7. Constructive ideas on how to improve the existing structure are stimulated. To the systems analyst the charts are instruments of study for analyzing and synthesizing organizations.
8. Clarification of organization and job titles is brought about.

The principal disadvantages and limitations of organization manuals and charts are the following:

1. They are not purposeful for small enterprises. One authority mentions an annual sales volume of $5 million as the minimum company size to warrant the use of a manual.*
2. They may make the organization too rigid, formal, and restrictive; may discourage informal contacts and teamwork among personnel.
3. They bring to light relationships that preferably should be left undisclosed, or at least not brought into prominence. Often charts give above-and-below status implications that arouse resentments. Although they are meant to convey only different degrees of authority and responsibility, some employees may attach unpleasant social implications to these gradations.
4. They convey only a limited concept of the organization structure and the network of inter-relationships that exists. They fail to show the many informal relationships that are so necessary to the smooth functioning of an enterprise. Neither do they reveal the quality of the leadership at the different levels.
5. They are troublesome and costly to maintain.

* Lyndall Urwick, *The Elements of Administration,* Pitman, London, 1943, p. 79.

Taking everything into account, it is generally recognized that organization manuals and charts can be useful instruments of management. This is especially true if they are correctly interpreted and their drawbacks and limitations are recognized.

FUNCTION-ALLOCATION CHARTS

As a means of graphically portraying and studying the apportionment of functions among various units the function-allocation chart is useful. Such a chart consists of organization boxes with a listing under each box of the functions and activities of that unit; it may be applied to the entire structure of firms, or merely to divisions, departments, or other organizational groupings. The function-allocation chart shown in Figure 4-4 is for a medium-size firm.

The illustration is moderately simple in order to demonstrate the techniques of organizational analysis and synthesis. The optimum breakdown of organization varies, even within the same industry. Regardless of size, composition, or complexity, the basic approach is to bring together significant facts about a structure so that they may be evaluated.

A function-allocation chart is only a grouping of activities, but it can be the starting place for penetrating examination. Nothing useful will emanate from it until we start asking questions and finding answers.

Organization Check List

In his evaluation of the allocation of functions, the systems analyst will find a check list helpful. The questions serve to stimulate the mind and to make certain that important aspects of the study are not overlooked.

Thought-Provoking Questions. The questions that could be included on a check list are numerous; among them are the following:

1. Is the organizational arrangement designed to facilitate management?
2. Is there appropriate centralization and decentralization of activities?
3. Are line-and-staff relationships established along satisfactory lines? Are they well understood by the individuals involved?
4. Have the span of control and chain of command been kept within proper limits?
5. Are the organizational units in balance? Is each designed to contribute its proportionate share in achieving the goals of the organization?

6. Is there undivided accountability where one unit is directly responsible to only one superior unit?

7. Does the organization have flexibility? Can it be conveniently expanded and contracted to meet changing conditions?

8. Is there a logical assignment of duties? Are functions allocated according to their similarities and related characteristics?

9. Are all necessary functions assigned?

10. Is the allocation of functions to specific units clear-cut?

11. Is there neglect of important functions or undue attention given to subordinate matters?

12. Do any executives appear to be overburdened with responsibilities?

13. Is there any overlapping of functions?

14. Does any unnecessary duplication of effort exist?

15. Is there a separation of control functions from the activities controlled?

WORK-DISTRIBUTION CHART

The work-distribution chart, also called a job-correlation chart or activity-analysis chart, is basically a spread sheet of summarized statements about the activities of an organization. Figure 4-5 illustrates the general layout of such a chart. The first left-hand column lists, roughly in decreasing order of importance, the major activities for which the unit is responsible. Each column to the right allocates the tasks as shown in the first column under the name of the employee performing that particular function. The amount of time spent on each task is recorded, as is the employee column totals.

The basic composition of a work-distribution chart shows work assignments in an order that forcibly brings attention to any inappropriate distribution of work. A properly constructed chart will show what share of work is performed by each employee with the time expended and relates this to the tasks and time spent by the other members of the group. This points out any misapportioned work, deficient utilization of skills, insufficient blending of tasks, and unnecessary duplication of work.

Analytical questions to ask in connection with the chart are presented later in the chapter.

Collecting Information on Tasks Performed by Employees

The mechanical procedures for gathering information on tasks performed by employees are greatly facilitated by the use of (*a*) the *time-task list* and (*b*) the *task summary list*.

WORK DISTRIBUTION CHART

DATE: ________________

NAME OF ORGANIZATION Purchasing Department			EMPLOYEE T. Thompson		EMPLOYEE M. Pine	
PREPARED BY T. Timberline			JOB TITLE Director of Purchasing		JOB TITLE Buyer	
APPROVED BY			CLASSIFICATION None		CLASSIFICATION GB-1	
NO.	ACTIVITY (KEY DESCRIPTION)	HOURS PER WEEK	TASKS	HOURS PER WEEK	TASKS	HOURS PER WEEK
1	Processing of purchase order requisitions	9.1	Authorize processing of requisitions for procurement of materials, equipment, and services Route requisition to buyer	6.5		
2	Making inquiries	42.9			Check sources of supply Telephone calls Prepare ''Request for Quotation'' form Interview salesmen	2.4 3.0 3.6 12.2
3	Procure items and services	63.5			Analyze quotations Select supplier Make draft of purchase order Obtain approval Place order Follow up	17.8
4	Handling correspondence	41.7	Prepare and dictate correspondence on special matters	12.4		
5	Maintain records	6.5				
6	Administration	15.0	General supervision: maintaining personnel records, answering questions, assisting on special projects Participate in the formulation of company procurement policies Seek new or alternative sources of supply to reduce cost, improve quality of end product, etc. Read and sign purchase orders Prepare reports Attend conferences	15.0		
7	Miscellaneous	15.7	Travel View vendor plants Attend national association meetings Study market conditions	4.3		
8	Allowance for nonmeasured work	5.6		1.8		1.0
	Hourly totals:	200.0		40.0		40.0

Figure 4-5. Work distribution chart.

ARRANGEMENT:

☐ EXISTING ☐ REVISED

EMPLOYEE C. Buttonwood		EMPLOYEE F. Cedar		EMPLOYEE S. Burns	
JOB TITLE Assistant Buyer		JOB TITLE Stenographer		JOB TITLE Clerk-Typist	
CLASSIFICATION AB-2		CLASSIFICATION PS-1		CLASSIFICATION CT-2	
TASKS	HOURS PER WEEK	TASKS	HOURS PER WEEK	TASKS	HOURS PER WEEK
				Time-stamp forms and correspondence	2.6
Assist buyer	13.3			Type forms and correspondence Distribute copies of ''Request for Quotation'' form Maintain purchase order requisition register	8.4
Proofread purchase orders Check prices, extensions, and grand totals Follow up	25.9			Type forms and correspondence Distribute copies of purchase order	17.6 2.2
		Take dictation Transcribe	8.5 20.8		
		Compare receiving report with purchase order Post receipts	1.2 2.1	Post data on sources of supply: quantities, prices, quality	3.2
		Filing Clean typewriter Get supplies Stamp mail	2.2 0.9 1.7 1.5	Filing Expediting: errands Clean typewriter	3.5 0.6 1.0
	0.8		1.1		0.9
	40.0		40.0		40.0

Time-Task Lists. A time-task list records each task performed by an employee in the course of one work day, with the amount of time devoted to each of these tasks. (See Figure 4-6.) If the period covered by a work-

TIME-TASK LIST			
NAME OF EMPLOYEE	JOB TITLE	CLASSIFICATION GRADE	REFERENCE
NAME OF ORGANIZATION	IMMEDIATE SUPERVISOR	DATE RECORDED	APPROVED BY

TIME		ACTIVITY LEGEND NO.	DESCRIPTION OF TASK	COMMENT
FROM	TO			
8:30	8:45	7	File correspondence	
8:45	9:30	4	Take dictation	
9:30	9:45	7	Get supplies	
9:45	10:00	4	Transcribe	
10:00	12:00	7	File Correspondence	
12:00	1:00		Lunch	
1:00	1:45	4	Take dictation	
1:45	2:20	5	Post receipts	

Figure 4-6. Time-task list.

distribution study is one week, five time-task lists per employee will be prepared.

Information on tasks performed and the time spent on each is usually obtained from the supervisor or the employees themselves. Another method is to observe each employee and record the information on his activities. Although this approach may be commended on the grounds of its potential objectiveness, it is relatively expensive and time consuming.

Two good reasons for having employees fill out their own time-task lists are the ease in gathering the information and the accuracy of the information (based on the assumption that only the employee performing the work knows all the tasks involved). Conversely, the strongest point against having employees complete their own time-task lists is that the information may not be wholly dependable. In practice, it is possible to use certain techniques to reduce this possibility, such as: (*a*) issuing complete instructions to employees on the proper preparation of time-task lists, (*b*) requiring each supervisor to check information submitted by his employees, and (*c*) comparing time-task-list information with standards of performance for each task.

Sampling and time study methods may be necessary to ascertain the reliability of time-task-list information, especially when very accurate time values are required. The usual approach involves the periodic observation and timing of selected tasks performed by each employee, and comparing results with the less certain information that has been submitted.

Task Summary Lists. Figure 4-7 shows a task summary list. This form is used to summarize the information recorded on the task lists of each

TASK SUMMARY LIST

Florence Cedar — NAME OF EMPLOYEE

9/7 — PERIOD COVERED

SUMMARIZED BY

REFERENCE

LEGEND NO.	DESCRIPTION OF TASK	HOURS
4	Take dictation	8.5
4	Transcribe	20.8
5	Compare receiving report with purchase order	1.2
5	Post receipts	2.1
7	Filing	2.2
7	Clean typewriter	0.9
7	Get supplies	1.7
7	Stamp mail	1.5
	Total	38.9

Figure 4-7. Task summary list.

employee for the period under study. The information is then transferred to the work-distribution chart.

Analyzing the Completed Work-Distribution Chart

The analysis of a work-distribution chart breaks down into three elements:

1. *The analysis of the activities of the unit.* Each activity listed in the first column is examined with the basic objectives of the organizational unit in mind.

2. *The analysis of the tasks that contribute to the fulfillment of each activity.* This information cuts across the chart, reading from left to right and right to left, and is sometimes termed the horizontal analysis. The as-

sociated tasks relating to the performance of each activity are studied jointly and severally.

3. *The analysis of employee duties.* This element is called the vertical analysis because the information reads up and down the chart. The tasks listed in each vertical column are also examined jointly and severally.

Check List for Analyzing the Work-Distribution Chart

Check lists are simple but effective aids to analytical and creative effort. A three-part check list that the systems analyst finds useful in the analysis and interpretation of a work-distribution chart is as follows:

Part I—Analysis of the Activities of the Unit

1. What are the objectives of the unit? Do the activities of the unit coincide with them?
2. Does each activity make a necessary or desirable contribution to the fulfillment of the objectives?
3. Do the activities constitute a logical arrangement?
4. Are all the activities performed best by the unit being studied? Would it improve matters if certain activities were assigned to another unit?
5. Does any other unit in the company perform one or more of essentially the same activities? If so, is the duplication necessary or desirable? Can it be discontinued without adversely affecting operations?
6. Would the assignment of additional activities permit the unit to attain its objectives more efficiently?
7. Is the total time apportioned in accordance with the relative importance of each activity? If not, how should the time be distributed?
8. Do any activities appear to be taking more time than necessary?
9. Are there any activities that indicate the need for additional study (e.g., cost study, motion study, or forms analysis)?

Part II—The Horizontal Analysis

1. In what manner is the activity separated into tasks? Is the arrangement logical?
2. Is there any unwarranted duplication of tasks in accomplishing the activity?
3. What tasks associated with the activity take the most time? Is this justified?
4. Are the tasks spread too thinly throughout the unit? Is performance impeded because too many persons are involved in performing the tasks? Should certain assignments be consolidated so that they can be performed by one person?

5. Are the tasks associated with the activity being performed by too few people? Is one person burdened with work whereas another does not have enough? Should the work load be distributed more evenly?

6. Is there an indication that some employees are working under undue pressure because they handle all of the important tasks whereas other employees work solely on less urgent matters?

Part III—The Vertical Analysis

1. Do the tasks assigned to the employee form an appropriate combination of duties: Do the tasks properly belong together? Are they in harmony with one another?

2. Does the employee perform any unrelated tasks? If so, is this a desirable situation?

3. Does the work load assigned to the employee constitute a monotonous combination of tasks? Is it of such a nature that it might be responsible for excessive fatigue and reduced efficiency on the part of the employee?

4. Do the requirements for accomplishing the tasks utilize the special skills possessed by the employee? Are these skills being used to best advantage?

5. What is the ratio of time spent by the employee on tasks requiring the use of special skills to time spent on other tasks? Does the ratio indicate a division of effort that satisfactorily employs special skills?

The work-distribution chart has its value in evaluating the work assignments within an organizational unit. It is, however, solely a visual aid. The chart will serve to point out situations, but it cannot be expected to interpret them.

Many facts are needed to render a true interpretation of a work-distribution chart. There may be valid reasons for continuing a particular working arrangement although it is reflected as being undesirable on the chart; for example, a work-distribution chart may show that several persons with different job titles are performing the same tasks. On the surface it appears that there is an unwarranted distribution of the work load or that skills are not being properly utilized. However, such an allocation of tasks may be advisable to diversify monotonous work or provide additional skills to employees.

In conclusion, a work-distribution chart does not replace the need for background knowledge and good judgment. The information on the chart must be well understood before it is safe to make unfavorable conclusions from apparent evidence of improper distribution of work. Best results are obtained if the chart is studied jointly by the supervisor of the unit and the systems analyst.

APPENDIX

SET OF ORGANIZATIONAL DOCUMENTS FOR CONDUCTING A MANAGEMENT AUDIT

Representative Corporation
Quite Large Division
Anytown, U.S.A.

To: Date: July 24, 19XX

From: Systems-and-Procedures Department

Subject: Survey of Organization

Attached are a set of work sheets for use in preparing a detailed statement of your organization's objectives, responsibilities, and functions.

The purpose of gathering this information is (1) to determine the areas of authority and responsibility associated with the Production Department organizations so as to provide information for evaluation, and (2) to establish facts concerning the flow and "feedback" of information for the purpose of updating written procedures.

Before completing the attached sheets, review them carefully; when filling them out, please keep the following instructions in mind:

Objectives: State the objectives in terms of the contribution they make to the Production Department's basic purpose.

Responsibilities: Set forth the extent to which you are held accountable for the performance of work. Include areas where joint responsibility for performance is shared with other groups. If responsibilities in specific areas of work are not clear, please state.

In making suggestions—on page 2—please state your views regarding the relationship between your responsibilities and your authority. Are they compatible? Do they allow you to accomplish your objectives effectively?

You are requested to submit the completed work sheets to the Systems-and-Procedures Department by August 15, 19XX. Your cooperation in complying with this due date will be appreciated.

P. D. Quick, Manager

Responsibility and Function Analysis

Strictly Private

Page 1 of 3

Organization Title: ______________________________

Organization Head's Name: ______________________________

Prepared by: ______________ Date: ______________

Briefly state the overall objective(s) of your organization's effort:

____

____

____

____

____

____

____

____

____

____

____

____

Concisely state the limits of your responsibilities to accomplish the stated objective(s):

____

____

____

____

____

____

____

____

____

____

____

____

____

____

Responsibility and Function Analysis

Strictly Private

Do members of your group receive work assignments from any other person besides you? If so, from whom? Regarding what?

Please state any comments or suggestions (with substantiation) that you feel would help you in the accomplishment of your group's overall objective(s) (functional realignments, authority, facilities, services, etc.):

Signed: ______________________

Organization Head

5

Placement and Organization of the Systems-and-Procedures Department

In the development of a plan for the administration of systems and procedures, it is best to keep the organization as simple as possible and to have different parts grow out of the actual needs of the business. A company may begin by having one person devote his entire time to systems-and-procedures work. He is often designated as a systems specialist and made directly responsible to the head of the firm. As the work load increases, a small-size staff agency develops. In time the company's growth may warrant a sizable agency at the corporate level and one or more at the departmental level.

The systems-and-procedures department must be designed to fit (*a*) the distinctive characteristics of the business (work processes, staff and service functions, geographical location of buildings and facilities, etc.) and (*b*) the size, policies, and organization of the company. Management must determine the extent to which systems-and-procedures functions should be centralized.

THE PLACE OF THE SYSTEMS FUNCTION IN THE ORGANIZATION

Organizational placement of systems-and-procedures agencies within company structures generally follow one of three types of arrangement. These are discussed below.

Placement at the Corporate Level. Under this plan the systems-and-procedures work is performed by a central agency whose manager reports directly to a member of top management. Figure 5-1 shows three arrangements of centralizing the systems-and-procedures function with reference to the entire enterprise.

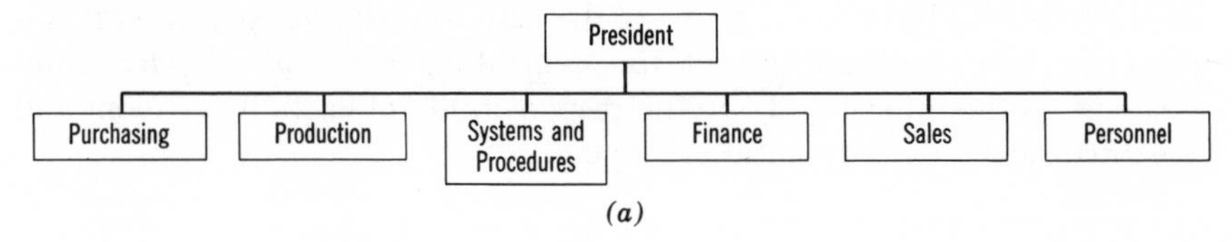

(a)

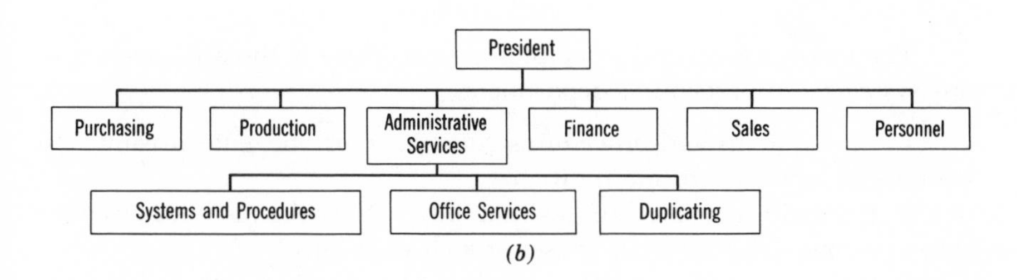

(b)

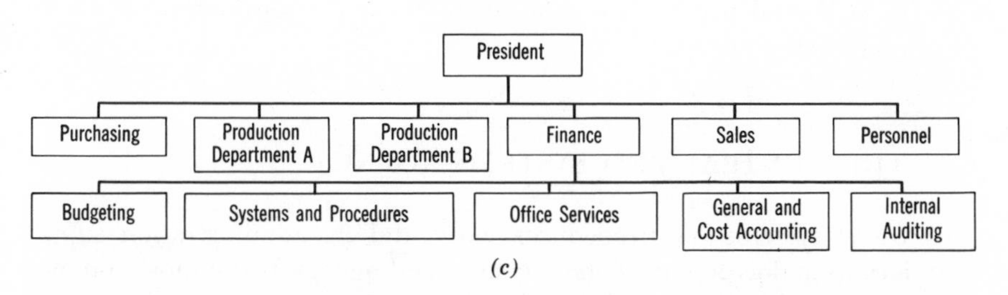

(c)

Figure 5-1. Three arrangements showing placement of the systems and procedures organization at the corporate level. In each case the systems and procedures agency services the whole enterprise.

Placement at the Departmental Level. In this case systems-and-procedures work is distributed among the various departments, such as the functional divisions of accounting, engineering adminstration, manufacturing, and marketing. Each unit may have its own systems staff responsible to the particular departmental supervisor. These staffs range in size from one-person units to organizations of considerable size.

A Combination of Corporate and Departmental Types. Here certain systems-and-procedures work is centralized under the direction and control of one systems-and-procedures manager, whereas the remaining work is decentralized and performed by the systems units of each major department. The question as to "What systems should be centralized?" is answerable on the basis of the scope of the activities: *Interdepartmental* systems are the responsibility of the central agency. *Intradepartmental* systems, on the other hand, are the responsibility of each department and are intrinsically departmental in scope.

ORGANIZATIONAL PRINCIPLES AND PRACTICES FOR SYSTEMS WORK

The following general principles are necessary in the organization of any systems-and-procedures department:

1. The systems-and-procedures function must be enterprising and continuous, not static or intermittent.
2. It should be geared to deal not only with problem situations after they arise but also with basic causes of such situations.
3. It should provide for the initiation of systems-and-procedures improvements.
4. It should be simple, effective, and economical.
5. It should provide for administration and coordination of all the work within the group.

THE CENTRALIZED SYSTEMS AGENCY

Since systems-and-procedures work and its administration supply services to major departments, a centralized agency is the most common type. Among the responsibilities of such agencies are the following:

1. Control over the data-processing programs of departmental systems staffs so that they are in line with company objectives.
2. Implementation of the "total systems concept;" in its true application, this term implies the master planning and development of integrated systems within the whole structure.
3. Maintenance of a minimum number of information systems and data-processing facilities to serve a maximum number of purposes.
4. Maximizing the value of information systems.

5. Promotion of data systems that transcend organizational lines and complement or encompass existing data systems.

6. Achievement and development of overall effectiveness by less duplication of effort, less equipment, and greater employee specialization.

7. Standardization of equipment, codes, formats, and various requirements for compatibility, without inhibiting development by premature standardization.

Factors that Influence the Design of the Centralized Agency

The following factors are essential to the design of all centralized systems agencies:

1. There is a need for master planning and development of integrated systems within the enterprise.
2. The systems organization must be formally structured and one person must be responsible for the administration of systems and procedures.
3. The systems manager must act in an advisory capacity to top management.
4. The department must be held responsible for systems that cut across departmental lines.
5. The systems function must be sufficiently independent of pressure from particular groups in the enterprise whose interest may conflict with the interest of the enterprise as a whole.
6. Emphasis must be placed on the design and control of systems rather than data-processing equipment.

Ordinarily, in a centralized organization the director of systems is responsible to one of the following executives: (*a*) president, (*b*) vice-president, and (*c*) comptroller.

The systems manager still often reports to the comptroller, but in recent years there has been a tendency to make the systems-and-procedures department directly responsible to the company president or a vice-president of administrative services.

ORGANIZATION OF THE SYSTEMS-AND-PROCEDURES DEPARTMENT

The organization of a systems-and-procedures staff depends on the scope of systems activities, the nature and complexity of assignments, the volume of the work load, and the background and capabilities of the indi-

vidual members. In small units, where the opportunity for work division is limited, there is apt to be little specialization among the members. Where this is the case, the job specification requirements are mostly for general practitioners—people with diversified knowledge and skills.
skills.

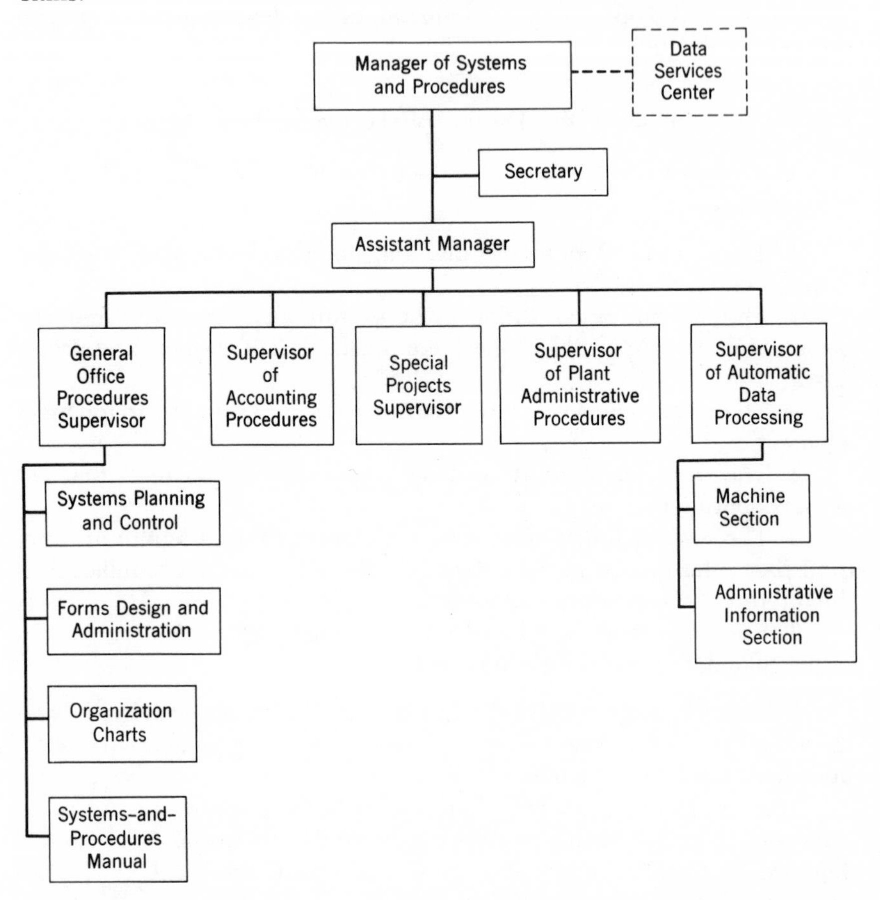

Figure 5-2. A representative organization chart of a systems department in a medium-sized industrial firm.

Specialization, as in other fields, requires large volumes of similar work or a cluster of closely related jobs. In large systems-and-procedures units the degree of specialization can be extensive.

Work divisions for systems-and-procedures organizational structures are generally made on the basis of functions. A list of such functions is as follows:

1. Planning and directing the installation and operation of data-processing systems for accounting, engineering administration, manufacturing, marketing, and quality control.
2. Preparation and maintenance of policy and procedures manuals.
3. Forms design and control.
4. Assisting in organization development and improvements by management audits.
5. Space planning—layout.
6. Automatic data processing (ADP). This includes data processing by conventional equipment (bookkeeping machines, duplicating and copying machines, cash registers, etc.); data processing by electric accounting machines (EAM)—tabulating equipment; integrated data processing (IDP)—common-language machines; and electronic data processing (EDP)—digital computer applications.

Figure 5-2 illustrates a typical organizational breakdown for a systems-and-procedures department. A different arrangement is shown in Figure 5-3. An explanation of the indicated organizational units follows the figure.

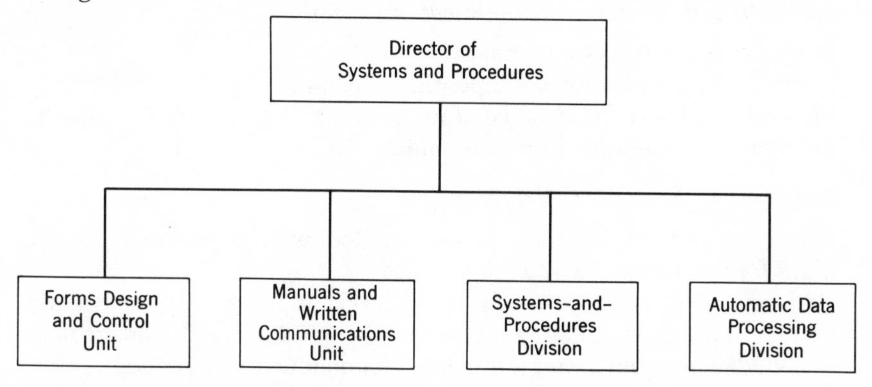

Figure 5-3. Organization chart of a systems and procedures agency. An explanation of the functions, duties, and responsibilities of the indicated organizational units is contained on the following sheets.

SYSTEMS-AND-PROCEDURES AGENCY

The function of the systems-and-procedures division is to provide management with pertinent information and control services for obtaining maximum effectiveness of the enterprise. The agency has the responsibility to conduct studies of systems and procedures, policies, organizations, functions, and performance. It is charged with the work-simplification program, the cost-consciousness program, and the forms-management program. It also must adapt appropriate procedural operations to automatic-data-processing machines and equipment.

Functions

Office of the Director

Sponsors improvements in the company's information-processing systems. Plans, directs, coordinates, and evaluates the overall activities of the systems-and-procedures division.
Exercises staff and technical supervision over the personnel of the systems-and-procedures division.

Forms Design and Control Unit

Provides forms-design services for staff and operating personnel.
Oversees the company's forms-management program.

Manuals and Written Communications Unit

Publishes documents for manuals.
Controls the distribution and updating of manuals.
Provides advice and assistance, if necessary, in the preparation of written reports, procedures, and communications.

Systems-and-Procedures Division

Develops and implements, in conjunction with department heads, plans for the improvement of operations.
Performs continuing research in scientific management, constantly seeking through imaginative ideas, creative thinking, and management knowledge new and better approaches for conducting activities.
Division responsibility includes the following:

Conducting management audits.
Developing and monitoring management reporting requirements.
Systems analysis and development.
Analysis and recommendations for the improvement of organizational structures.
Assisting in planning layouts.
Examination and improvement of reports, forms, and systems records.

Automatic-Data-Processing Department

Designs, installs, operates and maintains automatic-data-processing systems (EAM, IDP, and EDP installations).

Conducts feasibility and practicability studies.
Accomplishes machine programming of approved data-processing systems.
Operates automatic-data processing equipment.

6

Layout of Physical Facilities

The last four chapters have dealt with the organization of manpower. The following discussion deals with the organization or layout of physical facilities. The layout of a building area, in whole or in part, has as its objective the optimum utilization of the space with regard to the arrangement of its elements, such as machines, furniture and fixtures, equipment, and aisles.

Layouts are a proper concern of the systems analyst. In order to have effective systems and procedures, the arrangement of the physical facilities must make them possible. A poor layout will entangle operations, obstruct the flow of work, and cause many inefficiencies. The interdependence between layouts and procedural activities necessitates the consideration of both. The successful planning of a new or revised layout requires careful attention to the effect it will have on systems and procedures. On the other hand, a new or revised system must be in accord with the actual or planned layout in which it will function.

Layouts are of two principal types: office and plant. Our interest will center around the office layout, as this is the hub of most systems-and-procedures work. This does not mean that the systems analyst is without concern for the layout of manufacturing areas. Factory systems and procedures are strongly influenced by plant layout. In this connection, however, the establishment of suitable layouts for manufacturing areas usually requires the combined talents of systems analysts, production engineers, methods men, and other management personnel. Where conditions warrant it the professional services of industrial architects should also be engaged.

Two situations require a layout plan: First, when the initial arrangement of physical facilities must be made, and, second, when an existing layout must be changed (e.g., to accommodate new equipment or to bring about improvements). Despite certain problems that are peculiar to each, the same underlying aims and principles of planning apply in both cases.

AIMS TO BE ACHIEVED IN PLANNING LAYOUT

The general aims that are necessary in any successful layout are the following:

1. Optimum utilization of space;
2. Proper work flow;
3. Conditions conducive to good supervision;
4. Appropriate flexibility;
5. Sufficient open space;
6. Personnel safety and health;
7. Personnel comfort and conveniences; and
8. Pleasing appearance.

Optimum Utilization of Space

The optimum utilization of space is a fundamental consideration. Space is a major cost of operations. It requires expenditures for rent, taxes, heat, light, building maintenance, etc. All of these are allocable to each square foot of floor surface, whether utilized for office or plant facilities.

There is merit in thinking of space in terms of its cost per square foot; it places the entire subject of layout in its true utilitarian setting. In spite of the fact that space is expensive, many companies are wasteful of it. Making wasted space productive is a potential source of substantial savings, which can often be realized with moderate effort and reasonable expenditure. When additional space is necessary it is sometimes possible to reclaim enough unused or misused areas so that the entire expense of expansion or acquisition is saved.

Proper Work Flow

Good layouts are constructed in view of the effect they will have on work processes. Physical facilities should be arranged to provide for a smooth flow of work through the operational stages. Moreover, it is usually desirable to have the work move in as straightforward a direction as possible. The work should progress through successive work stations with a minimum amount of crisscrossing, backtracking, and handling.

The layout of an area exerts a considerable influence on schedules. Delays and interruptions are often attributable to poor layouts with inadequate space, with the result that equipment is hemmed in, working quarters are cramped, and passageways are obstructed. Conversely, abundant space can be a source of difficulties also. This condition may cause excessively long lines of communications that require needless travel on the part of employees, unnecessary and expensive material handling, and much lost time in trying to meet schedule requirements.

When an area is suitably laid out, work flow is facilitated in at least two ways. First, it provides for straight-line work flow. Work proceeds directly from one station to the next with a minimum of travel time, handling, and expense. Second, it also provides for the utilization of mechanical devices such as belt and chain conveyors, overhead monorail carriers, and power trucks.

Conditions Conducive to Good Supervision

Layout arrangements must take into consideration the need for supervisors to oversee and direct the activities of the persons under their supervision. In situations that require rigid surveillance it is necessary to make entire working areas clearly exposed to the view of the supervisor.

Supervision requires giving instructions and directions. Layouts should be conducive to these essential activities; for example, supervisors of professional employees should have offices located so that discussions can be conducted with convenience and dispatch.

Appropriate Flexibility

Layouts should provide for possible expansion needs in order to avoid unnecessary and costly modifications at some future date. Here are six specific ways to provide for flexibility in layouts:

1. Situate departments that are likely to expand next to areas that can be freed for that purpose with comparative ease. It may be planned so that a "growth" department will eventually spread out to encompass adjacent areas that have been allocated for storage, conference rooms—or small departments that may be relocated easily.

2. Place permanent installations such as heavy machinery and equipment in locations where they are least likely to interfere with any subsequent changes in layout.

3. Gear the layout to peak work-load requirements based on operating standards that reflect past performance, current needs, and anticipated trends. Plan ahead! Layouts established with merely the existing work load in mind may become outmoded in a short time.

4. Build as much "openness" into layouts as economically feasible, avoiding unnecessary divisions and encumbrances.

5. Use partitions that can be conveniently moved. In situations where they will be used and reused rather frequently, movable partitions afford great flexibility and deserve serious consideration. However, because their initial cost is high, an organization may find it more economical to use fixed partitions that can be rebuilt as required.

6. Provide wiring and plumbing fixtures that will accommodate a variety of changes in layout. It is much simpler to put in additional fixtures for future needs when the layout is being set up initially than to have to do this work at a later date.

Sufficient Open Space

A well-designed layout must have sufficient aisle and circulation space. Employees should have convenient routes to and from their places of work. Provision should be made for a network of passageways that provide means for both employees and material to reach the work stations readily and without hampering other activities.

In most cases it is advisable to have a combination of main aisles and subaisles. Main aisles are meant to carry medium and heavy traffic within the working area, whereas subaisles are for light and medium traffic.

In some locations it is absolutely essential that ready access be provided for material-handling equipment such as power trucks. This holds true for certain office locations as well as places on the production floor. Keep in mind that power trucks furnish an excellent means for moving office supplies from the receiving platform to a central storage point within the general office area.

Where material-handling vehicles, manual or powered will be used, the aisles and intersections must be sufficiently wide to accommodate them.

Personnel Safety and Health

Needs for the safety and health of employees must be considered. First, it is necessary to abide by local ordinances concerning fire prevention, sanitation, and so on. Most of these laws are sensible and work for the benefit of all concerned.

Safety and health considerations ordinarily extend beyond legal requirements and justly embrace humanitarian and economical considerations. Every company that contends to have the welfare of its employees at heart insists on surpassing minimal safety and health requirements. Moreover, suitable provisioning for safety and health increases productivity. Time and again it has been proven that employees do their best work in an environment that is orderly, clean, safe, and healthy. Cluttered aisles, lack of proper safeguards around moving parts of machinery and equipment, and crowded working conditions are typical causes of accidents that can be prevented through good layout planning.

Personnel Comfort and Conveniences

Closely allied to the aims of personnel safety and health are those of comfort and conveniences. High on the list of desirable features is proper air conditioning: heating, cooling, and ventilation. By maintaining a comfortable temperature and atmosphere, it is possible to enhance productivity substantially. Moreover, marginal or unsuitable floor space can sometimes be reclaimed through air conditioning.

Provision should also be made for facilities such as pleasant rest rooms, drinking fountains, water coolers, telephones, and even vending machines. These should be conveniently located. An employee should not have to walk a great distance to get a drink of water, wash his hands, and possibly obtain refreshments like soda and candy. Of course, whether or not items such as soda and candy will be made available is a matter of company policy; but if they are offered, they should be conveniently located.

Pleasing Appearance

Attractiveness in a layout is always desirable. This is particularly true if offices and rooms are frequently visited by outsiders such as customers and vendor representatives.

Also, appearance can make a significant difference in prestige, morale, and output.

BASIC PRINCIPLES FOR THE PREPARATION OF LAYOUTS

The following basic principles apply to the preparation of most layouts:

1. Design the layout around the dominant flows of work. The subordinate and auxiliary functions should be made tributaries to the main stream of activities.
2. Locate organizational units with similar or related functions in convenient proximity. Departments whose work is closely interwoven or

requires much personal coordination should be placed near each other.

3. Arrange the work stations so that the work flows from point to point in as direct a course as possible.

4. Place machines and equipment where they will be conveniently accessible to the persons who will be using them.

5. Map out space predominantly in rectangular patterns, because areas of irregular shape are not as effective. However, avoid an assembly-line pattern by using, for example, a number of non-right-angle arrangements where appropriate.

6. Provide work stations of sufficient size to enable the employees to conduct their work in comfort.

7. Situate units with considerable public contact near entrances. This provides convenience and allows business to be conducted without hampering the work of other sections.

8. Partition off unpleasant, confidential, and untidy operations from general view.

9. Set apart or insulate units that perform noisy and distracting functions.

10. Make aisles sufficiently wide so that there is plenty of room for the amount of traffic that they will bear. There should be clear passageways from working areas to facilities such as water coolers, lavatories, and vending machines.

11. Build safety into the layout. Provide protection against injury from exposed or loose wiring, hot pipes, and moving parts of machinery and equipment. Be certain to furnish ready and free approaches to exits and fire escapes.

12. Use furniture and equipment of a uniform type and size. This renders a neat appearance and permits flexibility.

13. Locate water coolers, lavatories, and other facilities so as to minimize the time away from work.

14. Provide work settings that are appropriately attractive. Use colors, illumination, and an overall decor that radiates cheerfulness.

MATTERS OF PARTICULAR CONSEQUENCE

Private Offices

Individuals may be assigned private offices for one or more of the following reasons: (*a*) to manifest prestige, (*b*) to conduct confidential work in seclusion, and (*c*) to provide a quiet atmosphere for work that requires a great deal of concentration. All three of these reasons are generally considered in designing private offices for top-level executives, staff specialists, department heads, and professional personnel. Private offices should make it possible for such persons to hold informal conferences without disrupting the work of adjacent personnel.

Positions of prestige often dictate the need for offices that are not only private but spacious and decorative as well. Society in general, and the business community in particular, expects that persons holding high managerial posts have offices appropriate to their positions.

Except where the need for private offices is clearly signified by official standing or convention, the determination of who should have them may be somewhat problematical. Conceivably, there may be greater justification for subordinates to have private offices than for those of higher rank; for example, a staff assistant to a department supervisor may need privacy for performing confidential work such as research, planning, or preparing certain reports; whereas the department head may be better situated in the outer office area within low railings where he can be near the people he supervises. Low railings give an impression of privacy. Such arrangements are particularly useful where first-line supervision is involved.

Company policy usually governs the extent to which employees at different levels are provided with private offices. In formulating this policy the obvious advantages of private offices should not obscure the drawbacks connected with their use. The more important of these are: Private offices use more floor space; they are a deterrent to flexibility, as it is difficult and expensive to effect future changes; they increase the cost of installing and operating facilities for heating, lighting, and air conditioning; and they impede the functions of communication and supervision. Perhaps the biggest drawback is the encumbrances they place on the flow of work.

As significant as these drawbacks may be, the advantages of private offices can outweigh them. In a private office the creative output and the productivity of a professional or technical person will be much greater than under an adverse environment. Thus a private office can be well justified economically.

It may be desirable to limit the number of private offices and provide conference rooms and/or partial enclosures as a substitute. Conference rooms can be used as private offices for functions of a sporadic kind such as interviewing and making confidential plans. However, they do have the inherent disadvantage of not always being readily available when needed.

Some benefits of the private office can be obtained by the use of partial enclosures; for example, railings and short partitions will furnish a degree of privacy and prestige. One widely used short partitioned enclosure stands from 4 to 6 feet high; it has an upper encasement of either translucent glass for privacy or transparent glass to allow for supervision.

Semiprivate Offices

Two- to four-person offices have become a popular form of accommodation for middle management and professional personnel. This popularity is attributable to the intermediate choice they afford between the private office and the open area. Semiprivate offices can provide advantages of private offices without many of the drawbacks that normally accompany the wide use of separate rooms for individuals.

Prospective candidates for semiprivate offices include groups such as accountants, engineers, and systems analysts. It is a good idea to establish and maintain such groupings along homogeneous lines, so that they will be able to assist one another at close range.

A drawback of semiprivate offices is that they can produce inefficiencies. Third parties stopping in to "visit" interrupt the work of all occupants instead of one. The resulting discussions may be long and drawn out in the privacy of the semiprivate room.

Partitions

Many different kinds of partitions are used in setting up office areas. The separations may be full length, reaching from the floor to the ceiling, or they may extend only part of the way.

In recent years movable partitions have become quite popular because they can be easily erected, dismantled, and relocated. These partitions are prefabricated and are available with a wide variety of features. Some come equipped with wiring and outlets contained in the baseboard. One full-length type is completely sound conditioned, providing a sound insulation of about 30 decibels. (Soft radio music measures in the vicinity of 35 decibels.)

The partitions may be made of metal, wood, glass, plastic, or a combination of these; they may be solid, broken, or perforated; high or low; transparent, translucent, or opaque. Moreover, they may be in the form of glazed panels, railings, short dividers, screens, accordion folds—or part of a modular arrangement. The types and styles from which a selection may be made are great indeed.

Not to be overlooked is the possibility of forming partitions by *natural* means. In some situations units of office furniture—for example, filing cabinets and shelf racks—may be positioned to separate areas.

Modular Office Units

Modular units have become a prevalent form of office arrangement. Typically, these units are L-shaped and combine a desk on one side with an extended working surface at a right angle, covering drawers or cabinets; plus movable, short partitions. They are designed to be efficient, flexible, and somewhat private.

Every single unit can be easily and economically supplied with all the services that characterize a modern office, including air conditioning, good lighting, and communication facilities. These services may be effected by such construction techniques as ducts, channels, and tubes.

Table 6-1. RECOMMENDED ALLOTMENTS OF SPACE

Function	Area per Unit (square ft)
Top executives	400
Middle management personnel	300
Supervisors, first line	200
Clerical workers	50 to 125
Passageways:	
"Feeder" corridors	5 to 8 *
Main aisles	4 to 5 *
Secondary aisles	3 to 4 *
Footways	1.5 to 2 *
Rooms:	
Conference rooms	20 per individual; to accommodate up to 25 persons
Reception rooms	500 to accommodate 10 persons
Interviewing rooms	200
Cloak rooms	1.5 per individual
Private offices	125, minimum
"Cubicle" offices	100
Active records storage	5 per file
Inactive records storage	3.5 per file

* Length in feet.

Table 6-2. ESTIMATE OF WORK AREA REQUIREMENTS

Description	Number	Unit Allowance (sq. ft.)	Nature of Space (square feet)			
			Private	Semi-private	Open	Other
Division manager	1	400	400			
Department heads	3	300	900			
Supervisors	5	200	600	400		
Clerks, I	48	100			4800	
Clerks, II	37	100			3700	
Letter files	22	5				110
Stock room	1	100				100
Rest rooms	2	250				500
Conference room	1	500				500
Equipment	4	10				40
Utility closets	7	35				245
Miscellaneous	1	200				200
Totals	—	—	1900	400	8500	1695

94 persons) 12,495

132.9 square feet per person

Space Requirements

One of the most important questions that must be answered in preparing a layout is just how much space to allot the various elements of the arrangement. There are no stringent rules because the allotment of space for the different elements of layouts is subject to considerable variation. In each case the individual amounts of space will be influenced by such considerations as the size and shape of the relevant area in its entirety; location of stable utilities such as structural columns, doors, windows, and elevators; size and configuration of the furniture and equipment to be used; and the nature of the work to be undertaken. Recommended allotments of space for most applications are given in Table 6-1.

A complete layout must possess additional elements such as stock rooms, utility closets, lavatories, switchboard rooms, mail rooms, cafeterias, and auditoriums. Estimates for total space requirements are given in Table 6-2.

CONSIDERATIONS PRELIMINARY TO THE PREPARATION OF A LAYOUT

Experience has shown that those succeed best who have definite ideas about what they are going to do before they start. So it is with layouts. A good layout is not the product of chance or casual preparation; it is rather the result of serious study and thoughtful attention to the needs of the situation.

Typical of the points to be covered in connection with the approach in achieving effective layouts are the following:

1. Study the existing arrangement.
2. Survey the work involved.
3. Determine the requirements for future expansion or contraction.
4. Acquire a knowledge of dominant work flows.
5. Consider structural traffic patterns.
6. Study what the employees need in their work.
7. Determine requirements for furniture and equipment.
8. Visit places with layouts of the same type.
9. Obtain recommendations of furniture and equipment vendors.
10. Determine total space requirements.
11. Provide for allocation of space.

Study of the Existing Arrangement. When planning to revise an existing layout it is wise to make a thorough study of the existing arrangement. An examination of the existing layout will reveal its strengths and weaknesses: problem spots, opportunities for improvement, things to be avoided, and good ideas that should be carried over into the new arrangement. In addition, the study will provide basic information about the type, designs, and properties of the furniture and equipment.

In many companies layout drawings or blueprints are available, and where this is the case they can be very serviceable. Care should be exercised, however, to insure that the information shown on all such drawings is accurate and complete.

Reference to an available layout drawing will provide a graphic view of the present arrangement and its facilities. This will usually be of much service in planning an improved layout. From the drawing background information may be gleaned about a number of important factors, such as the furniture and equipment in use, space requirements, clearances that are necessary, and templates that will be needed.

A drawing of the present layout serves another very useful purpose. It provides a measure against which to evaluate proposed changes.

Survey of the Work Involved. In order to prepare an effective layout an examination of the work itself is necessary. This involves becoming familiar with the objectives of the organization, the work processes, and the facilities utilized. An advisable approach is to talk with appropriate department heads and unit supervisors about the work of their respective organizations. These discussions ordinarily cover the work of the individual unit as an entity and also include an account of the separate responsibilities of each class of employee. As suits the occasion, talks with management personnel should be supplemented by such discussions with individual workers as may be necessary to find out their exact functions.

As has been suggested above, the purpose of any layout is to carry out activities or services. Before a layout can be improved it is necessary to know how the work is accomplished, because it is absolutely essential that the layout be based on the work processes involved. Though it is not possible to state precisely what should be done in each situation, the survey should ascertain whether the operations are essential, whether they are being done in an efficient sequence without undue handling and movement, and whether or not they could be improved by the addition or substitution of equipment. Where indicated, detailed studies should be made to acquire more information about particular elements of the work.

Requirements for Future Expansion or Contraction. To predict realistically space requirements for the future is not an easy task. In some cases there is little background on which to base a sound judgment. Nevertheless, wherever it is practical factors such as fluctuations in the total volume of work, trends in economic and political activities, and likely technological changes should be considered. Provision should thus be made in the layout for accommodating future expansion or contraction requirements.

Dominant Work Flows. Nearly every office-layout study will reveal that the bulk of the work is processed along routine lines, stopping at the same points repetitively and moving onward through recurring courses of action. An exact knowledge of these dominant work flows will provide satisfactory results.

The movement of work from one station to another and between departments is a basic consideration in determining the arrangement of the units of furniture and equipment.

In studying work flows it is often worthwhile to outline them directly on a copy of the floor plan. Lines of various colors or types should be used to delineate the different routes over which the work travels, so that it will be easy to note any crisscrossing, backtracking, or unusual movements of the work.

Structural Traffic Patterns. Floor areas almost invariably have structural features that establish basic traffic patterns and more or less set the boundaries for the flow of both people and work. Among the more common are walls, abutments, windows, pillars, stairways, and elevators. Such features are permanent. Consequently, a layout must be constructed in conformity with them in order for it to be successful.

Employee Needs. Best results are achieved when layouts are based on the needs of the individual employee. What does he require to perform his work well? Has he need for much privacy? Quiet? Plenty of room? Additional or special furniture? These and similar questions must be posed, and their answers will largely determine specific requirements.

Requirements for Furniture and Equipment. It is necessary to explore carefully the requirements for furniture and equipment. The individual items must be considered as to type, size, capacity, and overall appropriateness to handle the kind and amount of work that is involved. In this connection it is desirable to prepare a list of the existing furniture and equipment, and to append to it any new items that are needed.

Visits to Places with Layouts of the Same Type. It is frequently advisable to visit other companies to observe a particular kind of layout firsthand. Such visits can be very informative; they can disclose potential difficulties that ought to be avoided; they can furnish guidance with respect to making decisions (e.g., in the selection of facilities); and they can generate inspiration and creative ideas.

Recommendations of Furniture and Equipment Vendors. Many manufacturers and distributors of furniture and equipment offer services that include expert advice concerning the installation and arrangement of their facilities. Because these firms have had extensive experience in solving layout problems, they are in an ideal position to give opinions and make suggestions. Ordinarily the advice is proffered without obligation, and it is widely recognized as wise practice to seek it liberally.

Other recommendations and ideas on layouts may be obtained from the publications of furniture and equipment manufacturers. Advertisements, instruction booklets, and promotional pamphlets can be excellent sources of information.

Determination of the Total Area Necessary. Consideration of the preceding matters is a prelude to determining total floor-space requirements for specific areas. Every person and item of furniture and equipment takes up a certain amount of space. A person assigned a standard 60-by-34-inch desk, for instance, may be thought of as occupying 51 square feet. This is demonstrated in Figure 6-1.

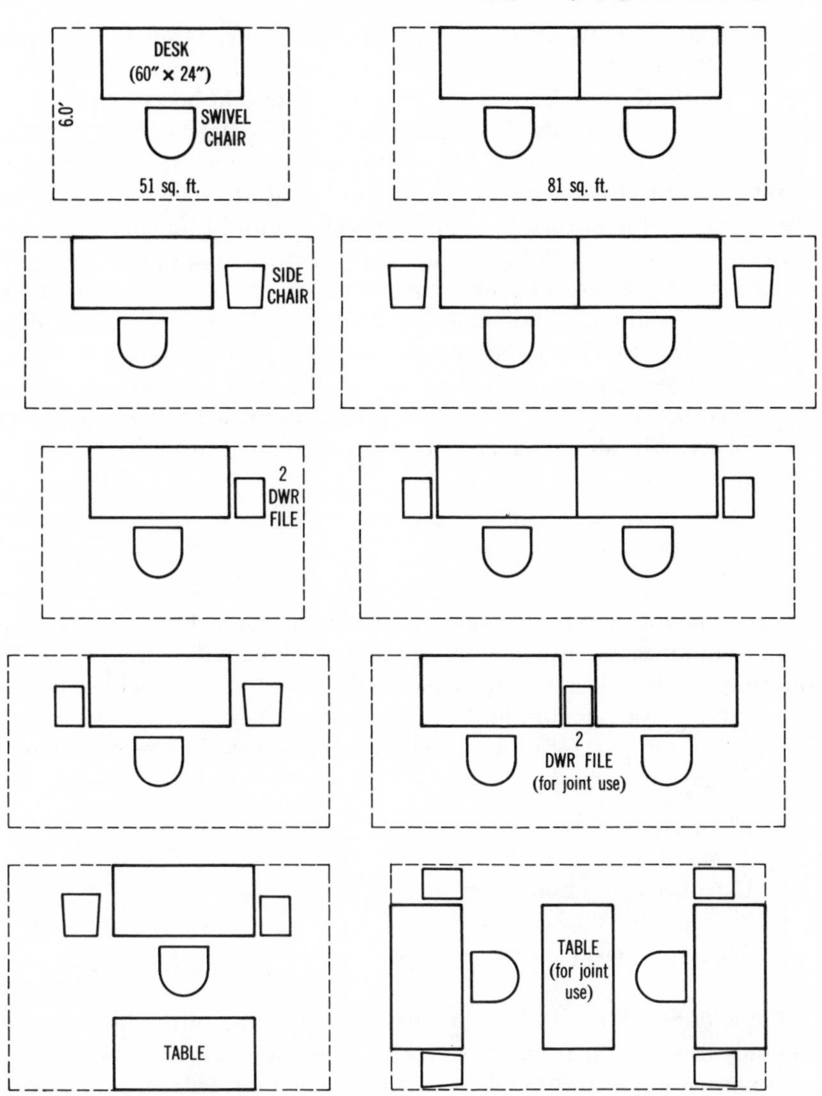

Figure 6-1. Various arrangements for desks and related office furniture.

It is necessary to look beyond the static space requirements; for example, the area allotted to a vertical file cabinet must allow for the extension of its drawers: approximately 28 inches for a unit with an 18-by-28-inch base. To gain access to a cabinet an additional area of at least 24 inches should be allowed.

As a basic approach, an average of 100 square feet per person (which includes space for aisles, drinking fountains, filing cabinets, and rest rooms) is often used to obtain a rough calculation of the aggregate amount of space required. The result is a composite figure that takes into account all the space requirements. This however should be used only as a guide.

Allocation of Space. Good office-space arrangement accomplishes organizational objectives with the least effort, greatest speed, desired quality of output, and least cost. The space provided must be adequate: neither too much nor too little. With this in mind, the space in question has to be allocated among whatever units or segments are involved. Factors influencing this allocation include such considerations as work processes, location with respect to other departments, and plans for the future.

In the development of a layout some adjustments are always necessary, and certain compromises must be made. As a rule an effective layout plan evolves only after trial-and-error experimentation with alternative arrangements.

MAKING THE LAYOUT

Let us now turn our attention to certain mechanical means and illustrative techniques, the use and observation of which will help to visualize arrangements, decide alternative locations, avoid actual problem situations, and suggest possible improvements—both corrective and productive. The more important of these matters are discussed under the following headings:

1. Floor plans
2. Templates
3. Use of cross-section paper
4. Showing the flow of work
5. Three-dimensional scale models

Floor Plans. A copy of an architectural drawing of the floor area is commonly used for the purpose of making a layout, since it shows the location of permanent features such as structural columns, entrances and exits, stationary walls, plumbing and electrical outlets, stairways, elevators, and rest rooms. The drawing that is used should be checked to insure that it reflects current conditions accurately. A typical floor plan for an office is shown in Figure 6-2.

Templates. One of the most practical means of designing a layout is by the use of templates in conjunction with the floor plan. Templates are patterns or guides that outline objects, such as the individual items of furniture and equipment; examples are shown in Figures 6-3 and 6-4. They

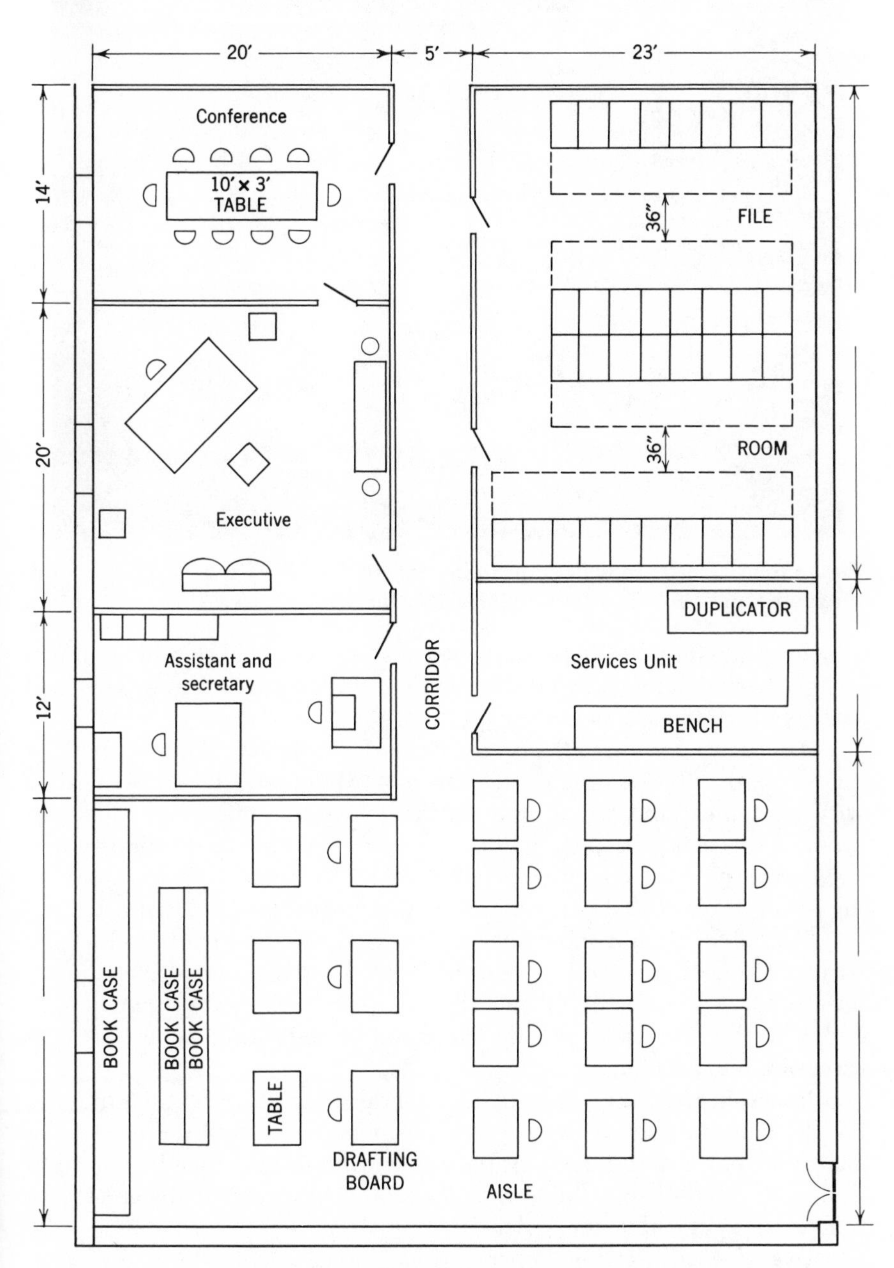

Figure 6-2. Typical floor plan of an office.

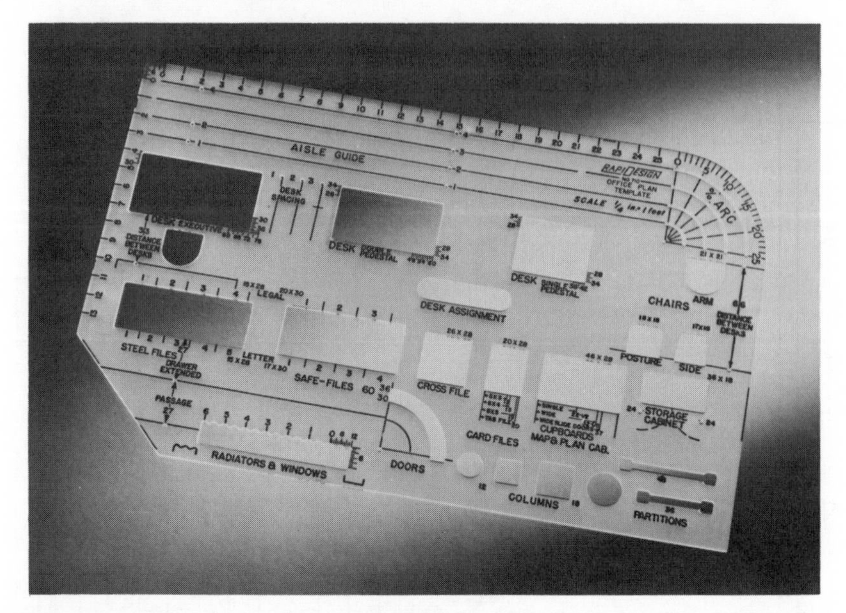

Figure 6-3. Plastic stencil-template for tracing outlines of standard office furniture and equipment. Courtesy: RapiDesign, Inc.

may be made from heavy paper, cardboard, or fiber, and should be drawn and cut to the same scale as the drawing itself. A suitable and widely used scale is ¼ inch to the foot.

An essential point in the preparation of templates is that they be adequately and fully identifiable. The face of a template should be marked with the name of the item it represents and any other information that will promote this cause: model identification, serial number, code number, and peculiar features. Recognition may further be enhanced by using contrasting colors or black-and-white hatching to indicate classes of items.

Templates should show not only the body of the item but also such parts of the item as extend beyond the body. Thus, for example, a template for a supply cabinet will indicate the sweep covered by the opening of its doors. In this way there is little likelihood of neglecting to take areas of extension into account.

Locating facilities is done by moving the templates over the surface of the drawing, in various experimental arrangements, until the most satisfactory layout is ascertained. Appropriate symbols may then be affixed to designate telephones, adding machines, and other small items such as time-card racks and water coolers.

Use of Cross-Section Paper. Instead of an architectural floor-plan drawing, cross-section paper may be used for making layouts. Much mea-

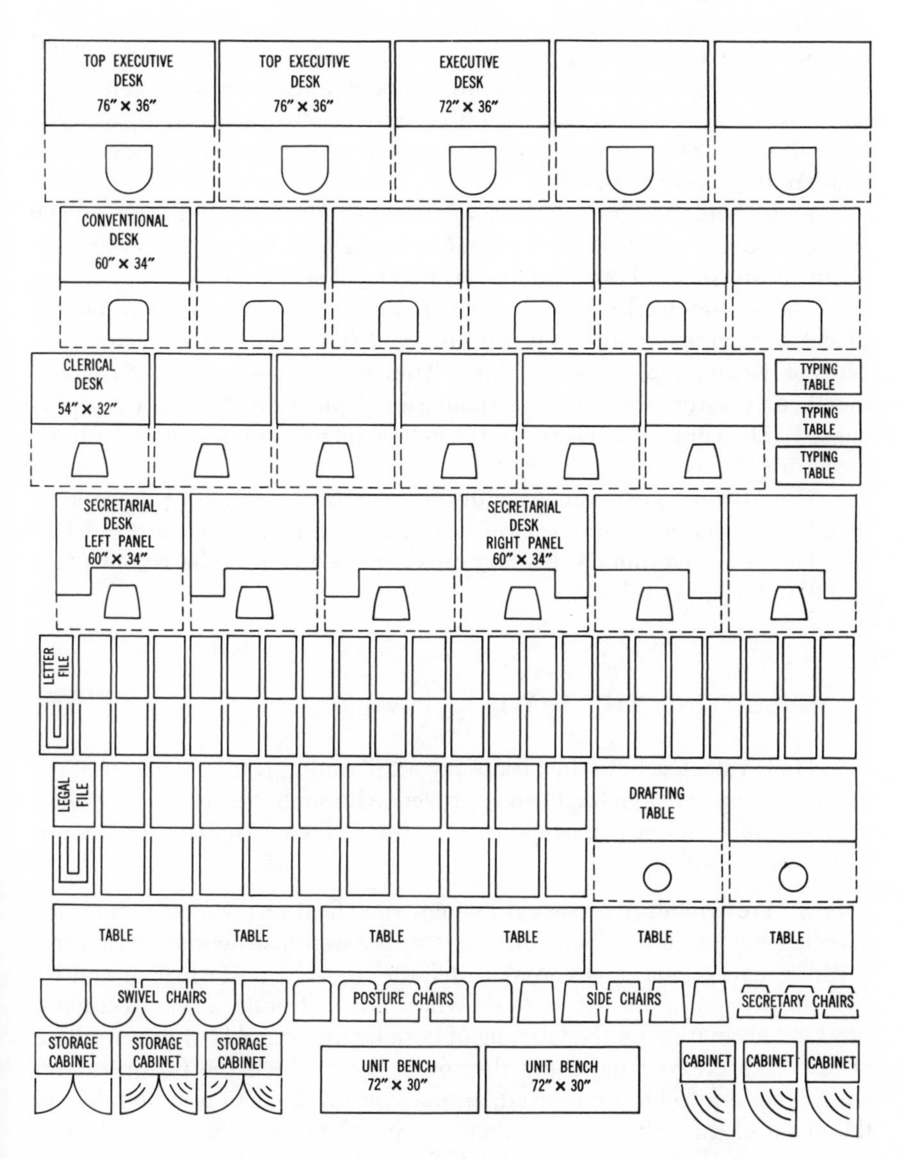

Figure 6-4. A sheet of office layout templates drawn to scale. These templates are usually printed on heavy paper or cardboard and are so arranged as to be easily cut out for use in making layouts.

suring is eliminated by its use, because the squares represent the scale. The lines of the paper furnish guides for making up the basic floor plan, walls and other structural features, and for shuffling the templates about until the optimum arrangement is found.

Showing the Flow of Work. As when studying the drawing of the existing arrangement, lines of various colors may also be traced on the new floor plan to illustrate the flow of work from place to place. Figure 6-5 portrays the flow of a sales order before and after improvements have been made in the layout of an office.

Scale Models. Three-dimensional models form an admirable means for developing layouts. The third dimension helps to give the viewer a realistic impression of how an actual arrangement will appear.

These models are simply scaled replicas of the items being used—pieces of furniture and equipment, units of floor space, and even duplicates of entire multistoried buildings. They may be made of plastic, wood, metal, or plaster. Kits of three-dimensional office models may be purchased. (Distributors are listed in the yellow pages of major-city telephone directories.)

The idea is to experiment with different arrangements of the models on a floor plan until, as in the case of shuffling templates, the most satisfactory layout is determined. Photographs may be taken of the layout plans and filed for future reference.

EXECUTING THE MOVE

Tied very closely to the task of layout planning is that of accomplishing a move once a plan has been approved. Although the problems of moving will differ in each particular case, the following specific points are generally helpful:

1. Provide all necessary drawings, specifications, and schedules for carrying out the move. Set forth the step-by-step instructions for accomplishing its various stages. Assign definite areas of performance and responsibility; point out who is to do what. The instructions should contain precise information on the labeling of both facilities and floor areas, notice of any restrictions concerning the routes over which the furniture and equipment are to be transported, and details on the action to be taken in the event of difficulties. Remember that problems are diminished if they are anticipated and provision is made for dealing with them.

2. Use unencumbered floor plans. This may be done through a series of drawings—one showing the layout of furniture and equipment, another the location of telephones, another the wiring for the area, and so on. The

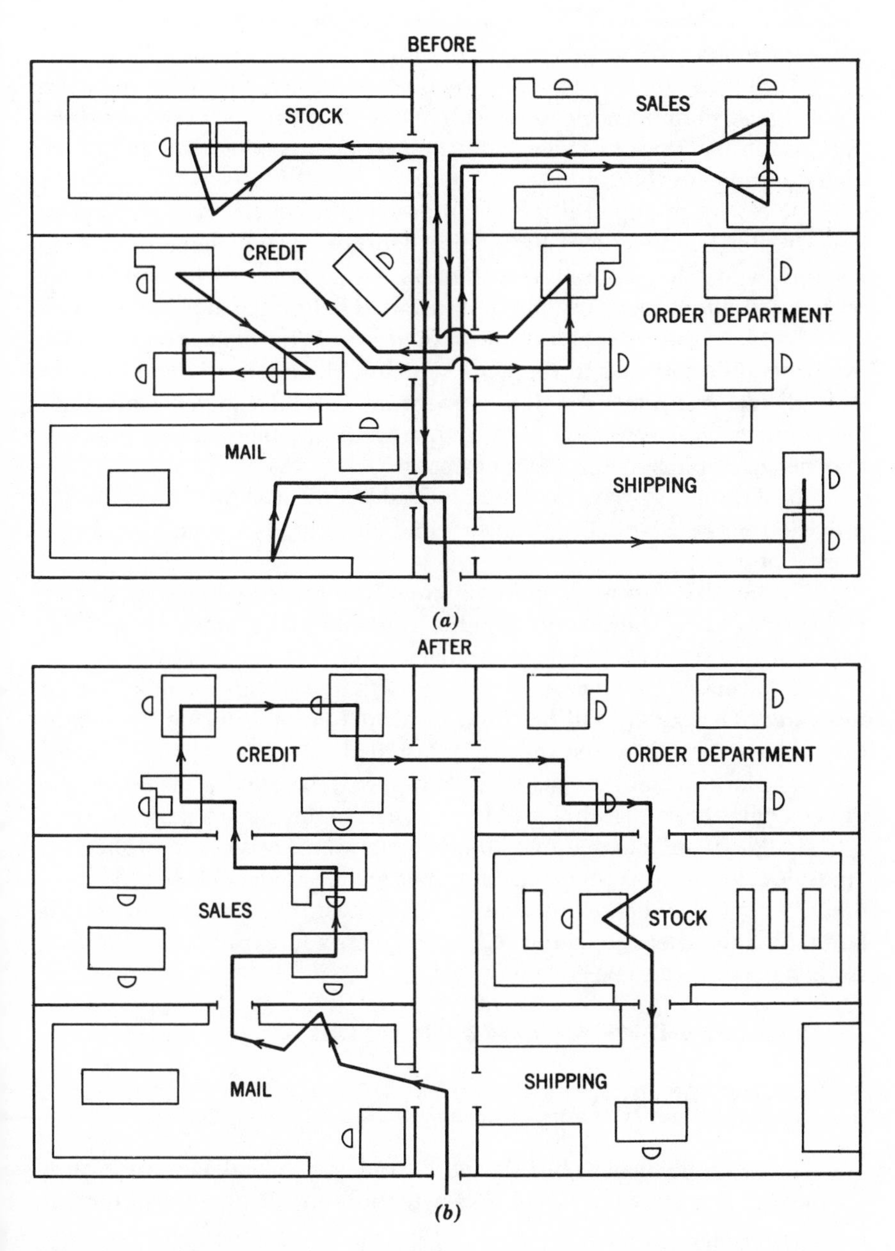

Figure 6-5. Flow of a sales order before and after improvements have been made in the overall layout.

furniture-and-equipment drawing should identify individual items, and the actual facilities should be tagged to correspond. It should show clearly where each item of equipment is to be situated in the new location.

3. Supply all persons concerned—supervisors, building engineers, millwrights, plumbers, electricians, etc.—with copies of pertinent drawings and instructions in order that necessary advanced work may be completed in time for the move.

4. Notify public utility firms of new requirements as soon as possible. Greater headway will be made when power, telephone, and water companies are given ample lead time in which to do their work. Rough wiring and plumbing is usually best installed before the move takes place.

5. Make the move at a strategic time so as to minimize confusion and avoid possible setbacks in the work schedule. Uninvolved moves may be carried out overnight; complex moves may be done in the course of a weekend or holiday period; moves that are both complex and extensive may be accomplished in a series of steps.

6. Arrange to have facilities moved on an authorized order. Make provision for substantiating transfers and checking the condition of each item of property.

7. See that furniture and equipment are protected against damage in transit. Where necessary, suitable use should be made of padding, packaging, and covers to safeguard items while they are en route.

8. Secure or remove parts of items that protrude or may become separated. Desks that will be tilted or turned on end during the moving should have their drawers emptied or fastened.

9. Mark portable items such as typewriters, adding machines, and desk calculators so that they will be returned to the person assigned.

10. Postpone making immediate changes in a newly established layout unless the need is pressing. Proposals for improvement should be cordially accepted and carefully evaluated, but basic changes should actually be made only after the layout has been given a fair trial and a thorough study has shown it to be advantageous.

DEVELOPMENT OF AN IMPROVED LAYOUT

A "BEFORE AND AFTER" CASE STUDY *

The general offices of Burton and Farraday,** a small accounting firm, are located in a 35-year old high-rise bank building. From time to time the

* Facts for this case study are presented through the courtesy of the General Fireproofing Company.

** Fictitious name.

firm gradually leased additional space as company personnel increased and extra space became available. Figure 6-6 shows the original floor plan resulting from this expansion. Observe that the public corridor extended to an entry adjacent to the offices of another company, Research Associates, Inc.*

The original layout included the following drawbacks:

1. To secure customer files it was necessary to cross a public corridor. This necessitated keeping the file room locked at all times.

2. Although the vault took up considerable space, it was not fireproof.

3. The traffic pattern, which grew as additional space was required, became extremely complex and was a distraction to all office employees. The partner consulted frequently with the supervisor in the office at the far left. The secretaries were constantly distracted by traffic to and from the staff room.

4. Although the space was included in the lease cost, many areas (such as the entry, anteroom, and closet) were not usable for office space.

5. As is true of most buildings of this age, lavatories were adjacent to every "wet" column, and many columns were evident.

6. The radiators, which were under the windows and exposed, were objectionable from an aesthetic standpoint. Venetian blinds in all states of disrepair were hanging at the windows. There was no attempt to standardize furniture, wall colors, or floor covering. The lighting fixtures were all on pendants from a 12-foot ceiling. All framework was dark mahogany with steel baseboard; doors were painted in grained mahogany. The whole atmosphere was chaotic and unattractive.

The layout of the old arrangement was studied with adequate consideration for the present and foreseeable needs of the firm. The study indicated that if Research Associates, Inc., could be moved to area 13 on the new plan (Figure 6-7), the public corridor could be eliminated and additional space could be acquired for Burton and Farraday. As it turned out, Research Associates, Inc., agreed, and the plan shown in Figure 6-7 was put into effect.

The revised plan provides many practical improvements, better utilization of space, and an attractive interior. Among the advantages of greatest significance are the following:

1. Direct access to the reception area from the main corridor leading from the elevators so the firm's entry can be seen from the elevator lobby.

2. Direct access to all offices with a minimum of foot travel, yet, all offices can be secluded by closing the door.

3. All columns have been recessed into walls.

* Fictitious name.

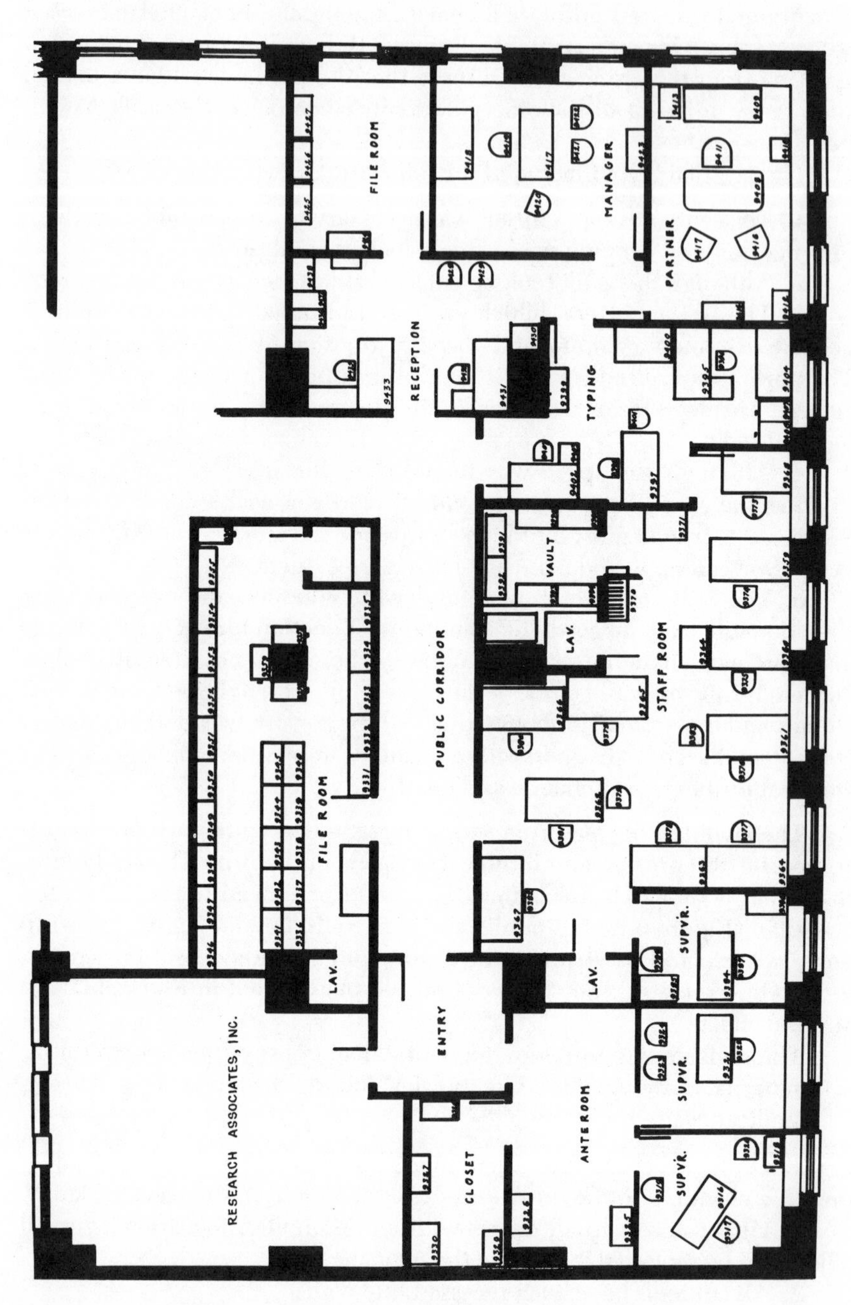

Figure 6-6. Original floor plan—Burton and Farraday.

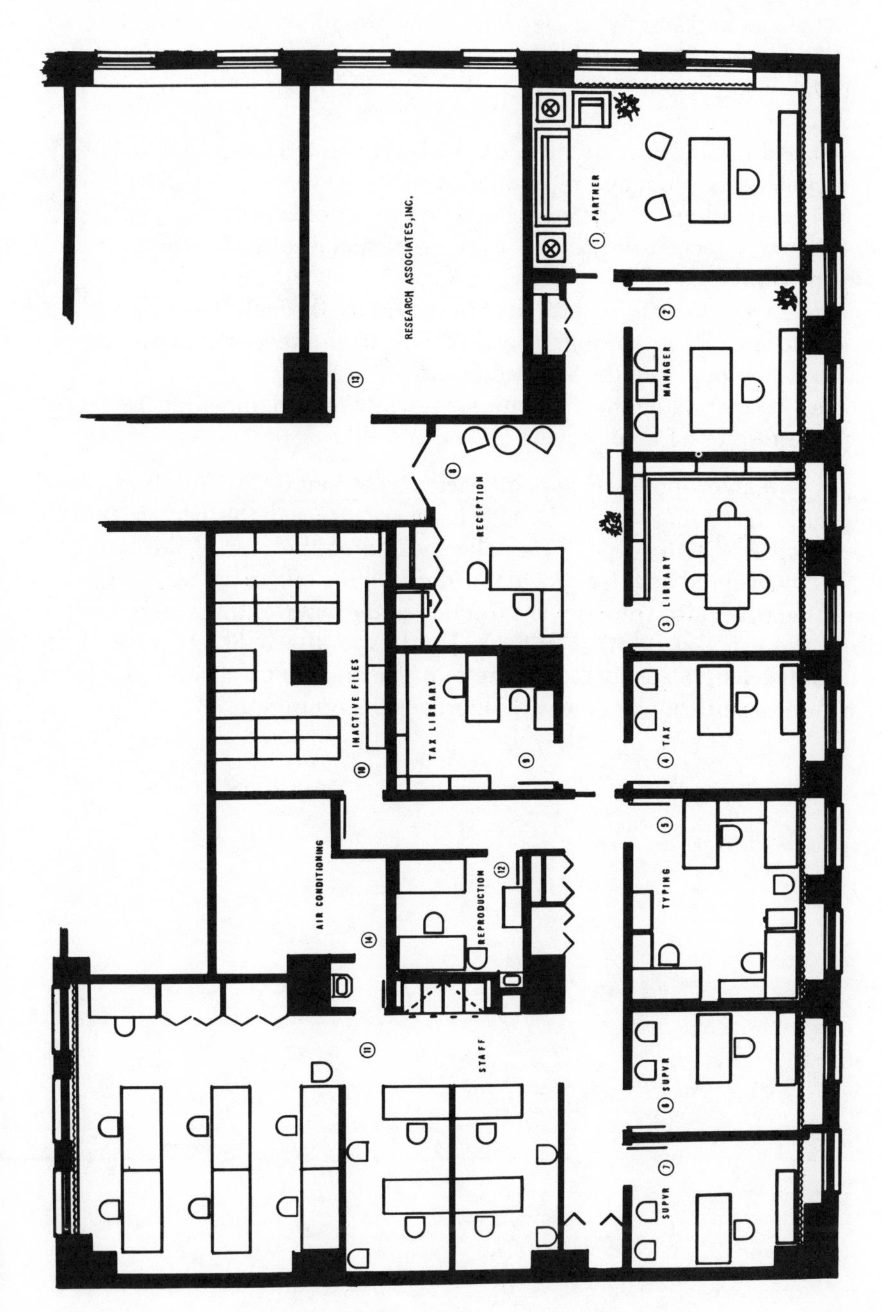

Figure 6-7. Final floor plan—Burton and Farraday.

4. Coat and supply storage have been provided where needed.

5. There is access to the inactive-file room and to all areas of the office without using a public corridor; thus the file room need no longer be locked.

6. Additional private offices have been created, including a library-and-conference room and reproduction room.

7. A thorough study was conducted of all office stations and record housing to select furniture and equipment specifically designed for each job and function.

8. Expansion can be provided in the future through the cupboard adjacent to area 13 or through the cupboard in the reception area into the inactive-file room and the suite adjacent.

9. Aesthetically, the new interior is warm and dignified; it creates the right atmosphere for employees and clients alike.

From an economic viewpoint each participant is satisfied. Burton and Farraday has gained an enclosed private area for a slight increase in rent. Research Associates, Inc., originally located at the rear of the building, quickly accepted smaller space in order to have offices at the front of the building (they also were given carpeting and draperies that were removed from the original partners' office). The lessor—the bank—is pleased because the change has increased the rentable area on this floor and created a beautiful suite of offices, which upgrades the building.

part 3

The Systems Study

7

Initiating a Systems Project

Systems surveys are diverse in kind and in scope, ranging in extent from comprehensive examinations that result in a wholly new system to special investigations that provide answers to questions or solve some particular problem. Our approach in this chapter and in the rest of Part III covers concepts and principles that are normally associated with a broad systems study.

PROJECT REQUESTS

Before presenting the steps in a systems survey it is necessary to discuss the sources that initiate systems-and-procedures assignments, the reasons for selecting a study area, and the methods employed in project requests.

Suggestions for conducting systems-and-procedures studies usually come from management personnel such as department heads and staff officials or from members of the systems organization. Occasionally suggestions come from outsiders, such as forms salesmen and data-processing-equipment salesmen. These people are systems oriented, alert to omissions and deficiencies in current operational practices, and in a position to propose their ideas or equipment for consideration.

Among a number of reasons why particular projects are initiated are complex data-processing problems, unmanageable backlogs, poor customer service, recognition of the need for better management-information systems, urgent time-scheduling considerations, or a desire to study benefits that might be derived from mechanization.

Depending on the degree of formality desired by management, project requests may be made verbally, by memoranda, or by means of a form especially designed for the purpose. It is the policy of many systems organizations to insist that all requests be made in writing; some units may act on a verbal request, but only with the proviso that it be followed up by a written one.

Having the request in writing serves to assure that the initiator gave the request serious and sufficient thought. This helps to minimize the number of impulsive-type projects that consume valuable time and lead nowhere. A written request also serves as a valuable record that states the general nature of the problem, the departments involved, the date, and so on. Without a written document the systems analyst may later find that an oral request had been misinterpreted or that the initiator had constructed his original verbal statements to shift responsibility to the systems analyst. A written request acts as a safeguard against possible misunderstandings that lead to bad feelings and a poor working relationship. Of importance, too, is the fact that it can protect the systems analyst from unwarranted attacks on his work.

BASIC STEPS IN PERFORMING SYSTEMS-AND-PROCEDURES STUDIES

There are 10 basic steps in the systems investigation, as follows:

The Preparatory Steps

1. Studying the assignment;
2. Performing the preliminary investigation;
3. Evaluating and selecting hypotheses; and
4. Planning the formal research activities.

The Formative Steps

5. Collecting the information;
6. Analyzing the information; and
7. Developing Solutions.

The Concluding Steps

8. Presenting recommendations;
9. Acting on approved recommendations; and
10. Following up.

A breakdown of the work into 10 steps does not mean that the systems analyst should always consciously follow the sequence of these steps. Many times a number of the processes are carried on nearly simultaneously and quite unconsciously. Often, too, there is the need to backtrack and retrace steps to acquire more information before proceeding with the study. The 10 steps are a formalized outline of a process that a systems analyst, with or without awareness, uses over and over again as he works.

The preparatory steps are the subject of this chapter. Steps 5 through 10 are described in later chapters.

Studying the Assignment

First, the systems analyst wants to make certain that he properly understands the assignment. For this purpose he will usually meet with the systems manager and/or initiator of the project. At this stage all that may be known is that a problem exists or that the nature of the problem is known but not well understood. Often the initiator knows what he wants but has no idea of how it may be obtained. As previously suggested, getting the request in writing can be a clarifying factor. In any event the systems analyst will want to make certain that the assignment is as clear and precise as possible at this point.

When the nature of an assignment is clearly understood, it is good practice to prepare a project-assignment sheet, such as that shown in Figure 7-1. Many systems organizations require that this be done. The systems analyst will record a statement on this form that (tentatively) defines the general purpose and scope of the study. This serves to focus attention on the problem, provide an indication of the work that is involved, and, in general, aid in guiding the systems analyst in the early phases of the investigation.

Because of the many unknown facts involved, it is not practical to complete the project-assignment form when it is initially set up. Nevertheless, from the beginning it is a valuable aid to the control and handling of systems projects. Information can be subsequently added or changed, because assignments often must undergo substantial modifications soon after the investigation gets under way. The approach to the project should therefore be sufficiently flexible to accommodate various contingencies as they develop.

The Preliminary Investigation—The Exploratory Survey

After the assignment is understood the systems analyst is ready to begin the "preliminary investigation," or, as it is sometimes called, the "exploratory survey." This serves as a preliminary to a complete and detailed study, and has for its purpose the acquisition of (*a*) first-hand familiarity with the problem factors and possible causes of difficulties, (*b*) some idea of the scope and complexity of the work that lies ahead, and (*c*) background information for evolving hypotheses.

PROJECT-ASSIGNMENT SHEET

Systems-and-Procedures Department

The Exemplifying Company, Inc. Project No.

Project title

Purpose and scope

Personnel assigned

Assignment objectives

Date received	Requested by
Assigned	
Target	Approved by
Completed	

Figure 7-1. Systems-and-procedures project-assignment form.

In conducting an exploratory survey a good starting point is to check the existence of earlier studies by referring to the "permanent" files of previous investigations. The systems analyst may find that the problem has been investigated previously and that any further investigation should be approached as a modification and extension of the previous work. Even previous work with major deficiencies can become useful for the current study, because sometimes the chief value derived from searching past records is in discovering what *not* to do.

The systems analyst should also confer with other staff members of the systems organization to find out what background information they may have on the problem. At this time other records—such as pertinent manuals, organization charts, correspondence, and forms files—should be reviewed.

Of particular importance in carrying out the exploratory survey is the interviewing of key executives and supervisors, especially those who have a direct interest in the project or who are in some way affected by it. One technique is to call these management people together in a meeting at which a member of top management expresses his interest in the study and thereby solicits their cooperation. If this approach is used, it is desirable to allow time for a discussion of the problem. Participation at this stage helps to consolidate the efforts of all levels of management in support of the project.

Techniques of Exploratory Interviewing. Whether or not a preliminary meeting is held, in practically all cases it will be necessary to directly confer with persons such as company executives, professional men, and other key management personnel. Since it is not unusual for such persons to have strict limitations imposed on their time, and since their advice and consent may be instrumental to the ultimate success of the project, interviews with these persons are usually prepared for carefully in advance so as to be most productive of time and effort. The following rules ordinarily bring good results:

1. Make a definite appointment and be punctual.
2. Have or obtain knowledge about the person involved prior to the interview. Know something about his personal characteristics and interests. This information can be invaluable when making introductory remarks and in knowing what to expect from him under various circumstances.
3. Possess a fundamental and practical knowledge of the subject matter. Interviews are most effective when the parties "talk the same language," for then the person being interviewed does not have to go into unnecessary explanatory detail.

4. Come to the interview prepared to lead the discussion and to draw out information; that is the responsibility of the systems analyst.

5. Be prepared to ask pertinent questions aimed at bringing out a progression of ideas on the subject.

6. Encourage the person to talk, and listen attentively until he is finished. Do not interrupt. Allow him ample time in which to assemble and express his thoughts; it may be that he is taking so many things into consideration that extra time is required to put his thoughts in order.

7. Hold some leading questions in reserve with which to stimulate and direct the conversation should it bog down or wander.

8. Be natural. Awkwardness and the inability to converse freely and explain matters will be eliminated.

9. Show an interest in his problems and needs. What information is he using in his work? What does he think he needs and is not getting?

10. Submit in advance information that is detailed and requires study, such as statistical tables, mathematical calculations, and involved statements.

Broad-Brush-Stroke Approach. The exploratory survey is intended to acquire information on a broad front, not to gather detailed facts and data. The systems analyst should find out just what is being done, why and how it is being done, and when and where it is being done—but not on an in-depth basis; that comes later. This means he must comprehend the problem or project as a whole and fill in only such details as are necessary to the purposes of the exploratory survey.

The work of the systems analyst at this stage is like that of an explorer in a land where there are few, or perhaps numerous, trails to follow. He is looking for suggestions or leads to provide clues, and he must proceed with care and deliberation. He must be inquisitive and alert to uncover—through interviews and observations—promising avenues of exploration. He should perform "up- and down-stream" investigations to uncover the possibility of the problem having its origin in organization at some point preceding or following the immediate area under consideration.

In comparison with the fact-finding step, yet to be discussed, the exploratory survey is really a small-scale inquiry. Depending on the scope and complexity of the project, a systems analyst may be able to complete this phase in a relatively short period, especially if he is familiar with the areas under study or has worked on similar projects in similar lines of business. On the other hand, the initial survey may range from several days to several weeks. Whatever time is reasonably spent on this phase, is time well spent, for if it is omitted or slighted, preparation of a meaningful project plan will be difficult or impossible.

Evaluating and Selecting Hypotheses

Most systems investigations are susceptible to one or more approaches. However, some approaches will be better than others, and the problem is to select the best. This step requires deliberation in weighing and examining reasons for and against a choice or measure. It is necessary to ferret out the pros and cons in the systems situation, evaluate them, and determine a course worthy of the time and effort required for a thorough research job.

The evaluation of pertinent hypotheses and an intelligent selection of those that merit study is an important phase of the systems investigation. The selected ones serve as leads or approaches to the solution.

Frequently the experimental survey will point up the desirability of pursuing only one hypothesis. This is advisable when all things have been considered in relation to their importance, and the weight of the evidence clearly favors the one; it saves aimless work and loss of time. Although one hypothesis is accepted for a starter, others will follow in its wake. One hypothesis inevitably leads to another, and in the sequence of events hopefully more precise "possibilities" will evolve.

Selecting the proper hypotheses serves a twofold purpose: they function to guide the study, and they help to establish its boundaries. With the selection accomplished, the systems analyst is ready to turn his attention to planning the formal research activities.

Planning the Formal Research Activities

Before assuming that there should be a full-scale examination of the area under study, including all processes, operating conditions, forms used, built-in controls, and personnel involved, a few questions should be asked. One is "Would such an elaborate undertaking have enough merit to be worth the effort expended?" In certain situations the answer will be an unqualified no. One reason might be that the solution to the systems problem is already revealed on the basis of the facts assembled. All that remains is to take action, as for example, to install or modify certain basic procedures.

Another question is whether there is time for gathering additional information. The urgency of the situation may call for stopgap measures. When, for example, customers are being lost because of faulty or improper billing procedures, immediate action must be taken. It may be necessary to quickly institute new procedures so as to expedite the processing of payments and thereby eliminate the alienation of customers.

A common situation restricting a wide survey is the lack of manpower or special talents. Demands on the systems staff's resources in other important areas may make it necessary to curtail or suspend a thorough and penetrating study.

Expedient measures that stop short of a complete systems study have given birth to the term "fire-fighting approach." Known also as the "trouble-shooting approach," it is geared to the solution of specific problems as they occur. Often the technique is subjected to adverse criticism, sometimes without justification. On the positive side, it is service oriented and practical; it is useful, as we have already seen, in emergency applications. On the negative side, systems staff members may spend so much time patching up difficulties that they have no time for studying systems as a whole. Consequently, in the course of time there arises a patchwork of ill-structured procedures for carrying out operations.

Shortcuts and interim solutions to problems have their place in systems-and-procedures work, but they should be no substitute for overall systems studies. Going on the assumption that a comprehensive study is to be made, our discussion will now proceed to the next stage; namely, planning.

On the Importance of Planning. Any event succeeds or fails in direct proportion to the preparation given it. Conducting systems-and-procedures work is no exception. Unplanned systems work is inefficient; that is, wasteful of time, money, and effort. There is a lack of coordination, and there is no clearly defined sequence of work. This points up the practical necessity for planning.

Planning is the laying out of a course of action to be followed. It provides measures necessary to influence and assure the fulfillment of systems work. It serves to organize an attack on the problem and helps the systems analyst know where he is going, minimizing backtracking and "taking off on tangents." By directing his efforts in a purposeful way, he concentrates on the important issues.

Attributes of Sound Planning. Sound plans are thorough, purposeful, realistic, strategic, and readily understandable.

A thorough plan is comprehensive; it reflects complete information on the composition of all significant factors. This does not mean that all details, side issues, and routines are embodied in the plan, but it does mean that provision has been made for fitting them into the framework of the plan. Thoroughness is achieved by making an adequate inquiry into the nature of the problem, by considering all sources of relevant information, by taking counsel from others, and by an aggressive search for truth.

A realistic plan conforms to actual and true expectations rather than to some ideal; for example, a schedule may require much effort to carry out, but the demands imposed by the schedule should be capable of accomplishment within the prescribed time.

A strategic plan is one in which it is possible to adjust for changing circumstances. Although the plan of approach should be thought out very carefully, no program of future action can be certain.

A good plan is designed so that it promptly conveys its meaning to all concerned. Such a plan is as clear and as simple as possible.

General Pattern of Planning. Thorough planning specifies the following:

1. *What work is to be done?* This involves (*a*) determining the problem area and boundaries, and (*b*) determining the type and sources of data needed.

On the basis of origin, information is of two types. When information is presented or made known by the same organization that did the fact gathering, it is called *primary*. When information is presented or made known by some organization other than the one that did the collecting, it is designated as *secondary*. The classification may be extended further. All information derived from printed sources is called *published information*. Information that is obtained from the originating source, without any intervening medium, is termed *direct*.

2. *How will it be done?* Should department heads and administrators be interviewed personally or should mail questionnaires be used?

3. *When will the work be done?* The priority of the project must be determined and the work scheduled. *Scheduling* is the preparation of a timetable within which the operations are to take place; it establishes a work-breakdown structure, so that the subdivisions of work become meaningful in terms of specific goals.

4. *Who will do it?* The complexity of the assignment, the time factor, and the required skills will enter into the selection of personnel for the study. In some cases a single systems analyst may handle the entire assignment. In other cases the work may be broken up into segments and distributed among a team of *generalists* and *specialists*. Team size normally ranges from two to six persons.

8

Fact Gathering

After clearing the project with the executives concerned, and performing a preliminary investigation, the systems analyst is ready to pursue the collection of data on an in-depth basis. Details of the present system are obtained in the following ways:

1. Interviewing supervisory and operating personnel.
2. Observing how the work is being done and studying the output.
3. Tracing specific data-flow networks from the point of inception to that of final disposition; for example, the documentary flow of a sales and cash-collection system might begin with the receipt of a customer's order as prepared by a salesman, or on a form provided by the customer, and be terminated with the remittance received in payment.
4. Examining records, forms, documents, mechanical equipment, and other media used for the recording and processing of data.
5. Sending out mail questionnaires.

INTERVIEWS

Contacting the Supervisor

In conducting the survey of a particular system courtesy and common sense dictate that the interviewing start with the supervisor in charge. Even though top management has authorized the study, the sanction and cooperation of the supervisor should be explicitly sought.

When meeting the department head for the first time it is advisable to obtain an introduction by the supervisor's superior or a high management official. This not only takes care of the amenities involved but demonstrates management's interest in the project more forcefully than would be the case if the systems analyst appeared on the scene by himself.

The seasoned systems analyst is considerate of the fact that operating management has regular duties to perform. In arranging for an interview he makes a definite appointment that fits into the working schedule of the supervisor.

An important factor is the building of good working relationships. Although an announcement may already have been issued describing the project and what it proposes to accomplish, it may be well to explain once again—giving it the personal touch—the objectives and techniques of the investigation. The whole conduct of the meeting should be along friendly and informal lines. The cooperation and suggestions of the supervisor should be invited, and any questions he may ask should be answered. Assurance may also be given that all findings and problems will be first discussed with him.

From the supervisor the systems analyst can expect to obtain a general understanding of the whole operation, how it ties in with the work of other departments or units, and the names of the employees to be interviewed to acquire the detailed information.

The supervisor should not normally be expected to supply full details on an operation. He cannot possibly keep abreast of all modifications that tend to be introduced on a job. When, as sometimes happens, the supervisor volunteers such information, the systems analyst later learns that there is a difference between the way the work is actually done and the way the supervisor thinks it is done. The systems analyst therefore must verify the facts for himself. He should trace the data flow from one work station to the next, in line with the movement of the paper work, discussing with the individual workers their contributions to the functioning of the whole system. Information obtained in this manner is verified by first-hand observations; that is, inspecting the actual procedures in action, examining the output, delving into the files and records, and so on. The systems analyst also obtains sample copies of forms and documents during each interview. In this regard he wants to know the circumstances that give rise to the preparation of the form or document, the manual or mechanical method by which it is prepared, the quantity generated, and the disposition.

Information that is sufficiently accurate and complete for the purpose is, of course, essential. If pertinent data under examination are not clear to the systems analyst, he should extend his investigation to the extent necessary to achieve this end; for example, if the purpose is to ascertain whether certain practices are followed as stated verbally or in a procedures manual, the tracing of a few well-chosen transactions should furnish the answer.

The fact gathering must be thorough. No significant data should be overlooked. Even seemingly unimportant operating practices may offer insight into problem situations or point up areas in need of improvement. The question as to just how far the systems analyst should probe will depend on each particular case. The general rule in practical survey inquiries is to stop at different levels, according to the interest, purpose, or "feel" of the systems analyst who is making the survey.

Within the scope of the project any unusual or irregular work should be critically investigated. One potential danger is the temptation to follow only routine work flows and to ignore other factors that may be of importance. To avoid falling into such traps the systems analyst should make a point of uncovering *exceptions,* such as, for example, special methods for handling out-of-the-ordinary work, shortcuts used by the employees but unknown to the supervisor, and repetitive tasks that are performed at indefinite times.

Interviewing the Employees

In approaching the individual employee on the first interview the systems analyst must realize the many threats he may seem to present to the status quo. Employees may be fearful that some change will upset the informal organization; that the introduction of new equipment will make their skills obsolete; or that, if the work is streamlined, there will be unemployment. Of greater importance, they may fear that a report of individual performance may be made secretively to supervisory management. These fears can be allayed if the systems analyst realizes the human-relations aspect of this work, avoids creating fear, and destroys it where found.

It should be made clear that the company is not interested in individual performance during the survey, but in the overall operations of the system. Employees will cooperate in a positive way if they are treated in a courteous and understanding manner. From the start the systems analyst should dispel any misconceptions about the project. He should make it known that the reason for the study is to learn enough about the system—the relationship of documents, the organizational elements engaged in processing such documents, operations presently mechanized, the sufficiency of information, etc.—to provide a factual basis for determining the best method of designing and installing an improved system.

An atmosphere of informality is important in conducting successful interviews. The systems analyst must not be impatient. Time is needed in which to develop rapport, to "warm up." People usually make explanations best when they are acquainted, relaxed, and unhurried. It is then, after tension has been lessened, that employees talk freely. As a result the person being interviewed is likely to volunteer information about operating problems, bring pertinent sidelights into the conversation, and offer opinions and recommendations. It is worth remembering, too, that significant facts often emerge after the formal part of the interview has been terminated, when people feel they can release their inhibitions and become plainspoken. In their approach systems analysts sit at the employee's work station in a comfortable and relaxed manner that is intended to put the person at ease and induce free response.

One of the first rules for all good interviewing is—use tact; do not criticize. The employee will not feel kindly disposed toward cooperating if he is being asked to justify every function he performs; besides, when all the facts are known, any initial criticism may prove to be unjust.

Something needs to be said about attitude. In seeking to make a favorable impression the systems analyst must guard against conveying the idea that he knows it all. Indeed, a certain amount of humility is in order; it is the person being interviewed who usually knows more about the practical details of the work, of the intricacies in it with which no one else is familiar and which sometimes can be of major significance to the study. The difficulty usually lies in securing his confidence to the extent that he will be willing to reveal them. To this end an unpretentious attitude of not knowing all the answers, but of being willing to learn, is most helpful. This condition of mind is as important as an intelligent choice of questions.

Remember throughout that the employee has his regular work to perform. Do not impose unnecessarily on his time. Although the interview should not be hurried, when the systems analyst has all the information he requires the interview should be terminated and not allowed to stretch out.

Advantages of the Interview over Other Fact-Gathering Methods

When the interview is used for fact gathering, the systems analyst is obtaining information directly from others in personal meetings, in contrast to obtaining information by means of mail questionnaires or the telephone. The chief virtue of the personal interview method is that it affords a direct approach to problems through face-to-face contacts. There are the following accompanying benefits:

1. The persons being interviewed may volunteer information of a personal or confidential nature that they would not ordinarily indicate on paper or relate over a telephone.

2. The interview accommodates persons who before imparting information like to see the interviewer (the systems analyst) and to receive guarantees as to how the facts will be used.

3. Personal contact permits the systems analyst to provide the stimulus necessary to "draw out" the answers to questions.

4. It enables the systems analyst to form an impression of the person who is being interviewed, to formulate some judgment as to the reliability of the answers, and to obtain information that may have been implied rather than explicit.

5. Interviewing permits the systems analyst to take advantage of clues and to follow up on leads.

6. It provides an opportunity for the systems analyst to impart on-the-spot information in give-and-take fashion, to render the personal touch, and to enlist the interest and effort of the respondent—an approach that is not possible in using a mailed questionnaire or in making a telephone call.

THE RECORDING OF INFORMATION

Note Taking

Generally, notes should be prepared by the systems analyst during the course of the survey, indicating briefly the results of his observations and inquiries on operations. Names of persons should be recorded as well as the dates of conversations. In the actual interviewing the systems analyst should not attempt to make an extensive recording of his findings, because this slows up the procedure. The notes should be made in the form of telegraphic paragraphs or rough charts; they should be clear enough to be intelligible when they are subsequently referenced. If, for reasons of expediency, the notes have been taken so rapidly that they are in danger of becoming quickly unintelligible, then they should be expanded and clarified as soon as possible. The guiding principle is to take notes with the idea in mind of accumulating adequate and accurate information for the preparation of a formalized chart or diagram, which in turn will be the basis for a penetrating study.

Note taking can involve the recording of very detailed and complex information. Typical recordings take into account the operations performed in a system; they show what is done, how each step is performed, where the work originates, where it goes, and why.

The on-the-spot recording of information pleases most people. They feel more inclined toward rendering details on a matter when the systems analyst is taking down notes rather than trusting to memory. The explanation is simple: Who wants to spend time presenting details when it is felt the information will not be remembered?

Charting the Notes

In order to get from the mere collection of data to their analysis, some way of organizing the mass of facts must be made. In general the process involves transcribing the notes obtained in interviews into flow process charts or flow diagrams. This the systems analyst does when he gets back to his desk after each interview or series of interviews.

Charts and diagrams are of several different kinds. Whatever the type, their purpose is to assist in organizing and visualizing the facts, and in designing improved procedures.

The different types of flow charts and diagrams with which the systems analyst should be familiar are discussed in the next chapter.

ASSEMBLING FINANCIAL DATA

As an integral part of a project assignment, it is often necessary to marshal financial information so as to appraise the economic benefits that are associated with contemplated changes. The typical elements to be considered in assembling the information are listed below.

Category	Primary Considerations
Personnel	Wages for each job classification and supervisory salaries.
Equipment	Depreciation and maintenance charges when company-owned; rental charges when leased.
Material	Forms and supplies.
Overhead costs	Insurance, taxes, administration expenses, etc., allocated by a suitable method such as a pro rata share of space costs.
Return on expenditure	Outlay of cash and other assets, savings that will accrue to the company, conversion and operating costs, and the time value of money.

The dollars-and-cents efficiency of a system in fulfilling its objectives and requirements is usually determinable with a reasonable degree of accuracy; unfortunately this is too often set aside in favor of "guesstimates". Figure 8-1 shows the savings to be derived as a result of installing an automated procurement system.

Indication of Savings To Be Attained by New System

	For the Period through the —				
	First Year	Second Year	Third Year	Fourth Year	Eighth Year
Savings (Reference: "Total Savings Resulting from Systematizing and Automating Procurement System" *)	$99,137.01	$198,274.02	$297,411.03	$396,548.04	$793,096.08
Cost of equipment	49,820.00	—	—	—	—
Depreciation	6,227.50	17,125.63	25,299.22	36,027.07	43,031.30
Savings over cost of equipment	49,317.01	148,454.02	248,591.03	346,728.04	743,276.08
Book value of equipment	43,592.50	32,694.37	24,520.78	13,792.93	6,788.70

This chart assumes a constant volume of business. An increase in business will proportionately increase the savings to be obtained.
* Referenced document not included in text.

Figure 8-1. Projected savings from the installation of a new procurement system using automated equipment—Flexowriters and Computypers.

For the purpose of determining potential savings cost data may be accumulated by departments, by jobs or orders, or in other ways suitable to the purpose at hand. In some cases the systems analyst will have access to existing factual data, historical cost data already compiled for another purpose; in other cases he will have to develop his own or make estimates. Very often, too, his results will encompass a combination of both factual data and estimates.

Factual data such as the following are basic and easily obtainable:

Average number of words in a typewritten line 12
Average number of lines to a typewritten page 26
Average typewriting speed—words per minute 60
Average time required to typewrite one page (minutes) 5

With this information, and knowing the hourly rate for a clerk-typist, it is possible to calculate the cost of typing a letter, page, or full report of the narrative type.

The various techniques of measuring basic units of work include the following:

1. Stop-watch study;
2. Sampling;
3. Mechanical or electrical counters;
4. Control numbers if the forms involved are prenumbered;
5. Time-and-output study; that is, by dividing the total time by the quantity of units processed.

Individual considerations determine which technique, or combination of techniques, to use in a particular case. Under all circumstances, however, allowances must be made for such factors as coffee breaks, down-time on equipment, and variations from the normal routine.

WRITTEN QUESTIONNAIRES

Occasionally the systems analyst will find it useful to request information for certain systems studies by means of mail questionnaires. When appropriately employed the questionnaire method of obtaining information offers a number of potential advantages, four of which are as follows:

1. The mailed questionnaire permits the systems analyst to work on other matters, related or unrelated, while acquiring information.
2. It gives the respondent an extended period of time in which to formulate replies. Moreover, supervisory personnel are given ample opportunity to consult with their subordinates before answering questions.
3. It is often a successful means of obtaining information from individuals who are difficult to contact by other means.
4. The method is relatively inexpensive if correctly executed.

On the other hand, several disadvantages arise in connection with the use of the questionnaire method. The most common of these are the following:

1. It is not suitable for obtaining certain types of information. Matters that are complex or deeply involved are better dealt with by the personal-interview method.
2. The list of questions must be kept short. Management personnel have an aversion to answering questionnaires that take a long time to complete.
3. Mail questionnaires have limitations with respect to obtaining impressions, criticisms, and suggestions.

4. The method is a comparatively slow means of obtaining information. Considerable time must generally be expended in the preparing, distributing, and, on the part of recipients, in filling out and returning questionnaires. To obtain replies it frequently becomes necessary for the systems analyst to personally contact individuals who fail to respond.

5. Many of the replies are apt to be unsatisfactory. This happens even when great care is exercised in framing the questions. Consequently, the systems analyst must also follow up with a personal interview to obtain the required information.

The subject of questionnaires was first discussed in Chapter IV with particular reference to obtaining organizational information. Most of the comments made regarding the construction and use of organizational questionnaires applies with equal significance to questionnaires used in other systems-and-procedures studies. The questions contained in questionnaires should be as easy to understand as possible and formulated so as to elicit answers that are free from bias. They should be clearly phrased to avoid misinterpretations. Leading and ambiguous questions must not be used. A well-written cover letter that is informative and makes an appeal for cooperation can be an effective means of getting the questionnaires completed properly and returned on time. The systems analyst should specify a realistic date for their return. From a practical viewpoint it is sometimes necessary to request that they be returned considerably in advance of the time when actually needed. This gives the systems analyst a period of grace in which to expedite replies that would otherwise be truly late. However, it is prudent to use this technique in moderation. If done in excess, asking for returns a week or two in advance loses its value, because the recipients sooner or later become aware of the practice and react accordingly.

9

Analyzing and Improving Systems

CRITERIA FOR SYSTEMS EVALUATION

A major consideration of systems analysis and design is the evaluation of a system's effectiveness while in operation. Criteria are applied while collecting the data and studying the design of the system. These measures of effectiveness, a model list of which is presented below, enable the systems analyst to make value judgments as he proceeds with his work.

1. A good system provides pertinent and timely information at the lowest possible cost.
2. Systems must be especially designed to meet the needs of the business and its managers.
3. To the extent practical systems should be designed to utilize management techniques developed to allocate available resources most effectively.
4. Systems need to be in accord with company policies.
5. To design an efficient system it is necessary to consider the human factor—the nature of man and his limitations—as well as the capabilities of any machine that the personnel will be required to use.
6. A system should be simple and workable.
7. A system should be able to withstand change by providing for ease of expansion and for extra output capacity.
8. In the design of a system consideration should be given to making it as automatic as possible.
9. Activities should be arranged to provide a straight-line flow from one step to the next, avoiding needless backtracking.
10. Where feasible data should be recorded at their source, or as close to the source as possible, so as to reduce or eliminate the need for recopying.

11. No information or service should be produced that is not justified by its cost to the enterprise.

12. Wherever practical, functions should be assigned to organizational units in such a way as to reduce the need for coordination, communication, and paper work.

13. System requirements should dictate the need for equipment. Because an operation can be performed mechanically does not mean that mechanization is the best or least expensive method.

14. To achieve its objectives a system must be formulated on the basis of accurate information.

FLOW CHARTING AND DIAGRAMMING

The assembling of facts during the system study usually involves charting or diagramming. The placement of information in graphic form is desirable for a number of reasons, which may be summarized as follows:

1. Facts about operations can be recorded systematically and compactly, thus making it possible for the systems analyst to gain an overall comprehension of the entire system and its interrelated procedures.
2. Completeness and accuracy are promoted. Graphic representations tend to bring to light omissions and errors, and to keep the study within its proper bounds.
3. Analysis is facilitated. Since the data are portrayed in a pictorial arrangement, it is easier to visualize the entire flow of data and thereby to locate the difficulties and weaknesses of a system; for example, bottlenecks, instances of backtracking, and unnecessary duplication of work.
4. A basis for synthesis is provided; that is, it is possible to present a graphic picture of operations that is capable of suggesting constructive ideas.
5. The explanation and comprehension of new proposals is facilitated.
6. Present and proposed operations can be contrasted dramatically.

Charts and diagrams make an important contribution as part of the permanent records of a systems study. The need for referring to them occurs frequently. For some time after a system is installed the systems analyst may expect to be questioned about its operations.

While the systems analyst is working on a system the operations are fresh in his mind. As time passes, however, small but important details are forgotten. It is then that recourse to a permanent file of charts, diagrams, and other data enables the systems analyst to refresh his memory of an assignment.

Not to be overlooked is the long-range value of such records. They are a prime source of information for future studies of the same system.

Types of Charts. Systems analysts utilize what might appear to be a large number of flow-charting techniques due to the many modifications and adaptations of the same basic chart types. Although these variations are important, discussion of the different kinds of charts is clearer when done on the basis of broad classifications. The types described and illustrated in this chapter are the most commonly used and fall into three major categories: (*a*) flow process charts, (*b*) flow diagrams, and (*c*) movement diagrams. Each of these types of charts serves a specific purpose by organizing and symbolically portraying information in a particular way, making it especially suited for analysis. Which one to use depends on the nature of the material to be charted.

FLOW PROCESS CHARTS

For identifying and studying the detailed steps of a process, the flow process chart is excellent. It records the step-by-step elements of an activity, differentiating between productive and nonproductive elements. It shows the activity broken down in terms of "do" operations, transportations, inspections, delays, and storages; it also includes information that is considered desirable for analysis, such as time notations and distances traveled. A flow process chart may be prepared for an entire activity, including many departments, or it may be limited to a part of an activity. Whatever its scope, the chart is used to determine how operations and steps might be eliminated, changed in sequence, combined, simplified, or subdivided for greater efficiency.

Because of its structural design, the flow process chart has definite limitations. In general it serves best for recording the facts for the analysis of procedures that are relatively simple or involve a single work flow. For most applications the flow process chart is used to follow only one document of flow of work at a time. If copies are distributed or if additional documents are generated, their flow is shown not on the same chart but on separate follow-on charts.

The preparation of a flow process chart entails depicting vertically, by brief statements and symbols, the sequence of steps that comprises an activity. The symbols represent summary descriptions of the steps. They serve not only to classify the information in a uniform pattern but also to make a contribution to its understanding. Although these symbols are sometimes designed in a variety of geometric patterns, the following are the ones most commonly used:

Operation

Transportation

Inspection

Delay

Storage

An *operation* occurs when an object is created, destroyed, or in some way changed (e.g., by addition or subtraction). The object itself may be a product, form, report, or even a plan.

A *transportation* takes place when an object (product, form, report, or the like) is moved from one place to another. The methods of transportation include mail delivery, messenger service, trucking, and conveyance by means of belts, chutes, or rollers.

An *inspection* takes place when an object (product, form, report, plan, or the like) is checked or verified.

A *delay* occurs when action on an object (product, form, report, plan, or the like) is temporarily postponed; for example, the recipient of form X from department A may have to await the receipt of form Y before he can proceed with the next operational step.

A *storage* occurs when an object (product, form, report, plan, or the like) is retained or filed. Storages are occasionally classified into two types, as follows:

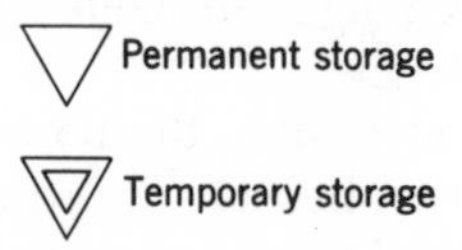

Numbers or letters are sometimes placed within or adjacent to symbols for the purpose of keying the processes to individual employees or organizational units.

To facilitate the drawing of the flow-process-chart symbols, a plastic

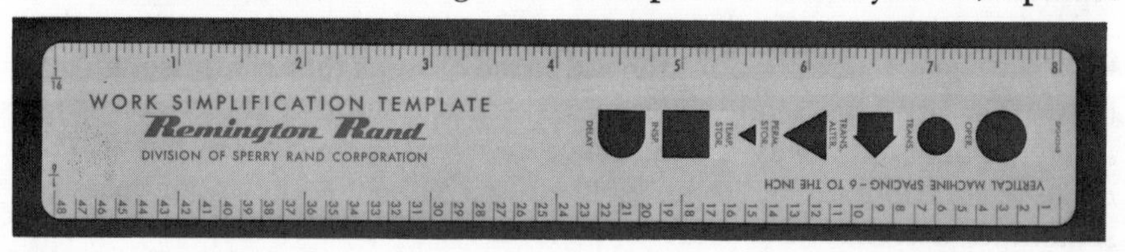

Figure 9-1. Multipurpose ruler with cut-out patterns for drawing flow-process-chart symbols.

ruler with cut-out symbols may be used, such as the one shown in Figure 9-1. Rulers and templates containing these and other additional symbols may be purchased from stationers and drafting-supply dealers.

Variations of the Flow Process Chart

Figure 9-2 illustrates the simplest form of flow process chart. On this type only the headings and captions are preprinted. The symbols are con-

PROCESS CHART

Name of Part or Product Requisition For Supplies - Rush Job

Chart Begins at Machine Shop Foreman's Desk, Ends on Typist's Desk in Pur. Dept. | Chart No.

Order No. | Lot Size | Dept. | Sheet 1

Charted by C.H.H. | Date Charted 7-28-45 | Bldg. M. E. Lab | of 1 Sheets

Travel in ft.	Time in min.	Symbol	Operations	Remarks
		1	Written longhand by foreman	
		1	On foreman's desk (awaiting messenger)	
1000		1	By messenger to secretary of head of department	
		2	On secretary's desk (awaiting typing)	
		2	Typed	
15		2	By messenger to head of department	
		3	On head of department's desk (awaiting approval)	
		O-3 INS-1	Examined, approved and coded (signed and code stamped)	
		4	On head of department's desk (awaiting messenger)	
2000		3	To Purchasing Department	
		5	On purchasing agent's desk (awaiting approval)	
		2	Examined and approved	
		6	On purchasing agent's desk (awaiting messenger)	
25		4	To typist's desk	
		7	On typist's desk (awaiting typing of purchase order)	

SUMMARY

Number of Operations	3
Number of Delays	7
Number of Inspections	2
Number of Transportations	4
Total Travel in Feet	3040

Figure 9-2. Flow process chart with preprinted headings and captions. Source: Asme Standard 101.

FLOW PROCESS CHART	Job Processing Customer Orders				Page __ of __ pp.		
	Department ______ Dept. Head ______				Charted by		
					Date		
Line Number	Operation / Transportation / Inspection / Delay / Storage	Step No.	Details of ☐ Present ☐ Proposed Method		Distance in feet	Quantity	Time in minutes
1	○⇨☐D▽	1	Receives order in mail room				
2	○⇨☐D▽	2	Stamps time and date				
3	○⇨☐D▽	3	To pricer		30		
4	○⇨☐D▽	4	Awaits pricing				
5	○⇨☐D▽	5	Prices order				
6	○⇨☐D▽	6	To sales department		1110		
7	○⇨☐D▽	7	Awaits checking				
8	○⇨☐D▽	8	Checks credit, price, etc.				
9	○⇨☐D▽	9	To biller		35		
10	○⇨☐D▽	10	Awaits action				
11	○⇨☐D▽	11	Inserts carbon in three-part form				
12	○⇨☐D▽	12	Types confirmation of order form				
13	○⇨☐D▽	13	Extensions are made				
14	○⇨☐D▽	14	Checks quantities and prices				
15	○⇨☐D▽	15	Confirmation copy and production copy placed in "out" box				
16	○⇨☐D▽	16	Files second copy by account name				
17	○⇨☐D▽						
18	○⇨☐D▽						
19	○⇨☐D▽						
20	○⇨☐D▽						
21	○⇨☐D▽						
22	○⇨☐D▽						
23	○⇨☐D▽						
24	○⇨☐D▽						
25	○⇨☐D▽						
26	○⇨☐D▽						
27	○⇨☐D▽						
28	○⇨☐D▽						
29	○⇨☐D▽						
30	○⇨☐D▽						
31	○⇨☐D▽						
32	○⇨☐D▽						
33	○⇨☐D▽						
34	○⇨☐D▽						

Figure 9-3. Flow process chart designed to show either *present* or *proposed* method, with preprinted headings, captions, and symbols.

FLOW PROCESS CHART

SUMMARY	Present		Proposed		Difference	
	No.	Time	No.	Time	No.	Time
○ Operation						
⇨ Transportation						
□ Inspection						
D Delay						
▽ Storage						
Distance Traveled		Ft.		Ft.		Ft.

Job ______________________

Department ______________________
Dept. Supvt. ______________________
Charted by ______________ Date __________
Note ______________________

Page___ of ___ pp.

Details of Present Operation	Operation Transportation Inspection Delay Storage	Distance in feet	Time in minutes	Line Number	Details of Proposed Operation	Operation Transportation Inspection Delay Storage	Distance in feet	Time in minutes
	○⇨□D▽			1		○⇨□D▽		
	○⇨□D▽			2		○⇨□D▽		
	○⇨□D▽			3		○⇨□D▽		
	○⇨□D▽			4		○⇨□D▽		
	○⇨□D▽			5		○⇨□D▽		
	○⇨□D▽			6		○⇨□D▽		
	○⇨□D▽			7		○⇨□D▽		
	○⇨□D▽			8		○⇨□D▽		
	○⇨□D▽			9		○⇨□D▽		
	○⇨□D▽			10		○⇨□D▽		
	○⇨□D▽			11		○⇨□D▽		
	○⇨□D▽			12		○⇨□D▽		
	○⇨□D▽			13		○⇨□D▽		
	○⇨□D▽			14		○⇨□D▽		
	○⇨□D▽			15		○⇨□D▽		
	○⇨□D▽			16		○⇨□D▽		
	○⇨□D▽			17		○⇨□D▽		
	○⇨□D▽			18		○⇨□D▽		
	○⇨□D▽			19		○⇨□D▽		
	○⇨□D▽			20		○⇨□D▽		
	○⇨□D▽			21		○⇨□D▽		
	○⇨□D▽			22		○⇨□D▽		
	○⇨□D▽			23		○⇨□D▽		
	○⇨□D▽			24		○⇨□D▽		
	○⇨□D▽			25		○⇨□D▽		
	○⇨□D▽			26		○⇨□D▽		
	○⇨□D▽			27		○⇨□D▽		
	○⇨□D▽			28		○⇨□D▽		
	○⇨□D▽			29		○⇨□D▽		
	○⇨□D▽			30		○⇨□D▽		

Figure 9-4. Flow process chart designed to show both *present* and *proposed* method, with preprinted headings, captions, and symbols.

tained in a single column and are written by hand by the analyst. A quicker way is to have the symbols preprinted; then a line is drawn to connect the symbols for each step in the process. Figures 9-3 and 9-4 show such charts.

On Charting the Present Steps of an Operation

In constructing a flow process chart it is necessary to remember the following points:

1. Determine the boundaries of the study and keep within them. Do not become involved in extraneous operations or side issues.
2. Follow the item or object from one end to the other. Actually tour the route in conjunction with the form, product, or whatever. Observe the actual operations and question the persons involved.
3. Register each step of the operation in the order of its occurrence. Care must be taken to see that no step is missed, because its importance may not be immediately evident.
4. Record other facts that are pertinent to the study, such as distance in feet, time expended, and special annotations.
5. Summarize the information. Total the number of times that each symbol appears, the transportation distances, and the time listings.

Evaluation of the Present Steps

Upon completion of the chart needed improvements become plainly evident by the disclosure of an excessive number of symbols of a particular kind, such as delays or transportations. If the problems are not readily apparent, careful observation and study will be necessary to disclose latent possibilities. To assist the analyst in making an evaluation, a checklist such as the following is useful:

1. Is each step actually a necessary part of the operation?
2. Are the steps arranged in the best possible sequence? Would efficiency be improved if the relative position of any step were changed?
3. Can any step be eliminated, simplified, or expedited?
4. Is it possible to combine certain steps into a single step that would improve overall performance?
5. Is there any duplication of steps? If so, is the duplication necessary or desirable?
6. Is the operation as free flowing as possible? If not, what step or

combination of steps causes congestion? Can the situation be alleviated by scheduling the work in another way?

7. Could certain machinery or equipment lighten the work load?

8. What are the causes of delays in performing the work? How can they be eliminated or reduced?

9. Would performance be improved if certain steps were broken down into finer steps?

10. Could a step or combination of steps be performed better by another person or organizational unit?

11. Could economies of time, transportation, or storage be realized by transferring certain steps to another person or organizational unit?

12. If a step or combination of steps is changed (e.g., by rearrangement or transference), how will the other steps be affected?

13. Are appropriate steps included in the operation for checking and controlling the work?

14. Is the work being checked for completeness and accuracy? If not, should such a check be done during this operation?

15. Are the frequency and manner of checking satisfactory?

16. Where 100-percent inspection of the work is being performed, could sampling be substituted? Should the work be sampled at additional points in the operation?

17. Are the requisites for accomplishing the work—facts, data, materials, forms, or whatever—being properly supplied? Are they being received on time and in good condition?

These questions indicate that a thorough analysis reaches into practically every element of an operation.

Developing an Improved Operation

The investigation, questioning, and critical evaluation of an operation will help evolve ideas for improvement. The flow-process-charting technique crystalizes the more promising ideas into one or more plans to improve the operation. This is done by listing and symbolizing the steps of a potential plan on a chart in the same way as was done for the existing method. These plans are then checked against the existing steps to determine their potential advantages. In practice one or more plans may be modified a number of times before the best plan is devised and instituted.

A rectangular box such as that shown in Figure 9-4 affords a convenient means for summarizing the various steps of both the present operation and any potential plan, and showing the cumulative differences be-

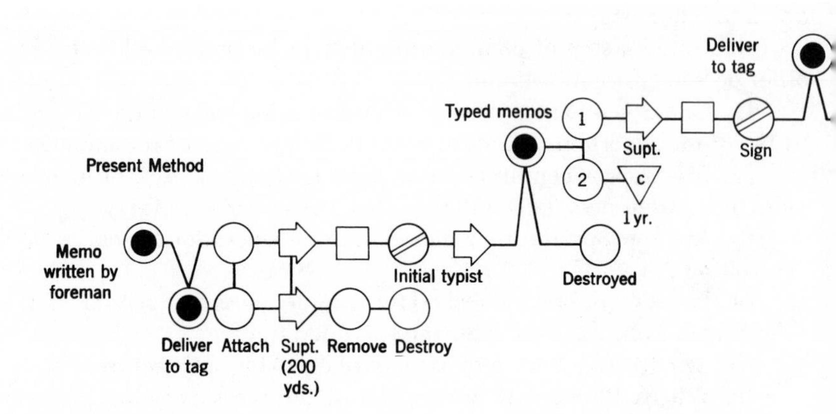

Elements	Symbols	Present Method	Proposal No. 1	Proposal No. 2
Origin	◉	8	2	1
Operation	○	15	0	0
Transportation	⇨	14	8	5
Inspection	□	6	4	2
Information added	⊘	3	2	2
Storage	▽	3	1	1
Delay	D	1	0	0
		50	17	11

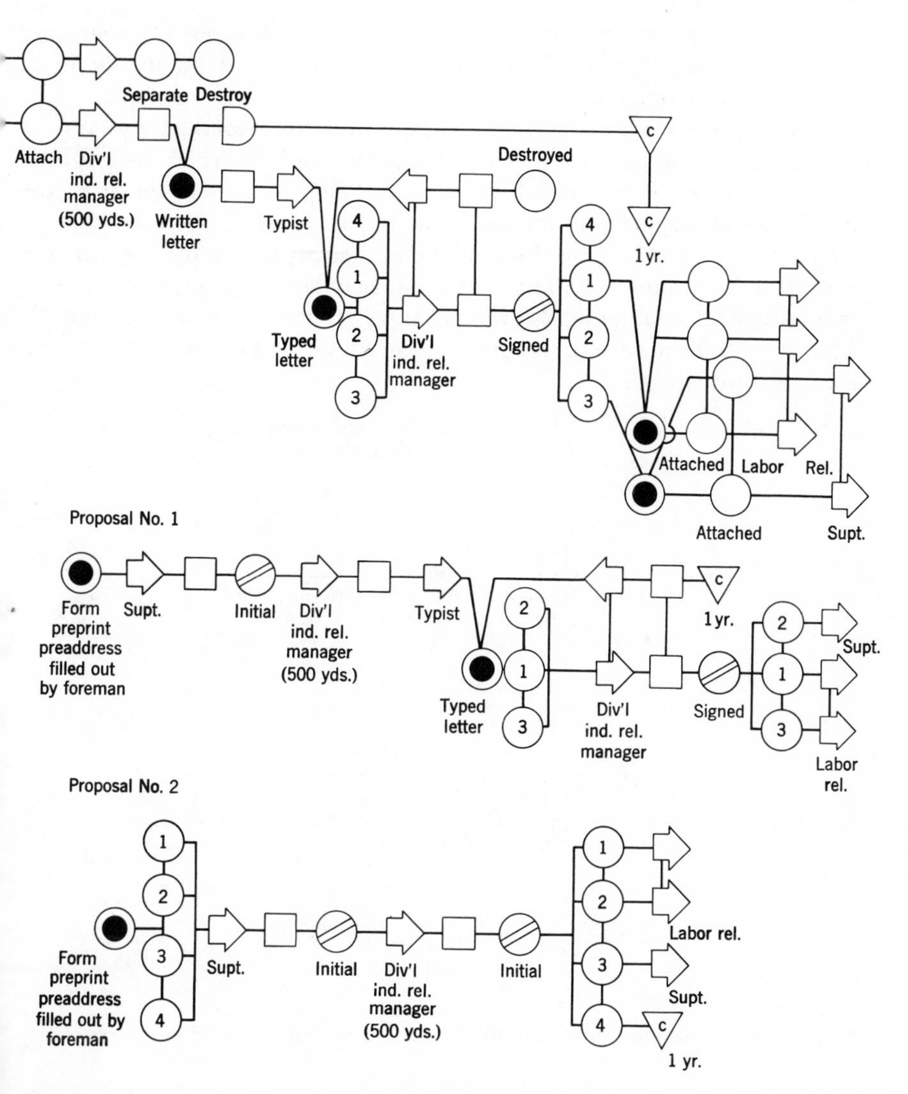

Figure 9-5. Flow diagram—process chart symbol type. Source: Canadian Office, Vol. 1.

tween the two methods. The advantages inherent in any revised plan are thus expressed in numerical terms that can be easily understood.

FLOW DIAGRAMS

The flow diagram is unexcelled as a graphical means for analyzing and synthesizing systems. Its importance as an interpretive medium can hardly be overemphasized.

A properly constructed flow diagram presents as perfect a picture of the operations of a system as the accuracy and reliability of the information from which it is derived will allow. It depicts a clear image of the intricacies of a system. It portrays what occurs to an object and other related objects as they flow through the system. It shows the sequence of events and it points out the relationship between a product and a form, one form with another, and different areas of work. Not too infrequently situations that are seemingly unrelated in diagram show correlations that are quite unsuspected.

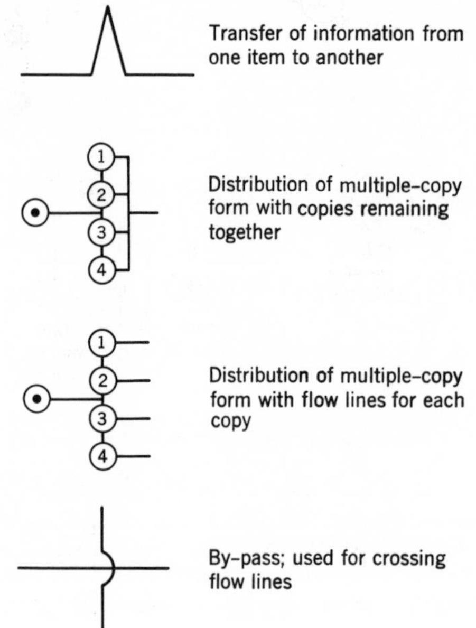

Figure 9-6. Symbols commonly used in the flow diagram of the process-chart-symbol type.

Based on the kind of symbols used, flow diagrams are of four basic types:

1. Process-chart symbol
2. Conventional
3. Automated
4. Schematic

Process-Chart-Symbol Type

Flow diagrams of this type contain symbols that are the same, or similar to, those used on flow process charts. Figure 9-5 illustrates this style of chart. Observe that in this case the symbols used for operations, transportations, inspections, storages, and delays are standard. Other commonly used symbols have been added. (See Figure 9-6.)

Conventional Type of Flow Diagram

For studying complex systems the conventional type of flow diagram is the most widely used of the different flow-charting techniques. Instead of using abstract symbolism, the conventional style of chart uses realistic portrayals; that is, symbols that resemble the particular objects they indicate. Examples are shown in Figure 9-7.

Flow lines are the connecting links on a diagram. Sometimes it is helpful to indicate these connections by different types and/or colors of

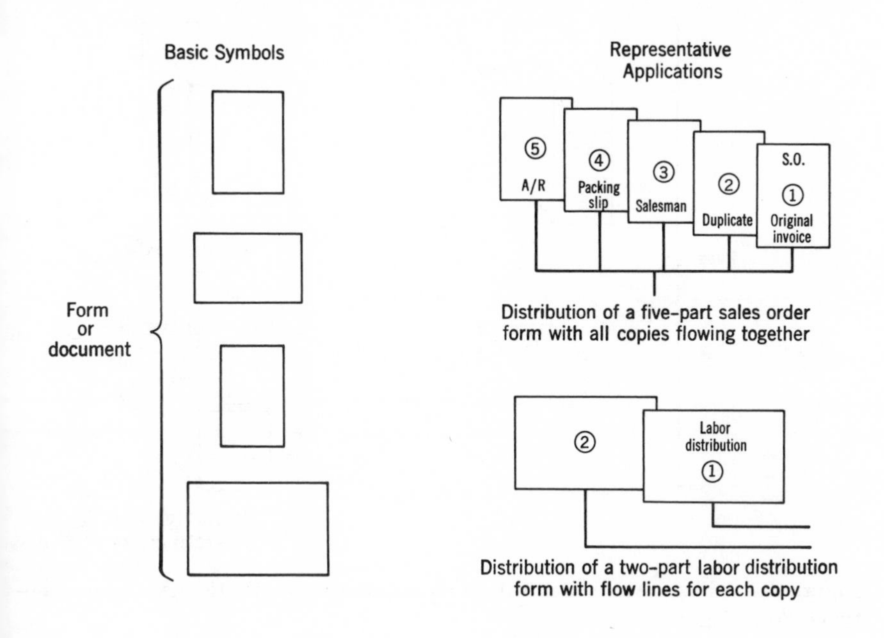

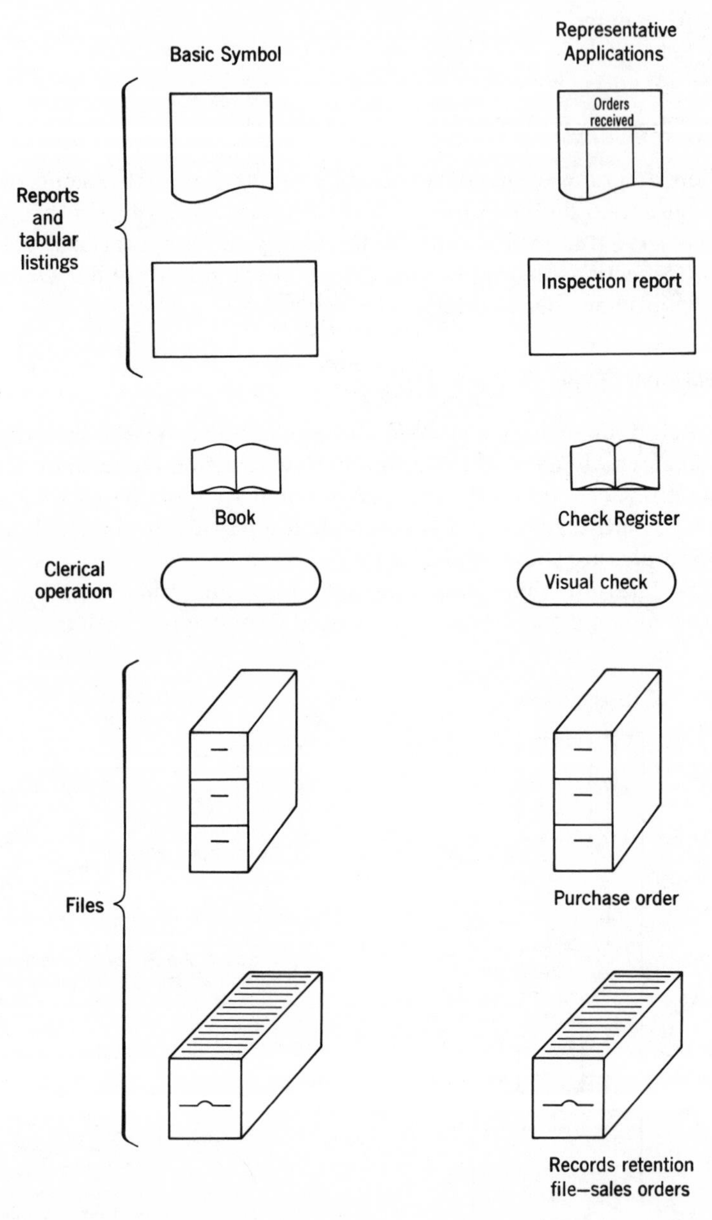

Figure 9-7. Basic symbols and their representative applications in the conventional flow diagram.

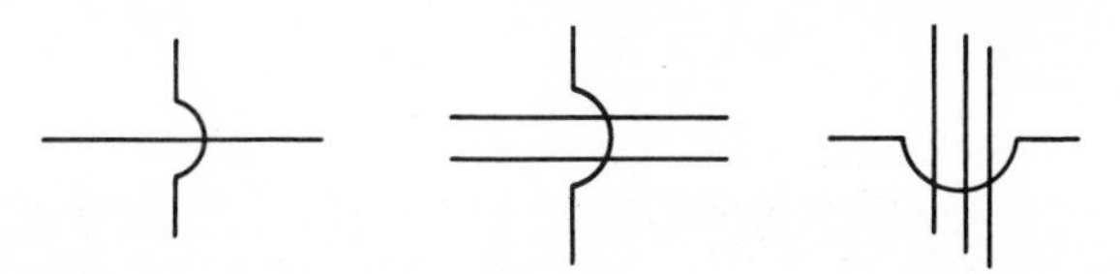

Figure 9-9. Examples illustrating the crossing of flow lines.

lines—solid, dotted, dash, etc.—that have been assigned specific meanings. Arrowheads are commonly drawn on the lines to clearly designate the direction of the flow. Three examples illustrating the crossing of flow lines are shown in Figure 9-8.

A combination vertical and horizontal flow diagram is shown in Figure 9-9, and a vertical flow diagram is shown in Figure 9-10.

Automated Data-Processing Diagrams

Charts of this classification are for the purpose of portraying mechanized operations, including those of electric accounting machines and electronic computers. Many of the same basic symbols may be contained on charts of this type as on the conventional flow diagram. A summary of the symbols used in automated data-processing diagrams is given in Figure 9-11. A set of IBM flow-chart symbols, which are widely used as alternatives, is given in Figure 9-12. An operational flow diagram of the automated type is shown in Figure 9-13, and a typical block diagram is shown in Figure 9-14. Figure 9-15 shows a set of IBM block diagram symbols, and other frequently used symbols are given in Figure 9-16.

Schematic Type of Flow Diagram

The schematic type of flow diagram (Figure 9-17) is primarily used for making presentations to management. Such charts are used to demonstrate systems broadly (e.g., in the presentation of a new systems proposal).

Two significant features distinguish the *schematic* from the conventional type of flow diagram. The first is that the schematic omits many details in order to focus attention on the highlights of a system. Secondly, the schematic frequently contains pictures along with symbols to emphasize and clarify cardinal points.

Fortunately, the systems analyst does not have to possess any special artistic talent in order to make use of pictorial representations. A number of companies produce prefabricated materials for constructing charts and diagrams, including artistic reproductions of various kinds. These are generally obtainable from stationery stores and dealers in artists' supplies.

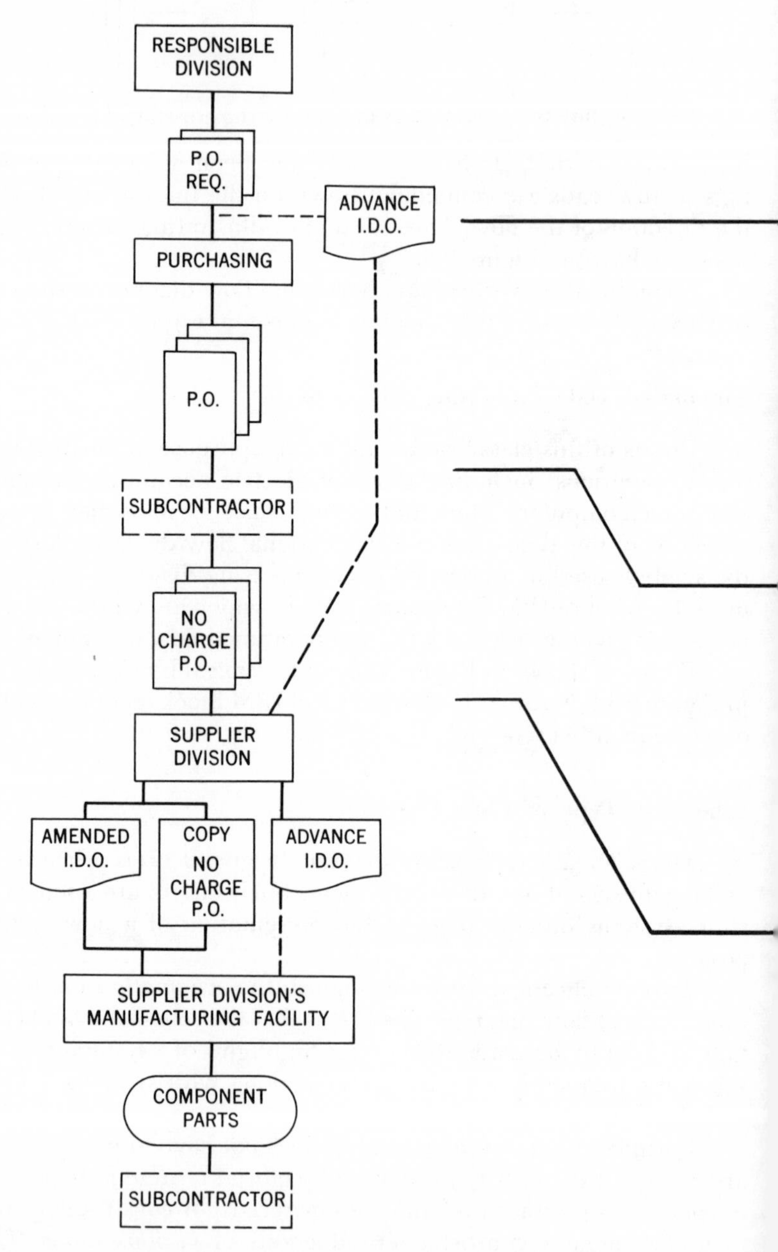

Figure 9-10. Vertical flow diagram—conventional type.

Exhibit A

FUNCTIONAL FLOW DIAGRAM
OF
END ITEM OR SUB-ASSEMBLY SUBCONTRACTING
WHEREIN DEARLEA SUPPLIES COMPONENTS

Functions
Responsible Division 1. Synthesize requirements on price, rate, and deliveries with Supplier Division. 2. Negotiate via Purchasing with subcontractor. 3. Issue Purchase Order Requisition, include repair information, I.D.O. No., 3 sets drawings, and specifications. 4. Direct "Advance" I.D.O. on Sup. Div. 5. Issue applicable internal orders.
Purchasing Department 1. Negotiate price, rate, delivery with subcontractor. 2. Issue P.O. on Subcontractor. 3. Include "Subcontractor's Responsibilities (Functions)" in P.O., i.e. issuance of No Charge P.O., noting when Government source inspection is required, etc. 4. Denote P.O. with I.D.O. No. on P.O.R. 5. Negotiate repair price with subcontractor.
Subcontractor 1. Comply with Dearlea P.O., including issuing No Charge P.O. to acquire components, denoting when Government source inspection is required, designating I.D.O. No. on P.O., returning rejected components to Supplier Division. 2. Place P.O. on Mkt. Div. for quantities in excess of those on Dearlea P.O. 3. Report progress & delinquencies, as required.
Supplier Division 1. Comply with "advance" I.D.O. 2. Fulfill complying No Charge P.O., & note all pertinent documents with No Charge P.O. Number. 3. Amend I.D.O. to include del. and ship instructions, and No Charge P.O. No. 4. Forward copy No Charge P.O. with amended I.D.O. to Dearlea Mfg. facility. 5. Estimate repair cost & affix responsibility on repairs. If Dearlea responsibility, repair at No Charge. If subcontractor's responsibility, advise Purchasing of cost; Purchasing will negotiate with subcontractor.

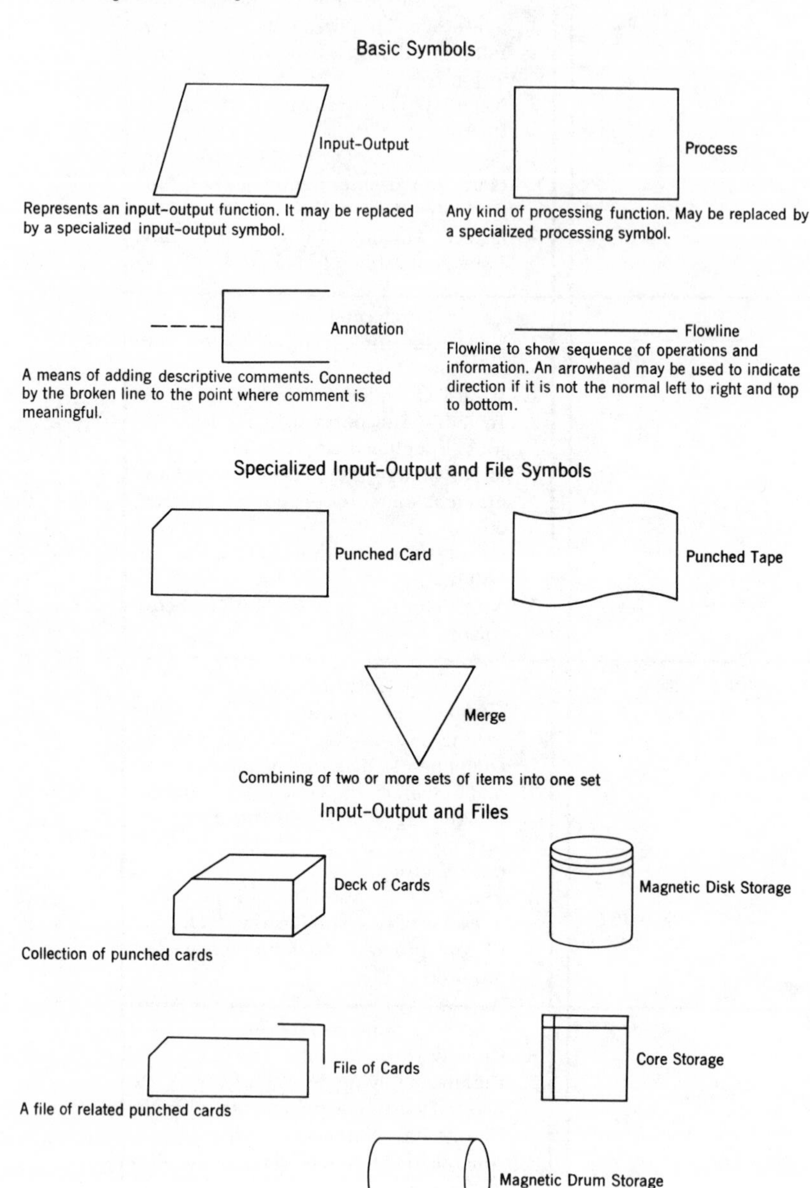

Figure 9-11. Summary of automated-data-processing symbols. Source: United States of America Standards Institute. For a complete listing see *United States Standard Flowchart Symbols for Information Processing*, USASI X3.5, 1966 (revised).

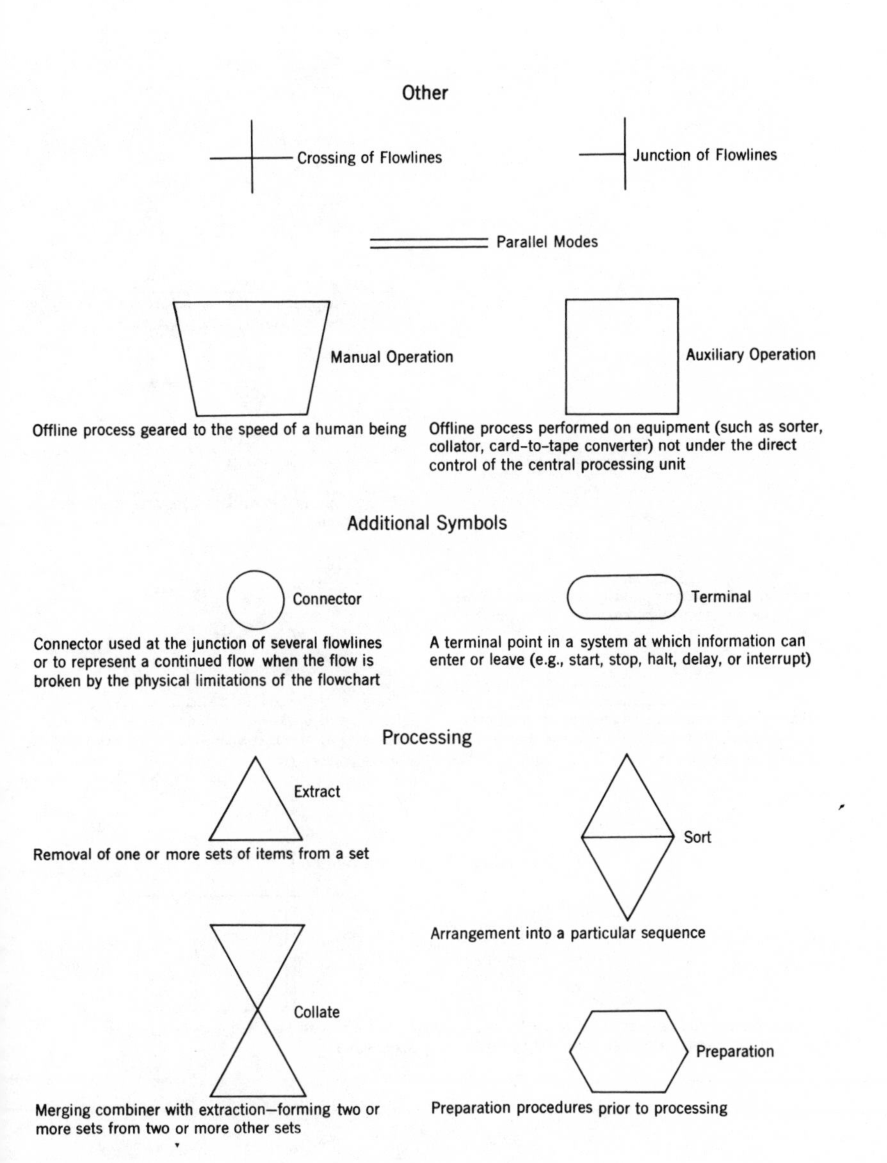

Figure 9-11. *Continued*

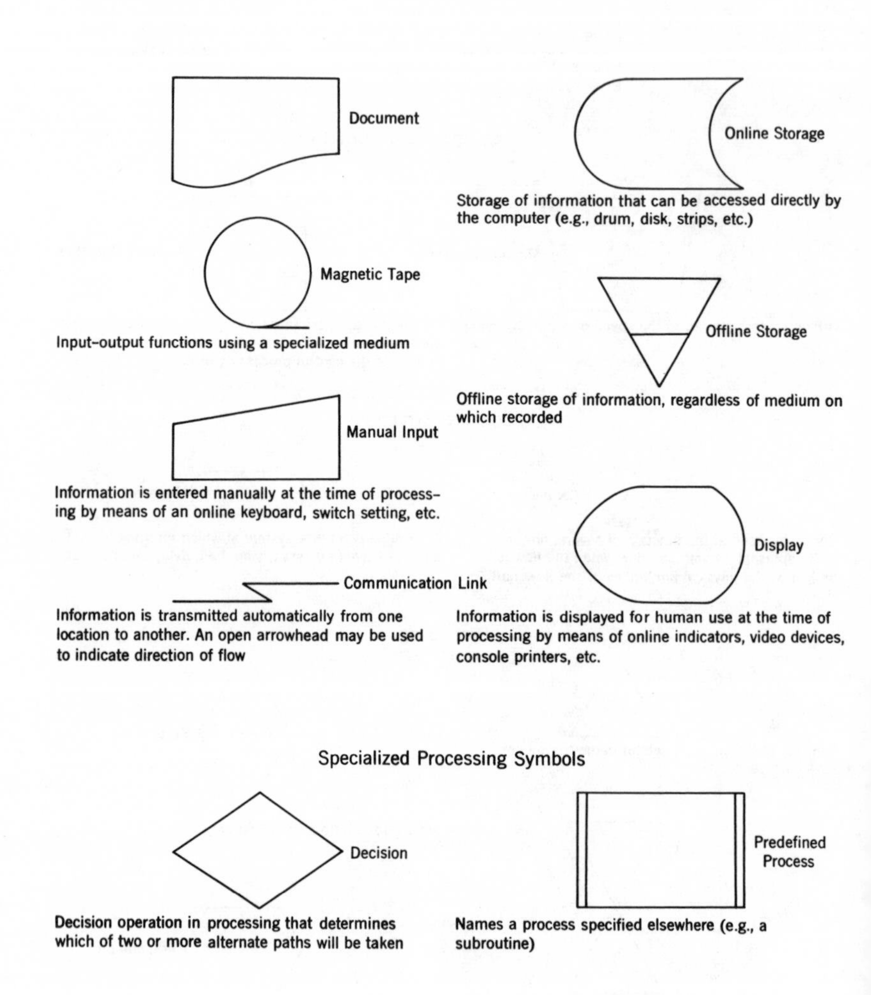

Figure 9-11. *Continued*

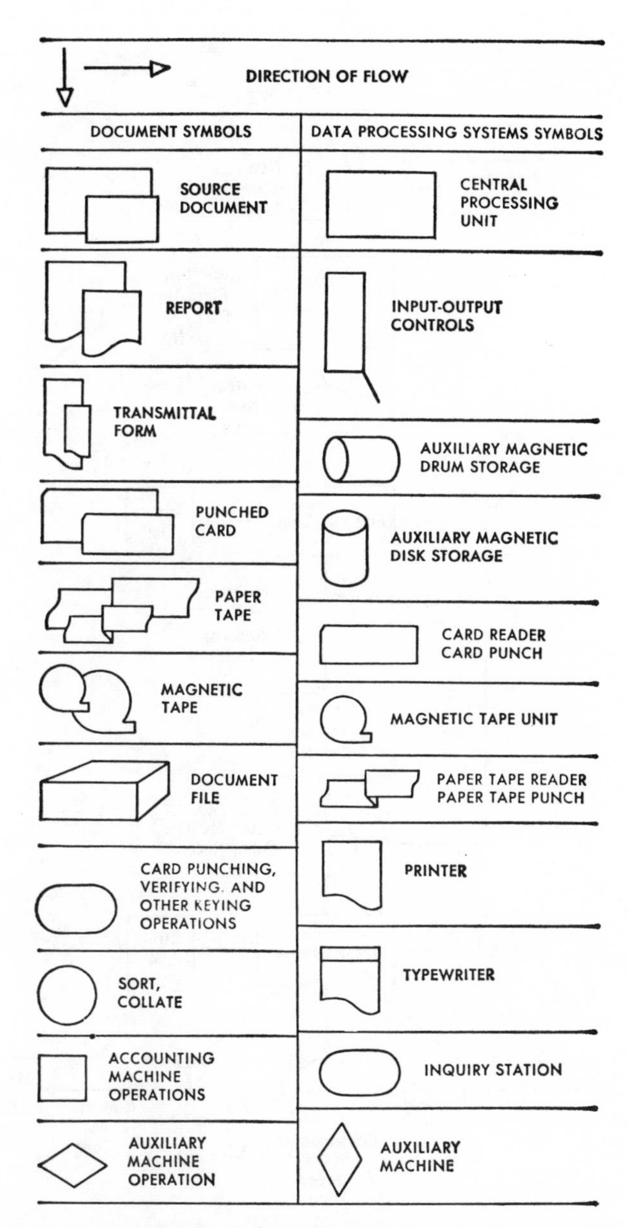

Figure 9-12. Set of IBM flow-diagram symbols, which are widely used as an alternative to those shown in Figure 9-11.

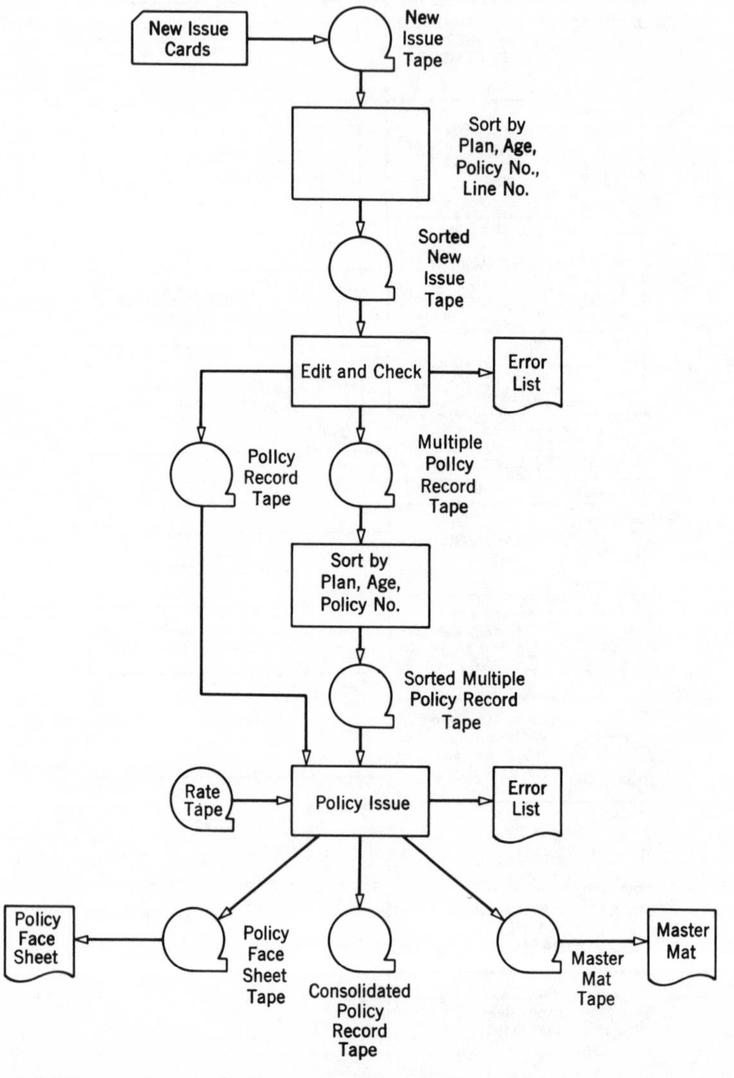

Figure 9-13. Automated type—operational flow diagram. Source: International Business Machines Corp.

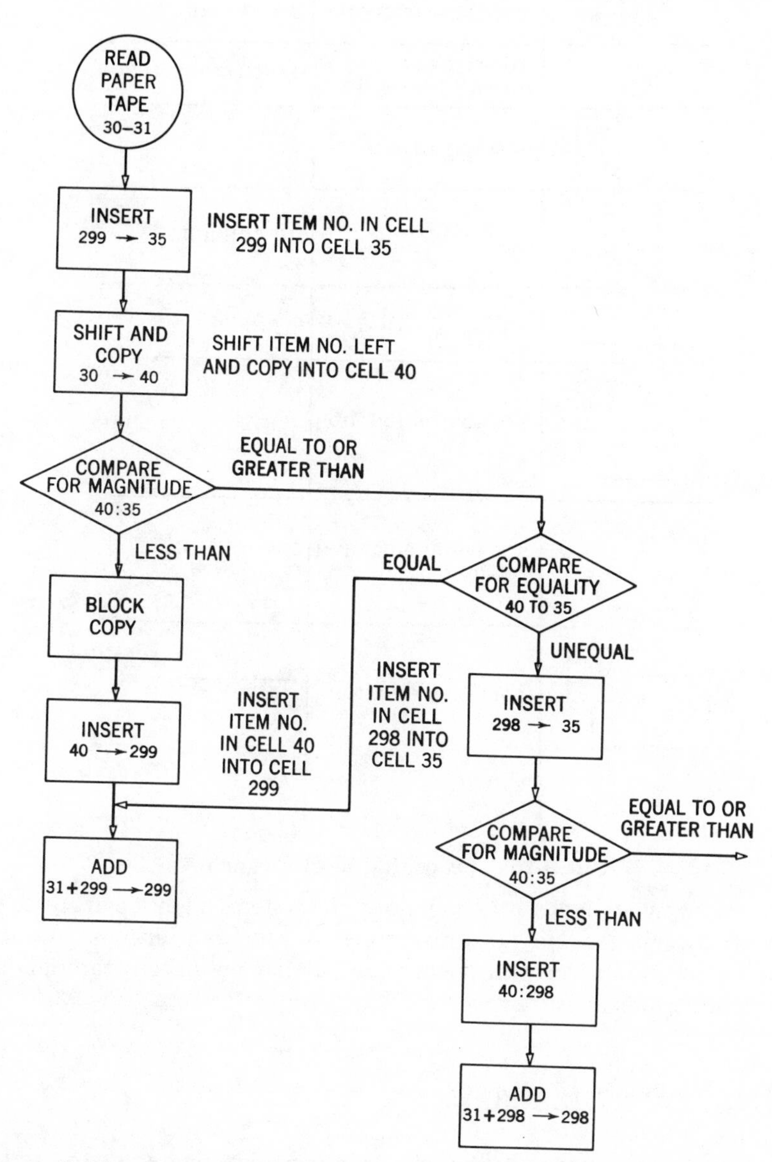

Figure 9-14. Automated type—block diagram. Source: The National Cash Register Co.

SYMBOL	STORED PROGRAM	STORED-WIRED PROGRAM
	DIRECTION OF FLOW	DIRECTION OF FLOW
	DECISION FUNCTION	PROGRAM EXIT
	CONNECTOR OR STEP IDENTIFICATION	BRANCH IDENTIFICATION
	TABLE LOOKUP (650)	
	CONSOLE OPERATION, HALT	STEP CONNECTOR
	PRIORITY ROUTINE IDENTIFICATION (7070)	COMMUNICATION
	INPUT-OUTPUT FUNCTION	
	PROGRAM MODIFICATION FUNCTION	CONTROL PANEL FUNCTION
	PROCESSING FUNCTION	STORED PROGRAM STEP

Figure 9-15. Set of IBM block-diagram symbols.

With the application of a little ingenuity the systems analyst can create his own artistic images. He can draw, trace, or cut from business journals, trade magazines, advertising media, and similar published materials the pictures he wants to use.

MOVEMENT DIAGRAMS

Also known as layout flow charts, movement diagrams are used to portray the movement of a person or object in connection with performing

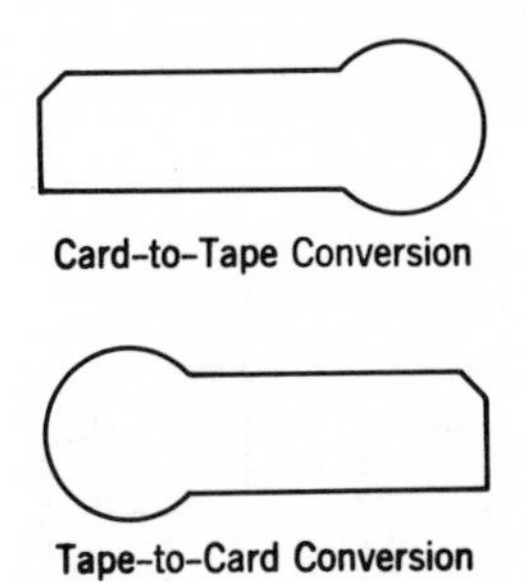

Figure 9-16. Other frequently used symbols.

work. Some are simply ordinary layout drawings with flow lines indicating the flow of work from one place to the next (see Figure 6-5). Others are symbolic, as shown in Figure 9-19.

ANALYZING AND EVALUATING A SYSTEM

After the factual information about the entire system has been diagrammed, usually on a conventional or automated flow chart, the next step is to analyze and evaluate these facts.

Every aspect of the system is studied and evaluated in terms of the principles set forth early in this chapter. Again, a questioning approach here, as in other analytical work, will be of value; for example, in posing the question, "Is the system sufficiently flexible?" the reply should conform with the principle, "A system should be able to withstand change."

Other questions are presented in the following check list. The specific questions to be posed will depend on the particular case and to a considerable extent on the insight and technical competence of the systems analyst.

Systems-Evaluation Check List

A systems-evaluation check list may be divided into two categories: (*a*) general, or key, questions and (*b*) specific, or pointed, questions. The key questions are as follows:

1. Why must the work be done?
2. Can any part of the work be eliminated?
3. Should the sequence of events relating to the processing of the work be changed?

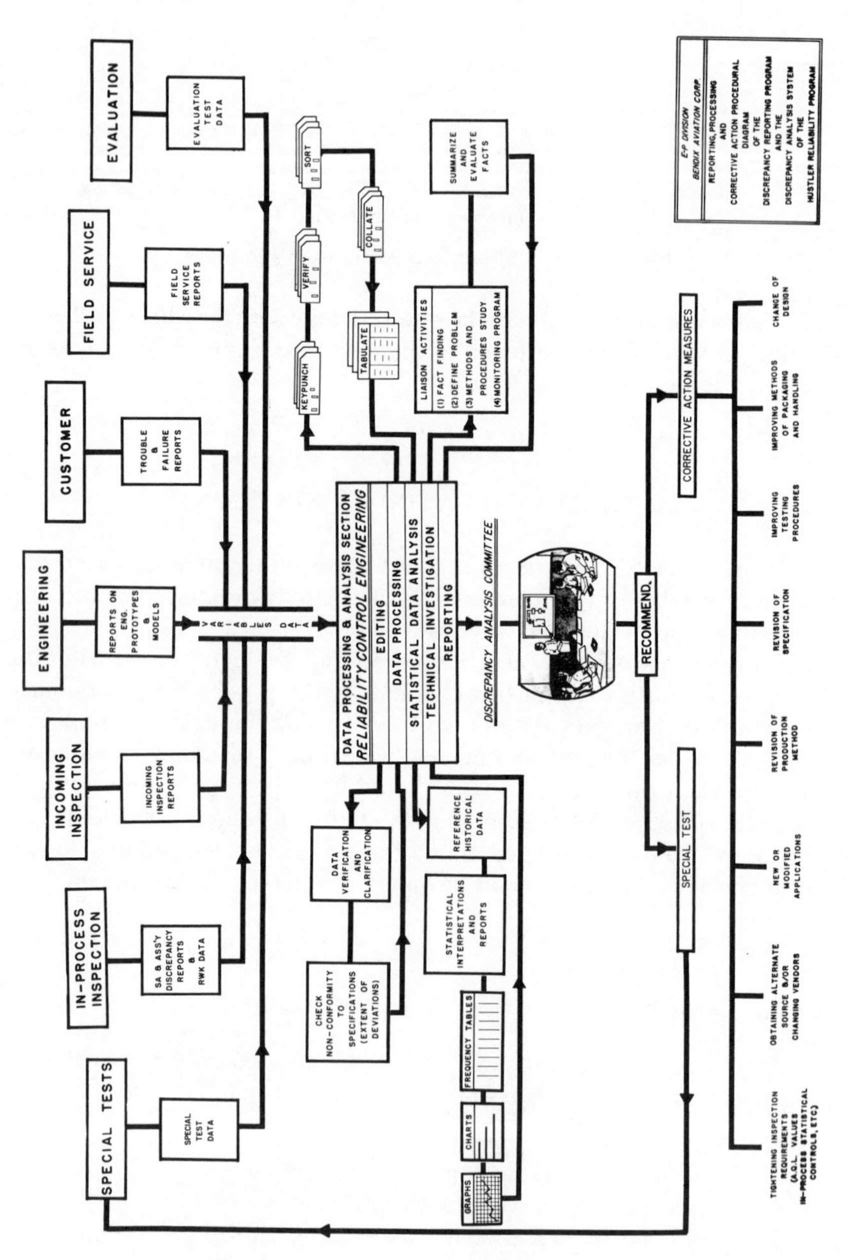

Figure 9-17. Schematic flow diagram.

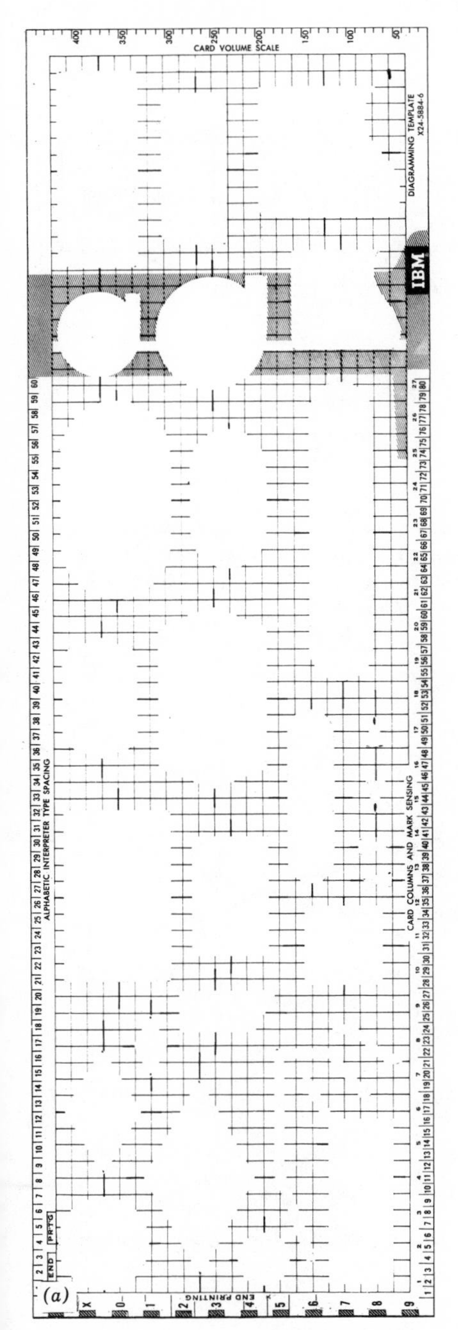

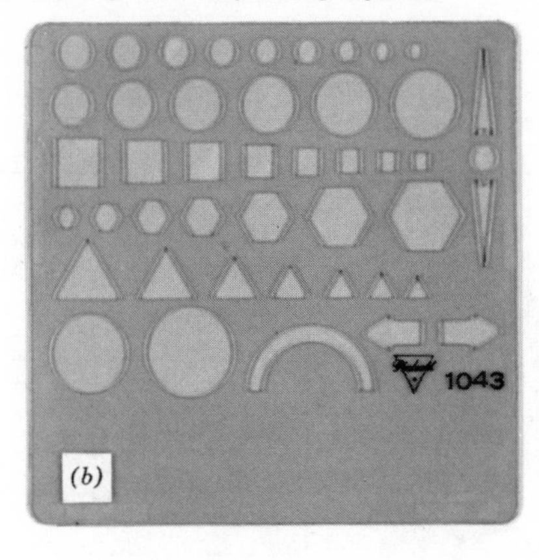

Figure 9-18. Templates for drawing flow diagram symbols.

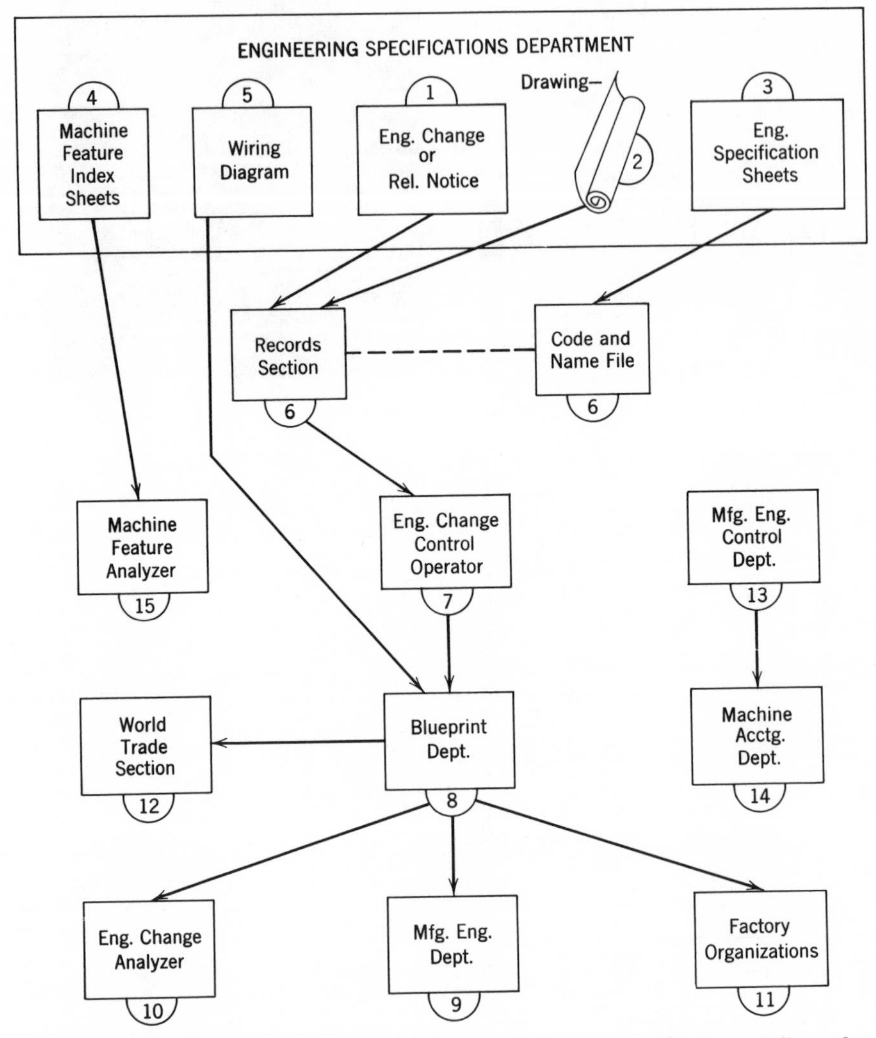

Figure 9-19. Movement diagram shows production procedures followed in processing an engineering change.

4. Can any activity be advantageously combined with some other activity?
5. Can any activity be simplified?

The specific, or pointed, questions of a systems-evaluation check list are broken down into four categories, as shown below.

Systems Operations

1. Does the system accomplish its intended purpose as efficiently and economically as possible?
2. Is the system sufficiently flexible? Can it handle normal variances in the work load?
3. Are all the operations arranged to permit a smooth flow from one activity, or group of activities, to the next?
4. Are the activities and responsibilities distributed among the various organizational units in a way that minimizes the need for coordination, communication, and data processing?
5. Are the activities grouped into workable, harmonious, and logical arrangements for accomplishing the desired results?
6. Are activities that require sequential performance placed close together?
7. Where appropriate, are similar activities grouped together? With regard to activities that complement one another, are they positioned within the framework of the system so as to provide the best means for achieving objectives?
8. Does each activity have a justifiable purpose?
9. Is the system as self-regulating as possible?
10. Have mechanization and automation devices been designed into the system wherever practical?
11. Is any operation a bottleneck in the system?
12. Is there adequate means throughout the system to accommodate exceptional situations?
13. Is there any unnecessary overlapping or backtracking of activities?

Design and Distribution of Forms

1. Are the forms used in the system suitably designed? Are they in accord with whatever requirements exist for their completion—entries made by hand, by machine, or by both of these methods?
2. What are the reasons for the existence of each form? Are the reasons valid?
3. Does the form satisfactorily fulfill its purpose?
4. Are there an adequate number of copies?
5. Are all copies necessary?
6. Would additional copies be of value?
7. Can one copy be forwarded in order to dispense with the necessity for one or more other copies?

8. Do the copies flow freely throughout the system so that work is not encumbered?

9. What ultimate disposition is made of the copies?

Machines and Equipment

1. Does the equipment have the ability and capacity to perform the work?
2. Would auxiliary units for the equipment be of value?
3. Where could the equipment be placed for better performance?
4. Is the equipment being utilized at maximum capacity?
5. If there is unutilized capacity beyond that which is required to provide for rush work and emergency situations, can the equipment be shared by more than one organizational unit?
6. Is the combination of equipment in reasonable balance, with no one unit having a capacity that is unduly excessive with respect to the others? If not, would such a combination of equipment be practicable?
7. Is the equipment standardized? If not, would it be advantageous to undertake a standardization program?
8. Is it possible to obtain new equipment that would result in savings in time and cost?
9. What effect will new equipment have on other operations? Will existing equipment be rendered obsolete? What cost will be incurred as a result of this obsolescence?
10. Would simple mechanical devices (e.g., check writer, time stamp, or accordion rack) afford an improvement?

Filing Considerations

1. Are forms, records, and documents filed carefully and accurately?
2. Is the filing system satisfactory?
3. Should any specialized type of files be used (e.g., motorized card files or multiwheel units)?
4. Are active and inactive materials filed together?
5. Is there a record-retention program?
6. Would it be advantageous to microfilm certain records?

DESIGNING AN IMPROVED SYSTEM

Analysis requires the gathering of data, observing things as they are, and collecting and recording facts. This reveals the nature of things. Of what are they composed? What is their make-up? What makes the system work? How is it controlled? and so on. The experienced systems analyst soon begins to see patterns in his data, a relationship between one event and another, and the relationship of the *whole* to its parts. As this occurs, the essentially *analytical* activity is transformed into one of *synthesis*. Synthesis is the combining of relative data made available by the analytical effort, developing hypotheses and conceptions, and devising systems. In a word, synthetic activity is creative activity. It involves ingenuity and inventiveness.

In designing a system consideration must be given to existing problems, the outputs desired, the inputs needed, the chain of procedures necessary between input and output, and the possible advantages of mechanization. The particular factors vary from project to project. Although no two systems are exactly alike, the following idea-producing check list is sufficiently general to help the systems analyst in any type of project.

What? Why? When? How? Where? Who?

> I keep six honest men serving me
> (They taught me all I know);
> Their names are What and Why and When
> And How and Where and Who.
>
> —Rudyard Kipling.

Idea-Producing Check List *

What Other Applications?

1. What *new* applications for *old things?*
2. In what new ways could *this* be used? Without changing *it?* By changing *it?*
3. To what other application could *this* be put?
4. What else could be *constructed* from *this?*
5. What to do with waste, rejects, and the like?
6. Can other principles, old or new, find another use for *this?*

What about Adaptations?

1. What is similar to *this?*
2. What other *thing* or situation does *it* bring to mind?
3. Does *it* suggest new ideas?
4. Is there a parallel to be drawn from the past?
5. What else could be adapted?

* Adapted from a listing by Alex Osborn.

6. Is there something similar that could be used?
7. What can *this* be made to look or act like?
8. What additional ideas can be included?
9. What two *things* might lend themselves to an adaptation?
10. What *things* might be borrowed and adapted from other systems, related or unrelated?

What about Modifications?

1. What changes can be made to improve *this?* In what way? In what other ways?
2. Can *something* be simplified or omitted?
3. How about a new location? Different position? Different configuration?
4. What change can be made in the process?
5. How about changing the style or type? The color? The form? The motion?
6. What changes would make it more functional? More understandable? More appealing?
7. How about ease of handling?

What about Substitutions?

1. How about substitutes? What other component? Process? Ingredient?
2. What other *part* might work better?
3. Who else could do this better?
4. Where else could it be done?
5. What other item would work as well?

What about Additions?

1. What might be added to make it better?
2. Should it have more strength?
3. How about making it larger?
4. What extra feature?
5. How about more time?
6. Should it be done more often?
7. How about using more than one?

What about Subtractions?

1. How about making it smaller?
2. What can be left out?
3. How can it be made more compact?
4. How about condensing it?

5. How about less length? Less width? Less depth?
6. Can it be made lighter?
7. How about less time? Less often?
8. Can it be done with fewer parts? Fewer activities? Less motion?

What about Rearrangement?

1. What can be done in the way of rearranging the established pattern?
2. Can it be transposed?
3. What if the sequence were changed?
4. What other placement might be better?
5. Can the timing be rearranged?
6. Would it be better to do this earlier or later?
7. What about interchangeability?
8. What about the relationship between cause and effect?

What about Reversals?

1. What are the opposites?
2. Why not a turnabout? How about *after* instead of *before? Up* instead of *down? Lengthwise* instead of *sidewise?*
3. How would it act in *reverse?*
4. Why not put it on top? On the *bottom?* In the *middle?* At the *end?*

What about Combinations?

1. What things can be combined?
2. What about combining purposes? Combining units? Combining ideas?
3. What can be done by way of putting things together in groups?
4. What about an assortment?
5. What about a consolidation?
6. What can be combined to multiply the purpose?
7. What could be done in combination with others?

This check list has application to a wide range of systems-and-procedures work. It may be used to deduce ideas relevant to forms, equipment, subsystems, and all sorts of procedural combinations.

Questions, of course, do not automatically elicit answers or develop solutions. They can, however, do much to stimulate thinking, and in the fulfillment of any creative work the thought is father of the deed.

10

The Creative and Synthetic Aspects of Systems-and-Procedures Work

Whether in systems-and-procedures work or in some other endeavor, thinking may be said to possess two modes: the conventional and the creative. Both are aimed at problem solving. The technique of the first is applicable to problems wherein there is a single, correct answer to be obtained, such as an analytical problem has—an answer to be arrived at by logical thinking, mathematics, or experiment. Where this is the case an approach similar to the following is used:

1. Study the problem with care.
2. Decide what the problem requires you to do.
3. Plan the solution.
4. Bring to bear on the problem the essential reasoning and/or formulas that apply (use of existing knowledge: general equations, principles, rules, laws, etc.).
5. Express the conclusion.
6. Check the result.

Although this approach may be fine for one-answer problems, its value is negligible in instances that require creativeness. The creative problem has no single, clear-cut answer; it has many lines of inquiry and many solutions, some of which are better than others. Always—ideally, at least—a better solution remains to be found in the future.

There is, fortunately, a charted course to follow—*stages of creative thought*—leading to the production of novel ideas; a course, interestingly, that all creative thinkers follow, whether they are composers, biologists, painters, authors, or systems analysts. These stages, identified somewhat differently by various students of the creative process, are shown below:

I. Preparation, Incubation, Illumination, and Verification.
—Wallace.[1]

II. Recognition, Decision, Preparation, Search for Clues, Frustration, Insight, Action.
—Sharp.[2]

III. Orientation, Preparation, Analysis, Hypothesis, Incubation, Synthesis, Verification.
—Osborn.[3]

IV. Preparation or Orientation, Frustration, Moment of Insight, Verification.
—Hutchinson.[4]

V. Problem, Preparation, Frustration, Incubation, Insight, Verification, Communication.
—Ott.[5]

Let us now, by way of example, take the Ott scheme and describe each of the steps in some detail.

STAGES IN THE CREATIVE THINKING PROCESS

Problem. The real problem must be clearly understood, with its environment and its true limitations.

Preparation. Obtain all existing knowledge bearing on the problem. Rearrange ideas, imagine further data, mentally construct possible solutions. Completely immerse yourself in the problem.

[1] Stein, Morris I., and Heinze, Shirley J., *Creativity and the Individual,* Free Press, Glencoe, Ill., 1960.
[2] Sharp, H. T., "Here's How to Get Ideas in a Hurry," *Chem. Eng.,* **63,** 218 (July) 1956.
[3] Osborn, Alex F., *Applied Imagination—Principles and Procedures of Creative Thinking,* Scribner's, New York, 1953.
[4] Hutchinson, Eliot Dole, *How to Think Creatively,* Abingdon-Cokesbury, Nashville, 1949.
[5] Ott, Emil, "Stimulating Creativity in Research," *Chem. Eng. News,* 33, 2318–21, 1955.

Frustration. This happens when no solution emerges. Do not be dismayed, because frustration is a prelude to insight.

Incubation. Set problem aside for other work or, better, relaxation. Recreation is good; sleep is even better.

Insight. In a "flash" the answer comes; we *know* it is right.

Verification. The solution is tested in a practical way.

Communication. The idea must be "sold" to others, to produce benefits.

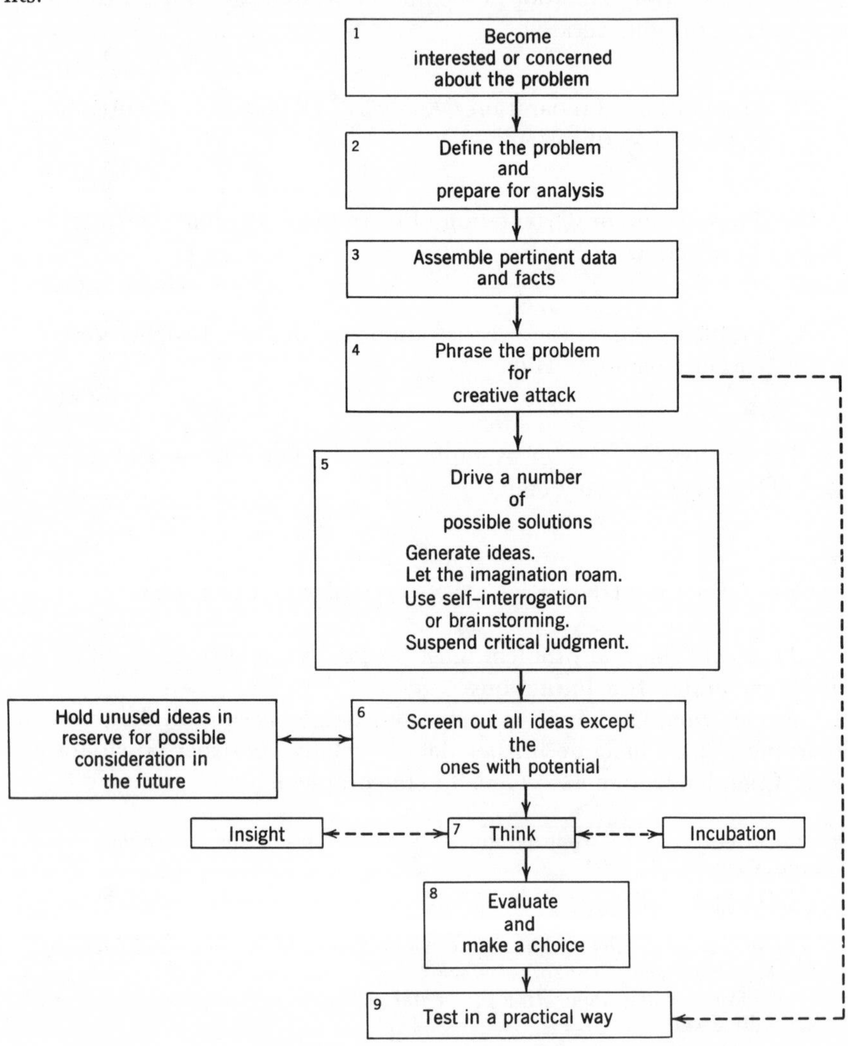

Figure 10-1. The abstract and practical use of the creative process.

These stages do not always follow in chronological order, and each stage need not always occur. There often is a skipping over and backtracking.

Having established a concise description of the creative process, we are in a position to give concrete form to its application. This is done in Figure 10-1. The first four steps that are shown in the chart are in the problem and preparation area. Facts are gathered and considered; in step 4 the problem may (or may not) need to be redefined on exposure to the facts.

In some cases the phrasing of the problem for creative attack may point the way directly to a solution or to a course of action. This possibility is portrayed by the dotted line going from step 4 to step 9. When the solution is not directly forthcoming (in step 4), the importance of building up a quantity of assorted solutions in step 5 can hardly be overemphasized. This may be done on an individual basis or as a team effort.

The most effective generation of new ideas is done by an individual working alone. The fact is that basic inventions, discoveries, and originations are the product of a single mind—group work in research laboratories notwithstanding. Edwin Land of the Polaroid Corporation, who has 300 American patents and 700 foreign patents to his credit, says: "There is no such thing as group originality or group creativity or group perspicacity." Even the ideas that emerge during the course of a brainstorming session—a type of idea-producing conference that we will presently discuss —must spring from the minds of individuals.

With a proper environment that is conducive to meditation and constructive thought, anyone can come up with a number of new ideas through discipline and hard work. Creative imagination is not, as is sometimes supposed, possessed solely by a few fortunate individuals. Actually, it is a trait that is widely distributed over the entire population.

As concisely stated by John E. Arnold of the Massachusetts Institute of Technology, the theory of creative thinking is based on the following hypotheses:

> First, that all men are born with a very definite, though limited, potential for creative work. Secondly, that this potential is at least partially independent of any other mental potential we may inherit. Dr. J. P. Guilford of Southern California seems to believe that there are three basic mental potentials: the first primarily associated with analytical or deductive type of thinking, and this is probably what we measure in our intelligence tests; the second associated with creative or synthetic thinking, and the third giving measure to our ability to judge and evaluate. His tests seem to indicate that they are partially, if not wholly, independent of each other, so that it is possible for one to be an analytical genius and a creative and judicial moron, or any one of an infinite number of combinations. The third hypothesis is that the creative process in itself is unique. It is the same whether it expresses itself in art, music, poetry, business, engineering, or in housekeeping. The fourth and last hypothesis is that the creative

potential can be realized through training and exercise, just as the development of our full capabilities along analytical lines can be obtained.[1]

Many systems analysts find it difficult to think creatively because they cannot find a starting point. Analyzing the background information is one good way of getting off the ground. To have the advantage of cumulative knowledge is a tremendous assist to inspiration and imagination.

Individuals working alone may improve their efforts by using check lists of the type used throughout this book. Questions stimulate the mind and draw out ideas. Professor Arnold went so far as to say, "The questioning spirit is the basic attitude, the first step toward being creative."

Another means of engendering ideas is through published information, especially through the reading of magazines and trade journals. Look also to the inspiration that may be obtained from allied and not-so-allied fields. "What," you may ask, "do other industries do when faced with a problem like this?"

WHEN ANSWERS DO NOT COME

Although finding a creative solution—"getting insight" as it is sometimes called—often occurs in a moment of concentrated effort, it need not. Sometimes, despite every deliberate technique and close mental application, nothing will occur. Practically everyone who has attempted to give a generalized picture of creative behavior says that whenever a problem remains unsolvable after much effort has been expended, it is advisable to turn to something else. The theory is that in doing so the strain is taken off the conscious mind and the unconscious mind is given an opportunity to mull the problem over, to come up with a solution, and to feed it back to the conscious. When the answer finally comes, it may come suddenly, like a bolt from the blue. Instances of this kind give rise to dramatic expressions such as, "Eureka, I've found it!" and "Wow, this is it!"

Many chemists, engineers, composers, authors, and poets—that is, scientists and artists in general—have described their experiences with ideas that have come in a flash, fully formed. Charles Darwin related in his autobiography, "I can remember the very spot in the road, whilst in my carriage, when to my joy the solution occurred to me." Similarly, Goethe tells us how still unorganized material became meaningful: "At that instant the plan of 'Werther' was found; the whole shot together from all directions, and became a solid mass, as water in a vase, which is just at the freezing point, is changed by the slightest concussion into ice." In a letter

[1] Arnold, John E., "The Creative Engineer," *Yale Scientific Magazine*, March 1956.

to a friend, Mozart writes, "What, you ask, is my method in writing and elaborating my large and lumbering things? I can in fact say nothing more about it than this: I do not myself know and can never find out. When I am in particularly good condition, perhaps riding in a carriage, or in a walk after a good meal, and in a sleepless night, then the thoughts come to me in a rush, and best of all." Modern creative thinkers give confirmation; for example, Dr. Arthur Glaser, 1960 recipient of the Nobel prize in physics, is said to have gotten his award-winning idea, his moment of insight, while staring at the foam in a pitcher of beer.

The gist of the matter is clear: We should not force our minds beyond a certain point. Otherwise we may become unduly frustrated, despondent, and fail in finding a solution. Dr. Eliot Dole Hutchinson, noted psychologist and author, describes the desired action in terms of a threefold pattern: "Deliberate effort, relaxation, and then effort again; consciousness, unconsciousness, consciousness; activity, passivity, activity . . ."

The realization that the conscious mind needs a diversion from what it has been concentrating on can be a great help in furthering creative effort. Sometimes merely turning to another task is sufficient. Edison made it a practice to have several projects under way simultaneously, and he would switch from one to the other as suited his mental state.

There are occasions, however, when the mind needs a diversion of another kind; it needs the relaxation that comes from sleep, recreation, or enjoying a pastime. A good night's sleep has long been cherished by eminent creative thinkers, because upon awakening the mind is refreshed and sparkling—ready for a new look at things. Other methods of relaxation that are recommended by many notably creative people include going for walks or a hike in the woods, listening to good music (classical, jazz, or popular—depending on one's taste), attending the theater or a concert, "light" reading such as mystery stories, and engaging in some sport or hobby.

At this point a fitting question is, "How long might it take before an answer comes?" Actually, no definite time can be given; it may take an hour, a day, several days, a week, or a few weeks—the time varies with the nature of the problem. It is not unusual for scientists who are working on difficult problems to set them aside for months or even years, working on them intermittently while pursuing other projects, until an answer is found. However, with problems of lesser complexity—the type met with daily—letting them lie dormant for a period ranging from a day to two weeks, making no strong effort at recall, may well prove to be all that is necessary. Hopefully, if urgency is involved, just "sleeping on the problem" overnight will be sufficient.

The aim of the previous discussion is not to fix a carefully formulated methodology on creativity (even if such were possible, which it is not), but to indicate in a general way what is known about it and to shed light on how new ideas are brought into existence. Given the concepts of conscious meaningful effort and imagination, we are in a position to court the Muse of creativity and to win her favors.

IDEA FINDING THROUGH TEAMWORK

Turning back to step 5 in Figure 10-1, we again note that this step is devoted to building up a quantity of leads as possible solutions to a problem. So far we have only covered self-priming methods for doing this. There is also another method, known as the team approach, which may take one of two forms: the co-worker method or the group method.

Co-Worker, or Partnership, Approach

Teaming up with another person can, under certain circumstances, be a boom to the development of novel ideas out of which something original and worthwhile ultimately emerges. History, past and modern, is peppered here and there with examples that support this contention. We are familiar with the delightful musical artistry of Gilbert and Sullivan, Rodgers and Hammerstein, and Lerner and Lowe; the distinguished historical works by the collaborators Dr. Charles A. and Mrs. Mary Ritter Beard; the scientific investigations of the physicists Pierre and Madame Marie Curie; and, in the field of scientific business management, the contributions of the Gilbreths, another husband-and-wife team.

Calling attention to the accomplishments of these personages serves solely to point out that collaboration between two individuals along creative lines can be effective. The success of the method may be attributed largely to the proper apportionment of the work, the fact that complementary talents are involved, the interchange and discussion of ideas, and the preservation of individuality.

Group Approach

The group approach is probably best exemplified by the research teams found in applied science and industry. This kind of mass assault on complex problems permits several persons to work under a group leader or research director on different aspects of the same problem. Nevertheless, new ideas are the product of the individual mind. Dr. Alfred W. Griswold, a former president of Yale University, expressed this view when he said in one of his commencement addresses, "Could 'Hamlet' have been written by a committee, or the Mona Lisa painted by a club? Could the New Testament have been composed as a conference report? Creative ideas do not spring from groups. They spring from individuals. The divine spark leaps from the finger of God to the finger of Adam, whether it takes ultimate shape in a law of physics or law of the land, a poem or a policy, a sonata or a mechanical computer."

Brainstorm Sessions

Developed by the late Alex Osborn of advertising industry fame, and introduced in 1939 into the advertising agency [1] of which he was a partner, "Brainstorm Sessions" describe conferences of a type used to elicit creative ideas. The technique itself was dubbed "brainstorming" to convey the idea of using the *brain* to *storm* a problem and to do so as a group effort with each participant focusing on the same problem.

Although the technique has its roots in the advertising field, we hasten to add that it "caught on" and has since been applied with success in other fields, both technical and administrative. Among those using it have been engineers and scientists doing research-and-development work, and, although not all who have tried it have expressed satisfaction, the method does have a considerable number of proponents.

Whether applied in advertising, engineering, business administration, or, as is our particular interest, in systems-and-procedures work, the procedure for conducting brainstorm sessions is essentially the same. The usual practice is to gather together a group, one with similar interests but diverse skills, and have the members turn their imagination free so that a quantity of ideas is produced. The emphasis is on quantity, without immediate concern for quality. This is done by invoking a noncritical rule and announcing that only positive suggestions are desired.

It is the responsibility of the chairman to call on the group for suggestions, to keep the talk on the subject, and to give everyone a chance to contribute. Here, as with individual self-interrogation, a check list may be used to spur ideas. The rules for such conferences are that there should be a clear statement of the problem, not just some vague generality, and that all criticism is out at the time.

As the suggestions are called to him the chairman writes them on a blackboard. He makes no effort to edit or evaluate, and someone in the group is assigned the job of recording the suggestions on paper so that they will be preserved. Easy talk and free-flowing ideas are wanted. Wildness is welcomed; the more far-fetched and seemingly silly the idea, the

[1] Batten, Barton, Durstine, & Osborn, Inc.

better. This encourages the participants to become less inhibited and to turn their imaginations loose; that is, to rain out ideas in torrents. The aim is quantity, in the hope that from a large enough number the one big idea will emerge. The underlying principle is the machine-gun concept: With a large number of shots going out, at least one will hit the target.

Inasmuch as *zany* ideas are not outlawed but encouraged, the chairman must be on guard against the comedian. Sometimes there will be comments like, "That's a crackpot idea," or "Let's fire the president." One way of handling such situations is to make it known that anyone who does not refrain from jesting must leave the session. The chairman must also be prepared to thwart critics, those ready and anxious to say, "That won't work because. . . ." It should be recognized that some persons are so accustomed to having logic and reason dominate their activities that they do not have room for the free play of ideas. Persons of this caliber are not best suited for brainstorming, but they can be extremely useful when the time comes to evaluate the ideas that are generated.

Brainstorm sessions are usually short, not more than an hour or so in length. Ordinarily, the number of participants should not exceed 12.

What happens to the list of ideas—sensible and otherwise? The preferred practice is to have it typed up and distributed to the participants, and to forward copies to the evaluation group for screening within a day or two. This evaluation group, at least partly composed of persons other than the brainstorm group, will first eliminate all obviously impractical suggestions and then decide on four or five that have potential value. Next, these select ideas undergo an evaluation process to determine the one best idea. Criteria for selecting the one best idea are the same as those used in making any sound decision.

Brainstorming rules, as well as some worthwhile suggestions both for individuals and groups, are given in Figure 10-2.

BRAINSTORMING RULES

1. Must keep these sessions entirely apart from other conferences.
2. Keep atmosphere informal, positive, permissive.
3. Judicial thinking and criticism are forbidden.
4. Use blackboard for visual as well as auditory stimulation.
5. Record what was said, in brief reportorial style; not who said it.
6. Eight to 12 people is best number; varying backgrounds desirable.
7. Use bell or red disk to signal infraction of Rule 3: "Three strikes and you're out!"
8. Announce subject before meeting—authorities differ on this.

HELPS FOR INDIVIDUAL USE

General

1. **Broaden your experience**—meet people; go to lunch with varying groups; attend meetings; take part in community activities; travel.
2. **Read—and write—a lot**—take notes of what you are reading or listening to; it makes no difference whether you ever read the notes, they will have sharpened the mental impression anyway.
3. **Play games** . . . bridge, chess, cryptograms—which require intense mental effort
   . . . **and have hobbies** . . . Handicrafts are better than collecting.

Specific

4. **Learn your most effective time of day,** or place or circumstances; early morning is often the time when the mind is sharpest and most liable to insight.
5. **Start now**—do not delay unnecessarily; if you need more time, get up an hour or two earlier.
6. **Assume a work attitude**—use pencil ("crowbar for moving the mind"—Osborn) and pad ("idea trap").
7. **State problem broadly** . . . as broadly as is consistent with the objective
   . . . **and write it down** . . . if you can't, it isn't clear enough.
8. **Ask important questions:** What about this? What if? Can this be changed? What is the *real* problem?
9. **Learn to withhold judgment**—be able to turn if off and on at will.

HELPS FOR GROUP USE

General

1. **Gain management approval.** This is not always easy to do; it is often hard to get the boss to admit that the thinking of all can be improved. Even among fellow workers there may be some suspicion and resistance in the beginning. It is advisable to move gradually and with a promotional campaign.
2. **Positive, permissive attitude.** Idea-forming stages must be free of criticism or evaluation, to overcome any reticence about expressing ideas; to encourage "green-light thinking."
3. **Encourage quantity of ideas.** The more ideas, the greater the probability of big ones.

Specific

4. **Morning meetings best.** The mind is freshest and most creative then.
5. **Comfortable room** . . . to go with the informal, permissive atmosphere. Some authors say to provide opportunity to smoke, have light refreshments.
6. **Avoid disparity in rank** . . . to keep lower ranking members of the group from being reluctant to express ideas freely.
7. **Avoid one-answer problems,** also those requiring a lot of calculations. Also avoid a search for possible fields for new-product research.

8. **Avoid breaking into little groups**—this defeats any meeting. Everything that is said should be directed to the whole group.
9. **Concentrate on the future**—this involves risking the status quo, but progress can be made in no other way.

Figure 10-2. Points worth remembering. Source: Kelley, Maurice J., "Understanding the Creative Process," *Chemical and Engineering News*, Vol. 35, No. 12 (March 25) 1957.

A NOT-SO-FINAL WORD ON CREATIVITY

As a creative person in business, the systems analyst must be possessed of many talents. Not only must he be a practical thinker, with his feet on the ground, but he must also be an investigator, a dreamer of sorts, an evaluator, and a person who accomplishes things of value.

part 4

Information Technology

11

Motion Study

THE CONSERVATION OF ENERGY, TIME, AND MONEY

In earlier chapters we have concerned ourselves primarily with the study of work flows having to do with entire systems, subsystems, and whole sets of interrelated procedures. Benefits from such studies can be beneficial, but they should not obscure the improvements that can be obtained by also studying the effectiveness of individual operations in considerable detail. This chapter focuses attention on the intensive study of certain single operations—operations that are repetitive, reasonably stable, and require substantial effort in terms of physical work.

Normally, the detailed study of an operation should be preceded by an overall study of the procedures of which it is a part. To conduct a detailed study in advance of a comprehensive survey could prove futile because a comprehensive survey may discontinue the operation studied.

Once the procedures related to the extensive operational practices have been evaluated, it is practical for the systems analyst to attend to the minute details that comprise the performance of individual operations.

MOTION STUDY AND ITS RELATIONSHIP TO TIME STUDY

Motion study and time study are two distinct implements of work simplification. They may be used jointly, or motion study may be used independently. As an approach to problems motion study has merit in its own right, leading to better performance based on the overall improvement of the motions themselves.

Time study, of necessity, must be founded on motion study. It would be impracticable to employ time study without motion study, because it is first necessary to know the elements of the work.

Motion study is an examination of work elements for the purpose of reducing fatigue, saving time, and reducing cost. It is directed at the attainment of higher productivity through the elimination of inefficiencies caused by unnecessary, cumbersome, and wasteful movements. A summary of the uses of motion-and-time study is given in Figure 11-1.

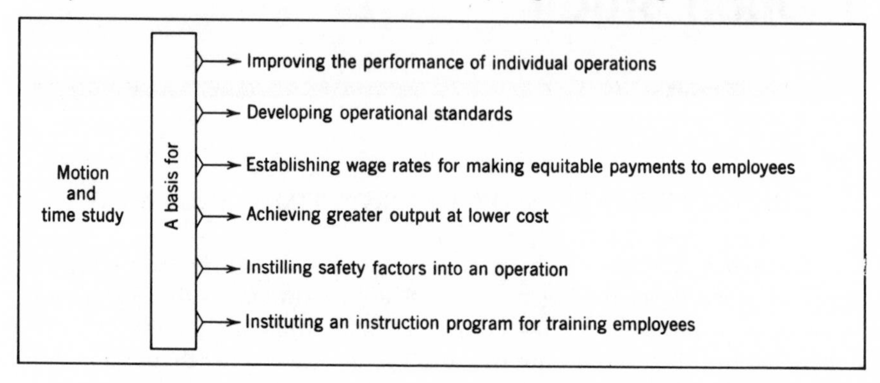

Figure 11-1. A summary of the uses of motion-and-time study.

The Historical Setting

Time study had its origin in 1881 at the machine shop of the Midvale Steel Company in Philadelphia.[1] It was nurtured by Frederick W. Taylor, who was later to become known as the father of scientific management. Taylor had this to say about time study:[2]

> Time study is the one element in scientific management beyond all others making possible the "transfer of skill from management to men" . . . "Time study" consists of two broad divisions, first, analytical work and second, constructive work.
>
> The analytical work of time study is as follows:
>
> a. Divide the work of a man performing any job into simple elementary movements.
>
> b. Pick out the useless movements and discard them.
>
> c. Study, one after another, just how each of several workmen makes each elementary movement, and with the aid of a stop watch select the quickest and best method of making each elementary movement known in the trade.
>
> d. Describe, record, and index each elementary movement, with its proper time, so that it can be quickly found.

[1] Subcommittee on Administration of the ASME, "The Present State of the Art of Industrial Management," *Trans. ASME,* **34,** 1197 (1912).

[2] Ibid., pp. 1199–1200.

e. Study and record the percentage which must be added to the actual working time of a good workman to cover unavoidable delays, interruptions, minor accidents, etc.

f. Study and record the percentage which must be added to cover the newness of a good workman to a job, the first few times that he does it. (This percentage is quite large on jobs made up of a large number of different elements composing a long sequence infrequently repeated. This factor grows smaller, however, as the work consists of a smaller number of different elements in a sequence that is more frequently repeated.)

g. Study and record the percentage of time that must be allowed for rest, and the intervals at which the rest must be taken, in order to offset physical fatigue.

The constructive work of time study is as follows:

h. Add together into various groups such combinations of elementary movements as are frequently used in the same sequence in the trade, and record and index these groups so that they can be readily found.

i. From these several records, it is comparatively easy to select the proper series of motions which should be used by a workman in making any particular article and, by summing the times of these movements and adding proper percentage allowances, to find the proper time for doing almost any class of work.

j. The analysis of a piece of work into its elements almost always reveals the fact that many of the conditions surrounding and accompanying the work are defective; for instance, that improper tools are used, that the machines used in connection with it need perfecting, and that the sanitary conditions are bad, etc. And knowledge so obtained leads frequently to constructive work of a high order, to the standardization of tools and conditions, to the invention of superior methods and machines.

Dr. Ralph M. Barnes in his book, *Motion and Time Study,* points out that ". . . it is apparent that Taylor made some use of motion study as a part of his time-study technique. However, he placed greater emphasis on materials, tools, and equipment in connection with the improvement of methods. It remained for the Gilbreths to develop motion study as we know it today." [1]

Frank B. Gilbreth and his wife, Lillian M. Gilbreth, were prime movers in advancing the science and art of motion study. To them goes credit for major contributions to our body of knowledge having to do with fatigue, monotony, and the transfer of skill. In addition, the Gilbreths developed process charts, originated micromotion study, and evolved the technique of motion study known as chronocyclegraph.[2]

[1] R. E. Barnes, *Motion and Time Study,* Wiley, London, 1915, pp. 7–8.

[2] All in all, the Gilbreths led a full and fascinating life. In 1949 the entire Gilbreth family came into prominence with the publication of *Cheaper by the Dozen* (Crowell, New York, 1948). This book was coauthored by two of their 12 children, Frank B. Gilbreth, Jr., and Ernestine Gilbreth Carey. It was made into a motion picture of the same name.

It is interesting to note that the Gilbreths were responsible for contributing three words to our technical language; namely, "micromotion," "chronocyclegraph," and "therblig." Each of these terms will be explained in context later on in this chapter.

Since the golden age of scientific management—the days of Taylor, Henry L. Gantt, and the Gilbreths—other very capable individuals have labored fruitfully on projects of motion-and-time study and in the broader field of work simplification. Two important names are Prof. David B. Porter and Allen H. Mogensen. Both of these men have succeeded in refining, expanding, and establishing many of the techniques we now have.

Research and education in motion-and-time study are expected to continue because the quest for the "one best way" is never ending. Nevertheless, thanks to those who have pioneered and developed the field, we presently have a useful collection of principles and practices at our command. The remainder of this chapter is concerned with these principles and practices.

It is interesting to observe how applications of the techniques of work simplification, including those of motion and time study, have spread from the plant to the office and now even into our homes. The principles and practices of motion economy have been applied with success to numerous household tasks. Entire work areas such as kitchens have been arranged to reduce fatigue, promote efficiency, and thereby increase output during the time in which the work is performed.

WHAT MOTION STUDY DOES NOT MEAN

One thing that motion study does not mean is to *hustle*. Taken by itself, *hustle* implies the indiscriminate speeding up of *all* motions regardless of their individual worth. On the contrary, the true intent of motion study is to seek out and eliminate motions that are unnecessary and to improve only upon those that are necessary.

Note also that the rightful use of motion study precludes any attempt to employ it as a means for exploiting workers. Motion study is a tool intended for the benefit of mankind; however, like many another tool of this sort, it may be employed for socially undesirable purposes. Whether the end result is beneficent or not depends on the understanding, judgment, and proficiency with which it is employed. Labor unions have long accepted it on the grounds that the more produced, the greater will be the amount of goods and services available for everyone. We cannot justifiably condemn the tool itself because there are those who would misuse it.

BENEFITS TO BE DERIVED FROM MOTION STUDY

When correctly applied, the methods of motion study serve the following purposes:

1. To improve the workplace and working conditions.
2. To eliminate aimless work and loss of time.
3. To establish correct movements and techniques.
4. To standardize the work.
5. To reduce unit cost.
6. To enhance productivity.
7. To overcome fatigue.
8. To assure worker satisfaction.
9. To promote safety and prevent accidents.
10. To minimize spoilage, equipment damage, and similar forms of waste.
11. To reduce errors in the work.
12. To improve the design and arrangement of equipment and machines.
13. To facilitate job training.
14. To remove bottlenecks in the flow of work.

MOTION-ECONOMY PRINCIPLES

Motion study can be of great service for both management and the worker if it is utilized wisely and judiciously. A wealth of past experience has produced certain fundamental principles that should be followed to obtain the most favorable productive results with the least amount of waste and fatigue. These principles are logically grouped under three headings:

1. Those that apply to the bodily motions of the worker.
2. Those that apply to the arrangement and environment of the workplace.
3. Those that apply to the design and use of tools and equipment.

Not every principle that is discussed below is applicable to each case; however, in the aggregate, these principles constitute a useful set of laws with general application.

Bodily Motions

Principles for the use of the human body are the following:

1. Use both hands in working; begin and end the movements of both hands at the same time. The motions should preferably be in opposite and symmetrical patterns. Only during rest periods should both hands be idle.
2. Use as few motions as possible to perform the task. Eliminate all motions that do not serve a purpose.
3. Use the minimum bodily motion that is necessary to get the work done. To the extent that it is possible, use finger motions instead of wrist motions, wrist motions instead of arm motions, and arm motions instead of body motions.
4. Establish motion patterns that are smooth, rhythmic, and somewhat curved. Avoid including any movements that entail abrupt changes in direction, because they cause delays and promote fatigue.
5. Relieve the hands of work that can be accomplished more advantageously by other parts of the body. By utilizing foot rather than hand motions it is sometimes possible to free the hands for other work. Do not overlook the possibility of employing other than terminal parts of the body; for example, a knee may be used to activate a control mechanism.
6. Endeavor to obtain a sequence of motions with built-in rhythm, continuity, and spontaneous flow. The sports field abounds in situations that illustrate this point; for example, the baseball pitcher's windup, the golfer's swing, and the swimmer's breast stroke.
7. Strive to attain balance in motions by giving each hand the same amount of work or movements to perform.
8. Establish the boundaries of normal and maximum reach (see Figure 11-2). For performing work on a horizontal plane, the maximum

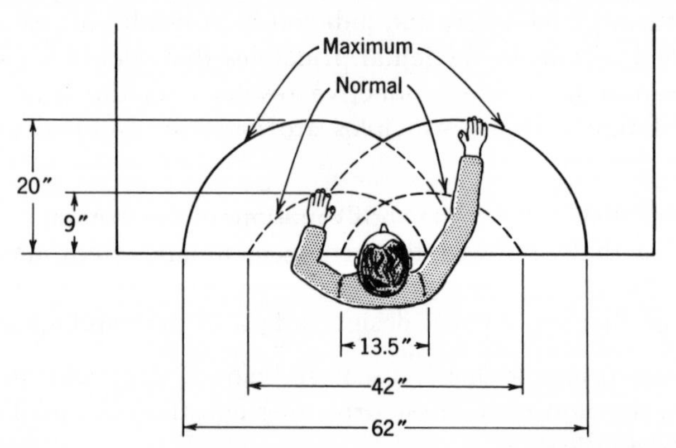

Figure 11-2. Normal and maximum working areas on a horizontal plane.

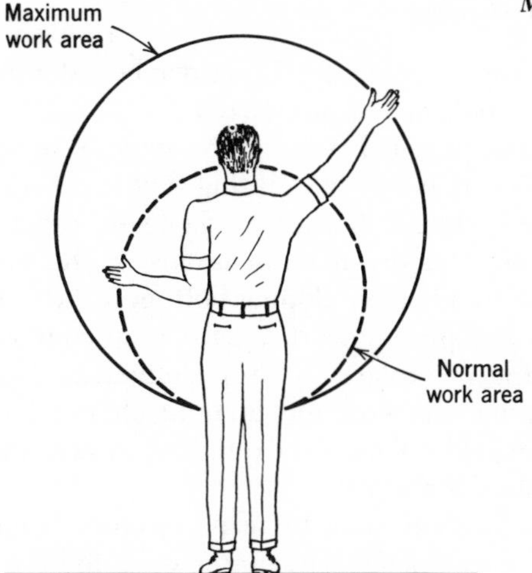

Figure 11-3. Normal and maximum working areas on a vertical plane.

working area for each arm is found by sitting (or standing) erect at the workplace (desk, bench, table, etc.) and swinging the forearm from the shoulder across the work area. Any movements outside this arc require additional effort, which encumbers the operation and induces fatigue. The normal working area for each arm is determined by holding the elbow near the body with the arm bent and employing the elbow as a pivot, establishing an arc across the work area. With both arms used similarly, the area where the arms cross in front of the body is the best place for performing operations that are common to both hands. There are also normal and maximum working areas for performing work on vertical planes. Figure 11-3 illustrates the boundaries of each.

Another important principle for the use of the human body is that the number of pauses and delays be reduced to a minimum.

Arrangement and Environment of the Workplace

Principles for the arrangement and environment of the workplace are as follows:

1. Locate materials, tools, and equipment within the normal and maximum work areas of either or both planes (horizontal and/or vertical), as suits the nature of the work.
2. Arrange the elements of the workplace to conform with the normal and maximum work areas. Preposition materials, tools, and equipment to permit the best sequence of operations and to eliminate or reduce such

activities as searching or selecting. Long reaches, stooping, and awkward motions should be abolished or minimized.

3. Employ basic physical laws to advantage. When practical, think in terms of sliding objects rather than lifting them up. Gravity-feed containers and dispensers should be used whenever practicable. The ideal gravity-feed unit deposits the material as close to the point of use as possible. Such units commonly have sloping bottom surfaces to allow the material to slide into the proper work area. Completed work is sometimes removable from the workplace by drop deliveries. This may be accomplished by "dropping" the work into a chute, channel, or through a bench hole and onto a conveyor. One of the quickest ways to dispose of an item is to just let it fall into a container.

4. Disengage finished items by using ejectors. A good example of an ejection device is a foot pedal that causes work to be released and thrust out from a holding position.

5. Make appropriate use of overhead space. Check the feasibility of having tools suspended in the air so that they can be conveniently used and automatically retracted.

6. Provide for alternate sitting and standing positions whenever the nature of the work permits.

7. Equip the operator with an adjustable posture chair for work that is performed in a sitting position. For work done in a standing position provide him with a floor pad made of rubber or some other spongy material.

8. See that proper illumination, ventilation, and temperature are provided. Eliminate hazards to health and safety, including polluted air, excessive noise, and unsafe tools.

Design and Use of Tools and Equipment

Principles for the design and use of tools and equipment are as follows:

1. Avoid the use of either hand as a mechanical device. Instead, provide physical fixtures to guide, support, and hold the work.

2. Employ tools, machines, and equipment that are designed to conform with the limitations and characteristics of the human body; for example, hand tools should be fashioned so that they can be easily and efficiently grasped. Handles on devices such as screwdrivers, cranks, and rollers need to be made large enough to provide sufficient leverage and comfort.

3. Use lightweight tools in order to reduce fatigue and sustain energy.

4. Consider the feasibility of combining tools. The typewriter eraser with a brush on the end is a simple example of a dual-purpose tool. Along more elaborate lines, it is sometimes possible to combine tools on a machine and have it perform two or more operations simultaneously.

5. Check into the possibility of employing tools and equipment of advanced design; for example, consider the use of such items as power tools, automatic conveyors (belts and pneumatic tubes), electric typewriters, and auxiliary units of various kinds.

6. Study the practicability of designing tools, machines, and equipment especially for the job. Sometimes a device that will cut in half the amount of energy and time necessary for the job can be made in the company machine shop.

MECHANICS OF CONDUCTING MOTION STUDIES

The carrying out of motion studies ordinarily falls into seven major steps. These are listed below.

1. Observe the present method.
2. Record the elements of the work. This is usually done on a chart.
3. Examine and evaluate the existing method and its elements in terms of the principles and practices of motion study.
4. Develop a better method by making desirable changes through eliminations, combinations, adaptations, modifications, and so on.
5. Record the improved method. As with the second step, this also is usually done on a chart.
6. Acquire whatever approvals are necessary.
7. Install the new method and oversee its performance. Included in this step is the training of workers and the necessary follow-up surveillance to assure proper conformance.

MOTION-CYCLE CHARTS FOR STUDYING OPERATIONS

Two graphic tools that are useful for the close study of operations are *operator charts* that depict the motions of the hands and other parts of the body in performing a cycle of work, and *operator-and-machine charts* that show the movements of an operator in conjunction with those of a machine.

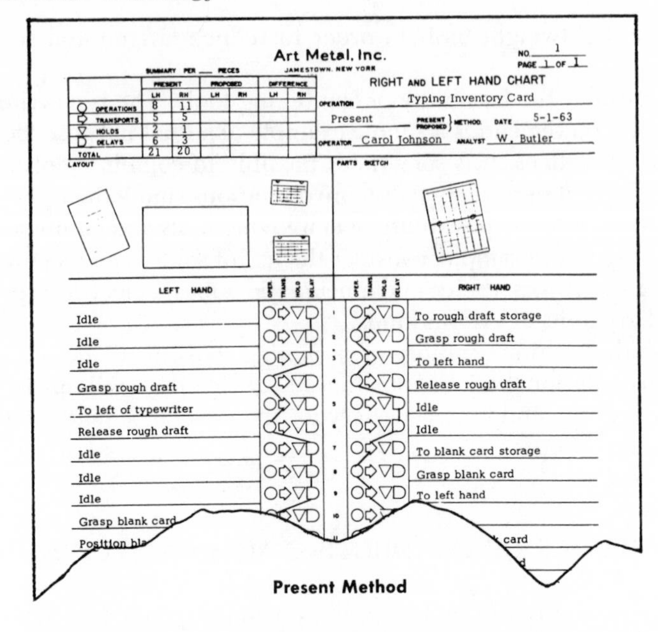

Art Metal, Inc.
JAMESTOWN, NEW YORK

NO. 1
PAGE 1 OF 1

RIGHT AND LEFT HAND CHART

OPERATION Typing Inventory Card
Present METHOD DATE 5-1-63
OPERATOR Carol Johnson ANALYST W. Butler

SUMMARY PER ___ PIECES

	PRESENT LH	PRESENT RH	PROPOSED LH	PROPOSED RH	DIFFERENCE LH	DIFFERENCE RH
OPERATIONS	8	11				
TRANSPORTS	5	5				
HOLDS	2	1				
DELAYS	6	3				
TOTAL	21	20				

LAYOUT

PARTS SKETCH

LEFT HAND		RIGHT HAND
Idle	1	To rough draft storage
Idle	2	Grasp rough draft
Idle	3	To left hand
Grasp rough draft	4	Release rough draft
To left of typewriter	5	Idle
Release rough draft	6	Idle
Idle	7	To blank card storage
Idle	8	Grasp blank card
Idle	9	To left hand
Grasp blank car	10	
Position bl	11	card

Present Method

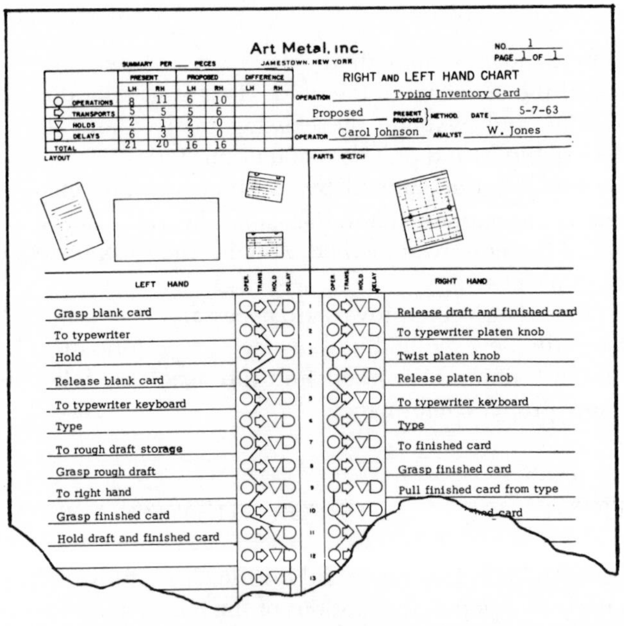

Art Metal, Inc.
JAMESTOWN, NEW YORK

NO. 1
PAGE 1 OF 1

RIGHT AND LEFT HAND CHART

OPERATION Typing Inventory Card
Proposed METHOD DATE 5-7-63
OPERATOR Carol Johnson ANALYST W. Jones

SUMMARY PER ___ PIECES

	PRESENT LH	PRESENT RH	PROPOSED LH	PROPOSED RH	DIFFERENCE LH	DIFFERENCE RH
OPERATIONS	8	11	6	10		
TRANSPORTS	5	5	5	6		
HOLDS	2	1	2	0		
DELAYS	6	3	3	0		
TOTAL	21	20	16	16		

LAYOUT

PARTS SKETCH

LEFT HAND		RIGHT HAND
Grasp blank card	1	Release draft and finished card
To typewriter	2	To typewriter platen knob
Hold	3	Twist platen knob
Release blank card	4	Release platen knob
To typewriter keyboard	5	To typewriter keyboard
Type	6	Type
To rough draft storage	7	To finished card
Grasp rough draft	8	Grasp finished card
To right hand	9	Pull finished card from type
Grasp finished card	10	card
Hold draft and finished card	11	
	12	
	13	

Proposed Method

Figure 11-4. Right-and-left hand charts of present and proposed methods of typing inventory cards. Source: Art Metal, Inc., *Office Standards and Planning.*

Both kinds of charts are also known as *workplace charts* because they relate to the detailed study of interlocking activities at single locations such as workbenches, desks, and machines. When an operator chart is applied solely to hand motions it is frequently termed a *right- and left-hand chart.* Figure 11-4 illustrates this type of chart.

It is not inferred that a motion study invariably means the use of some kind of chart. Experienced systems analysts often suggest improved methods of performance without having to resort to the recording of operational details. The charts are nevertheless beneficial both as visual aids for showing the need for particular changes and as educational instruments for training. In nearly every business there are certain operations that depend to a large extent on the use of charting to uncover ways of making improvements. In this case the use of charts is valuable.

Features of Construction

Motion-cycle chart designs frequently resemble those of flow process charts. Many of the same symbols may be used. As with flow process charts, the symbols that are used vary. The most commonly used symbols are shown in Figure 11-5.

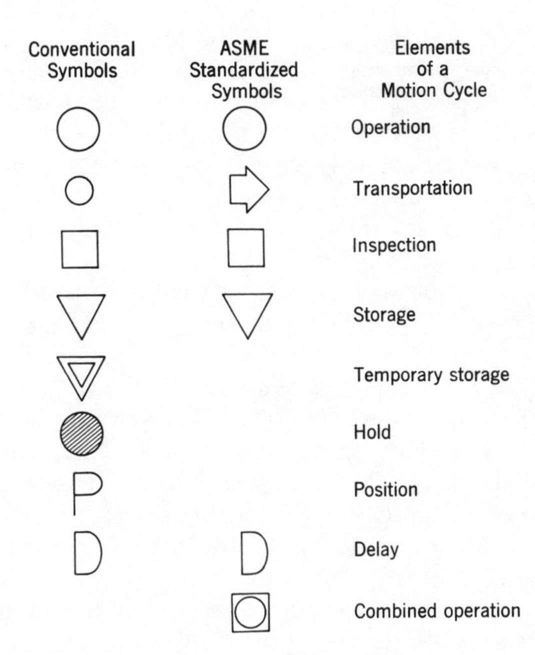

Figure 11-5. Symbols commonly used in motion-cycle charts.

Many motion-cycle charts are constructed with the use of only two basic symbols: one for designating operations and another for designating transportations. A transportation, as used here, refers to the movement of a body member toward or away from some object. Together the operation and transportation categories provide the most elemental of breakdowns. For many purposes only the symbols for these two categories need be employed in charting. Where more detail is desirable the use of one or more of the other symbols will provide the necessary refinement.

Various approaches are available to the systems analyst in recording motions on charts; for example, in constructing right- and left-hand charts some systems analysts prefer to observe the operations of each hand independently and to record the results on separate charts. The two charts are then merged into one.

Other systems analysts consider it best to simultaneously observe the motions of both hands and to record the findings on a single chart, as they occur. Whatever the approach, the idea is to construct a chart that accurately reflects the motions of an operation and thus lends itself as an aid to intensive study.

Therbligs

In performing their studies the Gilbreths evolved a highly refined schema for classifying and identifying the elements of a motion cycle. To basic divisions of operational tasks they applied the term "therblig," an anagram of the name Gilbreth. Each of the 18 therbligs was given a precise definition and assigned a mnemonic symbol, a letter abbreviation, and a color code. These are shown in Figure 11-6.

"Therblig" is a term that has been defined in many ways. Perhaps the most exact and most satisfying meaning is the one adopted by the Society for the Advancement of Management and the American Society of Mechanical Engineers. It is simple and to the point: Therbligs are "Gilbreth basic elements."

In theory, the breaking down of an operation into therbligs is a technique that is applicable to every motion-study situation. It is recognized, however, that motion-study applications in general warrant the employment of only relatively simple techniques. Many motion-study experts feel that the Gilbreths' list of therbligs is too cumbersome a tool for practical use in most applications. Ironically, its chief impediment lies in the refinement of the breakdown that it provides. Studies of numerous operations reveal that the time spent in the performance of certain of these therbligs is of such short duration that it is practically impossible to manually record them with an acceptable degree of accuracy and consistency. The

Name of Symbol	Therblig Symbol		Explanation	Color	Color Symbol	Dixon Pencil Number	Eagle Pencil Number
Search		Sh.	Eye turned as if searching	Black		331	747
Find		F.	Eye straight as if fixed on object	Gray		399	$747\frac{1}{2}$
Select		St.	Reaching for object	Gray, light		399	$734\frac{1}{2}$
Grasp		G.	Hand open for grasping object	Lake red		369	745
Transport loaded		T.L.	A hand with something in it	Green		375	738
Position		P.	Object being placed by hand	Blue		376	741
Assemble		A.	Several things put together	Violet, heavy		377	742
Use		U.	Word "use"	Purple		396	$742\frac{1}{2}$
Disassemble		D.A.	One part of an assembly removed	Violet, light		377	742
Inspect		I.	Magnifying lens	Burnt ochre		398	$745\frac{1}{2}$
Preposition		P.P.	A nine-pin as used in bowling	Sky blue		394	$740\frac{1}{2}$
Release load		R.L.	Dropping content from hand	Carmine red		370	744
Transport empty		T.E.	Empty hand	Olive green		391	$739\frac{1}{2}$
Rest for overcoming fatigue		R.	Man seated	Orange		372	737
Unavoidable delay		U.D.	Man bumping nose	Yellow ochre		373	736
Avoidable delay		A.D.	Man lying down on job	Lemon yellow		374	735
Plan		Pn.	Man thinking with fingers at brow	Brown		378	746
Hold		H.	Magnet holding iron bar	Gold ochre		388	$736\frac{1}{2}$

Figure 11-6. Table of symbols and colors for therbligs. Color symbols are used in this book as an expedient substitute for true coloring. In practice, these symbols are not used in the construction of simo charts. Instead, actual colors are applied to the charts by means of colored pencils.

mistakes thus introduced are apt to totally negate their usefulness. Their use, then, in normal practice has undeniable limitations.

Therbligs actually come into their own when used in conjunction with the performance of micromotion studies.

MICROMOTION STUDY

Literally, a micromotion is one of a number of small or microscopic movements. Being small, these movements are frequently performed with such rapidity as to defy study by ordinary observation; where this is the case the techniques of micromotion study find their real usefulness.

Micromotion study may be described simply as the study of detailed bodily movements and events by means of motion pictures. Its ultimate objective is the improvement of a work cycle. The nucleus of all micromotion studies is the motion pictures themselves.

In micromotion study the initial phase calls for photographing the entire operation with a motion-picture camera. For this purpose a 16mm camera is preferred over an 8mm one because the film allows greater detail, and is easier to edit and process. Other desirable features of the camera include a fast lens (F/1.9 is recommended) and the capability of operating at certain constant speeds. Today most cameras, including the spring-driven types, are designed to maintain a consistent pace while set at a given speed. The speed is the number of frames per second. Common speed ranges on cameras are 8, 16, 24, and 32; and 16, 24, 32, 48, and 64. The normal speed is 16; it is for general use. The half-speed, 8, doubles the rate of action in the projected picture; since it gives twice as much exposure on each frame, it is useful when the light is insufficient for fully exposed pictures at normal speed. Speed 24 slows down the rate of action to two-thirds normal; speed 32, to one-half normal; speed 48, to one-third normal; and speed 64, to one-quarter normal. The results produced at 64 speed are known as semi-slow-motion pictures. All speeds above 16 are for filming fast-moving events.

When an operation is filmed at a constant speed, time measurements are readily obtainable by merely counting the frames on the film between the beginning and the end of a movement or event. Time measurements, however, are usually obtained by placing a microchronometer, called a "wink" clock, in the field of vision. This is a fast-moving timepiece that will measure an element or therblig to $1/2000$ of a minute; that is, a unit of time known as a "wink." The clock is normally large enough to be easily seen when the film is projected.

A good projector is important. It should be capable of operating at varying speeds, including slow motion. The type that can be stopped to concentrate on one frame at a time is preferred. A reverse-motion feature is practical to permit portions of the film to be shown over again with ease.

Analyzing the Filmed Operation

In general practice the same operation is photographed several times. The film is then projected onto a screen, and the cycles are examined and considered with a view toward selecting a representative one for detailed study. The cycle thus chosen is subjected to a thorough analysis and evaluation.

The selected cycle itself may have to be projected a number of times. While it is being shown a micromotion worksheet can be prepared. The micromotion worksheet (see Figure 11-11) is a form used to record the facts concerning each basic element of the work cycle.

In a dual-hand operation the cycle should begin and end at approximately the same location for the two hands. The customary analytical approach is to record the motions of the left hand first and those of the right hand second. For information on the motions of the left hand the film is run through the projector and the observations are recorded on the work sheet. In the "Description—Left Hand" column goes a brief delineation of the movement or event. To the left of that, in the first column, is placed the clock reading. The third column is for recording the applicable therblig symbol. Computation for the "Subtracted Time" column follows the breakdown of the right-hand motions, after which the clock readings are subtracted to get the elapsed time for each therblig of both hands. The film is run again through the projector, and the information pertaining to the right hand is similarly reduced to writing.

Upon completion of the chart it is quite easy to visualize the entire cycle in action. Hand motions can be studied independently and in combination with one another.

Simultaneous Motion-Cyçle Charts

The simultaneous motion-cycle chart, commonly known as the "simo chart," provides a more sophisticated breakdown of motions and events than does the micromotion worksheet. It may be used in lieu of the worksheet or, as is more often the case, as an auxiliary instrument for facilitating detailed studies. When both forms are used the simo chart can be constructed from the information contained on the worksheet. Otherwise the simo chart is prepared in much the same manner as the worksheet; that is, directly from the film presentation.

The prime advantage of the simo chart rests on the breakdown of the elements it provides and the accompanying time scale. For these reasons it is more easily understood by persons who are not trained in the reading of such charts. It follows that simo charts make good training aids and are valuable in making people mindful of the advantages of motion economy.

Application of the Micromotion Technique

Micromotion studies are normally an expensive and time-consuming undertaking that must be employed with discretion. The technique is most suitable for situations where many highly skilled workers perform exactly the same tasks. The improvements made in one case may then be extended to the others. It will be found in general that the overall results are economical and truly satisfying.

THE CHRONOCYCLEGRAPH

This technique for studying motions is another of Frank Gilbreth's contributions to the field of motion study. There are variations of the method, but they have their roots in a basic idea.

In this technique motions of the operator are photographed with a still camera. By attaching a tracer light to a bodily member such as a hand it is possible to capture a record of the movement in a time-exposure photograph. The view thus produced is termed a cyclegraph.

To obtain measurements an interrupter is placed in the electric circuit accompanying the tracer bulb. When the light is turned on and off the path of the bulb appears as a series of elongated dots, the shape of which indicates the direction of a movement. The speed of motion affects the spacing of the dots; that is, slow movements produce marks of light that are close together; fast movements result in widely separated marks. It thus becomes possible to measure accurately the duration, speed, and acceleration of motions. The resultant record is known as a chronocyclegraph.

One further step is sometimes taken and that is to construct a wire model of the paths of motion. Such three-dimensional models serve the dual purpose of aiding the improvement of methods and instructing operators in the performance of correct motions.

EXAMINING AND EVALUATING THE FACTS

It is one thing to obtain facts and another to use them wisely. The objective is the improvement of the operation. Here, again, the questioning approach will be found useful.

Pertinent questions on bodily motions are as follows:

1. What motion elements, if any, can be eliminated?
2. Can any elements be simplified?
3. How about shortening the distance traveled or the reach?
4. Could certain elements be performed better if they were combined?
5. Are the most suitable members of the body being used?
6. Are any tasks being done by hand that could be done better by some other bodily member?
7. Have the physical capabilities and limitations of each member been properly considered?
8. Do the motions constitute a rhythmic and harmonious pattern?
9. Is the sequence of movements the best one?
10. Can either hand be relieved of holding material by the use of a fixture?
11. Would it be better to slide an item rather than to pick it up?
12. Are there any awkward motions involved? If so, how might they be adjusted in order to make them smooth and effective?
13. Do the hands begin and end their motions simultaneously? Are the movements symmetrical and in opposite directions wherever practicable?
14. What abrupt changes in direction take place? Can they be rectified?
15. Can a task being done by certain bodily members be more efficiently performed with the aid of a mechanical device?
16. Is it feasible to have the operator alternately stand and sit so as to reduce fatigue?

Questions pertaining to the arrangement and environment of the workplace are as follows:

1. Does the arrangement of the workplace permit the most efficient motions?
2. Are materials, tools, and equipment properly located?
3. Should certain tools be prepositioned?

4. Is there fatigue that can be eliminated by the introduction of a new tool? By rearranging the layout? By changing the working conditions?
5. Would it be practicable to employ chutes, conveyors, "drop deliveries," or the like?
6. Can any foot-pedal or knee-operated device be used to advantage?
7. Can material be deposited by means of gravity-feed dispensers?
8. Has the need for muscular effort been reduced to a minimum?
9. Is the workplace orderly and safe throughout?
10. Has excessive noise been eliminated?
11. Is there proper light, heat, and ventilation?

Questions pertaining to the design of tools and equipment are as follows:

1. Can any movements be made freer or easier by employing tools of lighter weight?
2. Are the tools properly designed for the work?
3. How about using a dual-purpose tool (such as a unit with a screwdriver on one end and a wrench on the other)?
4. Are hand tools easy to hold?
5. Would a specially designed tool, machine, or piece of equipment help to facilitate the performance of the operation?

ASSEMBLING BALL-POINT PENS

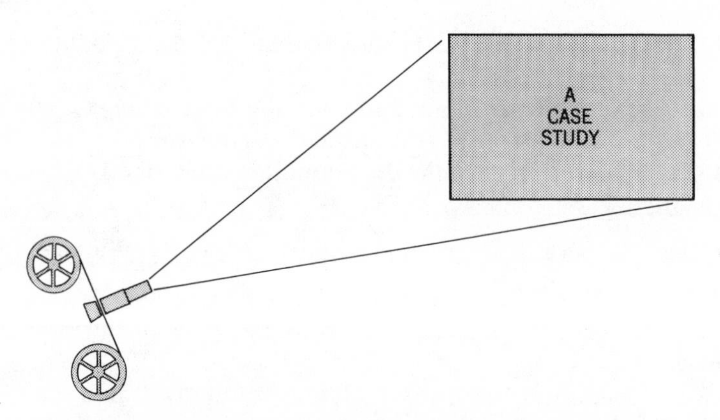

A CASE STUDY INVOLVING THE MICROMOTION TECHNIQUE

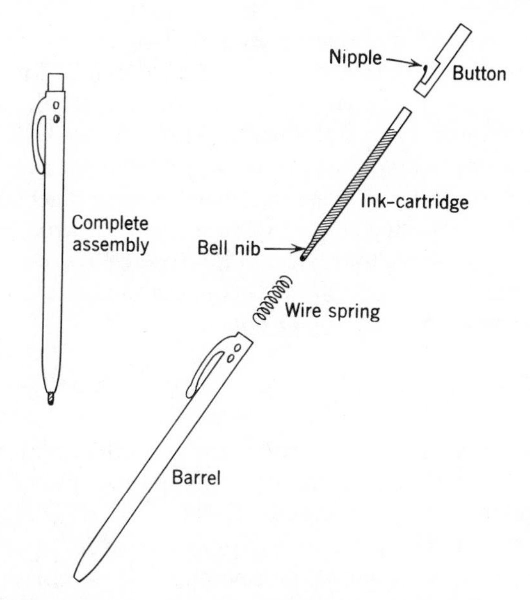

Figure 11-7. The final assembly and its parts.

Nature of the Operation

As shown in Figure 11-7, in assembling ball-point pens four parts must be combined; namely, the plastic housing, or barrel; the wire spring; the ink cartridge; and the plastic-molded knob called a button. The putting together of the parts is accomplished with the barrel held in an upright position. Three successive manual insertions take place: the dropping of the spring into the barrel, the slipping of the ink cartridge into the spring, and the subsequent positioning of the button.

Certain features of the pen should be noted. The barrel is tapered at its holding end and possesses a molded-in plastic clip at the other end. A quarter of a turn away from the clip are two ⅛ inch holes that extend to the core of the barrel. These holes are located one under the other and serve the purpose of engaging the nipple portion of the button.

The spring is of simple construction with two uniform ends, either of which may be dropped into the barrel first. In the case of the ink cartridge, however, the end with the ball nib must be placed foremost in the barrel chamber.

Topping the other components is the button. This unit is the binding force that unites all the parts. In assembly the button is introduced into the chamber with the nipple aligned with the two holes in the barrel. The button is then pressed until the nipple reaches the uppermost hole, where spring tension causes it to become lodged. This action completes the assembly, and it will not come apart without purposeful effort.

With the nipple in the upper hole of the barrel, the entire ink cartridge remains within the barrel. Pressing down on the button releases the nipple from the upper hole and causes it to become engaged in the lower hole. As a result, the ball nib is extended and held in place for writing. The nib is retracted by applying slight pressure on the protruding nipple. Consequently the wire spring that was compressed by the downward motion is released, forcing the cartridge to become withdrawn and the nipple to become again engaged in the upper hole.

Charting the Existing Method

For the purpose of studying the existing method with a view toward making improvements, five forms were completed. Each of these is discussed and shown below.

Operator Chart. The right- and left-hand chart in Figure 11-8 shows a conventionalized breakdown of the operation. It was made by acute observation, but without the aid of photographic equipment. For an interesting comparison note the difference between the motion elements contained thereon and the more elaborate listing shown on the micromotion worksheet.

Filming Specification Sheet. Take notice that the filming specification sheet (Figure 11-9) shows the location of the camera and floodlights at the time of filming the operation. Sufficient information regarding the photographic equipment is given to make it possible to verify the results should that ever be desirable; such recordings are in keeping with good scientific practice.

Cycle-Selection Chart. As mentioned earlier, in making micromotion studies the film is run through the projector in order to select a suitable cycle for analysis. Figure 11-10 presents a commentary on each of 13 cycles and relates that the fifth one was chosen for intensive study because of its representative characteristics.

Micromotion Worksheet. Figure 11-11 shows a micromotion worksheet that was made from viewing the selected cycle as it was being projected. When it was considered expedient—for the purpose of inspection—the projector was stopped and individual pictures were studied frame by frame.

Simultaneous Motion-Cycle Chart. This sheet (Figure 11-12) was prepared from the micromotion worksheet; its principal value lies in graphically portraying the motions and elements of motions in terms of a scale. From a study of the chart it is possible to visualize the entire operation.

Shortcomings of the Existing Method

A study of the existing method with the assistance of the foregoing charts and sheets revealed the following shortcomings:

1. Extensive holding of the barrel by the left hand.
2. Although it is preferable to have both hands end as well as begin their motions at the same time, the left hand is still holding during the time when the right hand completes its motions.
3. The right hand performs most of the motions, whereas the left hand is idle.
4. Parts are not sufficiently close to operation.
5. Location of the material makes it necessary to use shoulder motions in addition to those of lesser bodily members.
6. Stack of mixed pen parts does not permit smooth and rhythmic motions of the hands.
7. Movements are jerky because the material is too far removed from the operator.
8. No use is made of gravity-feed bins.
9. Disposition of finished assemblies by "drop deliveries" (or some other convenient method) is not made.
10. Failure to make use of a jig with a foot pedal when such a device would substantially reduce the work load.
11. Chair is too high for table and not of the proper type.
12. Materials are not prepositioned.

Features of an Improved Method

Out of a study of the existing situation grew the idea of designing a jig so that both hands could be put to maximum use. Referring to Figure 11-13, notice the construction of the jig. It will be seen that it is a simple device with two adjacent holes into which each hand can first insert a pen barrel and then assemble the pen parts.

The jig is made to be set into the table directly in front of the operator. It has one movable panel that is operated by foot to release the assemblies and let them fall into a basket under the table. In this way the jig is readily and easily cleared for assembling the next two pens.

One other innovation is of major importance; that is, the utilization of

RIGHT- AND LEFT-HAND CHART

OPERATION Assembling ball-point pens DATE 9-3-XX

OPERATOR Ray O. Dawn ANALYST U. James Roberts

DEPARTMENT 1410 DIVISION Manufacturing APPROVED

METHOD: [X] Old [ ] Proposed

Sketch of work place and indicated motion paths

Table
Storage bin
Stack of mixed components
Right-hand path
Left-hand path
Operator

LEFT-HAND ELEMENTS			RIGHT-HAND ELEMENTS		
DISTANCE (INCHES)	DESCRIPTION	SYMBOL	SYMBOL	DESCRIPTION	DISTANCE (INCHES)
10	Carry assembly to bin	→	→	Reach for barrel	18
	Place assembly in bin	0	0	Pick up barrel	
10	Return to work position	→	→	Carry barrel to work position	18
	Grasp barrel	0	0	Place barrel in left hand	
	Hold barrel	0	→	Reach for spring	18
	" "	0	0	Pick up spring	
	" "	0	→	Carry spring to work position	18
	" "	0	0	Assemble	
	" "	0	→	Reach for cartridge	18
	" "	0	0	Pick up cartridge	
	" "	0	→	Carry cartridge to work position	18
	" "	0	0	Assemble	
	" "	0	→	Reach for button	18
	" "	0	0	Pick up button	
	" "	0	→	Carry button to work position	18
	" "	0	0	Assemble	

Sheet 1 of 2 Sheets

Figure 11-8 *a*. Operator chart for old method of assembling ball-point pens.

RIGHT AND LEFT-HAND CHART
CONTINUATION SHEET

OPERATION Assembling ball-point pens DATE 9-3-XX

LEFT-HAND ELEMENTS			RIGHT-HAND ELEMENTS		
DISTANCE (INCHES)	DESCRIPTION	SYMBOL	SYMBOL	DESCRIPTION	DISTANCE (INCHES)
10	Carry assembly to bin				
	Place assembly in bin	0			
30	Total distance traveled			Total distance traveled	144

Sheet 2 of 2 Sheets

Figure 11-8 *b*. *Continued*

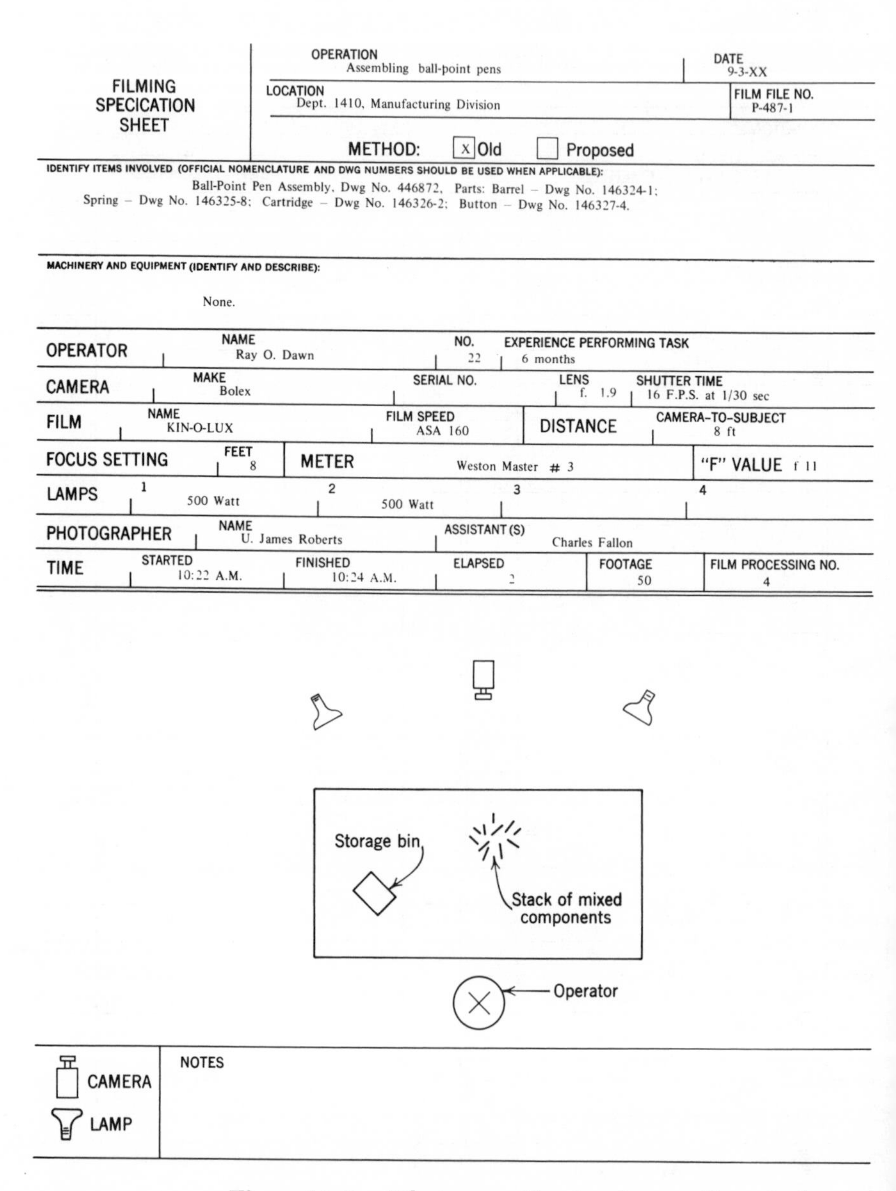

FILMING SPECICATION SHEET

OPERATION: Assembling ball-point pens | DATE: 9-3-XX

LOCATION: Dept. 1410, Manufacturing Division | FILM FILE NO.: P-487-1

METHOD: [X] Old [ ] Proposed

IDENTIFY ITEMS INVOLVED (OFFICIAL NOMENCLATURE AND DWG NUMBERS SHOULD BE USED WHEN APPLICABLE):

Ball-Point Pen Assembly, Dwg No. 446872, Parts: Barrel – Dwg No. 146324-1; Spring – Dwg No. 146325-8; Cartridge – Dwg No. 146326-2; Button – Dwg No. 146327-4.

MACHINERY AND EQUIPMENT (IDENTIFY AND DESCRIBE):

None.

OPERATOR	NAME: Ray O. Dawn	NO.: 22	EXPERIENCE PERFORMING TASK: 6 months	
CAMERA	MAKE: Bolex	SERIAL NO.:	LENS: f. 1.9	SHUTTER TIME: 16 F.P.S. at 1/30 sec
FILM	NAME: KIN-O-LUX	FILM SPEED: ASA 160	DISTANCE	CAMERA-TO-SUBJECT: 8 ft
FOCUS SETTING	FEET: 8	METER: Weston Master # 3	"F" VALUE: f 11	
LAMPS	1: 500 Watt	2: 500 Watt	3:	4:
PHOTOGRAPHER	NAME: U. James Roberts	ASSISTANT (S): Charles Fallon		
TIME	STARTED: 10:22 A.M.	FINISHED: 10:24 A.M.	ELAPSED: 2	FOOTAGE: 50 / FILM PROCESSING NO.: 4

Figure 11-9. Filming specification sheet—old method.

CYCLE-SELECTION CHART
for
MICROMOTION STUDIES

OPERATION Assembling ball-point pens DATE 9-3-XX

OPERATOR Ray O. Dawn ANALYST U. James Roberts

LOCATION Dept. 1410, Manufacturing Division FILM FILE NO. P-487-1

CYCLE NO.	TIME		COMMENTARY
	CLOCK	ELAPSED	
1	30	30	Incomplete cycle
2	283	253	Good cycle
3	572	289	Fair cycle
4	870	298	Fumble occurred during cycle
5	1128	258	Selected for analysis on the basis of its being the most representative cycle
6	1650	522	Two fumbles took place
7	1941	291	Fair cycle
8	2204	263	Good cycle
9	2520	316	Searching caused undue delay
10	2764	244	Best cycle – not representative, however
11	3117	353	Needless delay
12	3404	287	Fair cycle
13	3720	316	Delay caused by fumble and search

Figure 11-10. Cycle selection chart—old method.

MICROMOTION WORKSHEET

OPERATION Assembling ball-point pens

FILM FILE NO. P-487-1

OPERATOR Ray O. Dawn

ANALYST U. James Roberts

LOCATION Dept. 1410, Manufacturing Division

DATE THIS SHEET IS COMPLETED 9-3-XX

METHOD: [x] Old [ ] Proposed

Sheet 1 of 1 Sheets

CLOCK READING	SUBTRACTED TIME	THERBLIG SYMBOL	DESCRIPTION LEFT HAND	CLOCK READING	SUBTRACTED TIME	THERBLIG SYMBOL	DESCRIPTION RIGHT HAND
891	21	TL	Carries assembly to bin	888	18	TE	Reaches for barrel
895	4	RL	Releases assembly	910	22	St	Selects barrel
930	35	TE	Returns to work position	913	3	G	Grasps barrel
935	5	G	Grasps barrel	928	15	TL	Carries barrel to work position
975	40	P	Positions barrel	933	5	P&RL	Positions barrel and releases
1123	148	H	Holds barrel	945	12	TE	Reaches for spring
1143	20	TL	Carries assembly to bin	947	2	St	Selects spring
1153	10	RL	Releases assembly	951	4	G	Grasps spring
				972	21	TL	Carries spring to work position
				974	2	P	Positions spring
				994	20	A&RL	Assembles spring and releases
				1005	11	TE	Reaches for cartridge
				1009	4	St	Selects cartridge
				1012	3	G	Grasps cartridge
				1019	7	TL	Carries cartridge to wk. position
				1023	4	P	Positions cartridge
				1036	13	A&RL	Assembles cartridge and releases
				1054	18	TE	Reaches for button
				1056	2	St	Selects button
				1059	3	G	Grasps button
				1072	13	TL	Carries button to work position
				1080	8	P	Positions button
				1128	48	A&RL	Assembles button and releases

Figure 11-11. Micromotion work sheet—old method.

MICROMOTION STUDY
SIMO CHART

OPERATION Assembling ball-point pens FILM FILE NO. P-487-1

OPERATOR Ray O. Dawn ANALYST U. James Roberts

LOCATION Dept. 1410, Manufacturing Division DATE THIS SHEET IS COMPLETED 9-3-XX

METHOD: [X] Old [ ] Proposed Sheet 1 of 2 sheets

DESCRIPTION LEFT HAND	THERBLIG SYMBOL	TIME	CLOCK OR METER READING	TIME	THERBLIG SYMBOL	DESCRIPTION RIGHT HAND
Carries assembly to bin	TL	21		18	TE	Reaches for barrel
Releases assembly	RL	4		22	St	Selects barrel
Returns to work position	TE	35				
				3	G	Grasps barrel
				15	TL	Carries barrel to work position
				5	P&RL	Positions barrel and releases
Grasps barrel	G	5		12	TE	Reaches for spring
Positions barrel	P	40		2	St	Selects spring
				4	G	Grasps spring
				21	TL	Carries spring to work position
				2	P	Positions spring
				20	A&RL	Assembles spring and releases
Holds barrel	H	148		11	TE	Reaches for cartridge
				4	St	Selects cartridge
				3	G	Grasps cartridge
				7	TL	Carries cartridge to work position

Figure 11-12 *a*. Simultaneous motion-cycle chart showing old method of assembling ball-point pens.

MICROMOTION STUDY
SIMO CHART

OPERATION Assembling ball-point pens FILM FILE NO. P-487-1

Sheet 2 of 2 Sheets

DESCRIPTION LEFT HAND	THERBLIG SYMBOL	TIME	CLOCK OR METER READING	TIME	THERBLIG SYMBOL	DESCRIPTION RIGHT HAND
				4	P	Positions cartridge
				13	A RL	Assembles cartridge and releases
				18	TE	Reaches for button
				2	St	Selects button
				3	G	Grasps button
				13	TL	Carries button to work position
				8	P	Positions button
				48	A&RL	Assembles button and releases
Carries assembly to bin	TL	20				
Releases assembly	RL	10				

Figure 11-12 *b*. Simultaneous motion-cycle chart showing old method of assembling ball-point pens—continued.

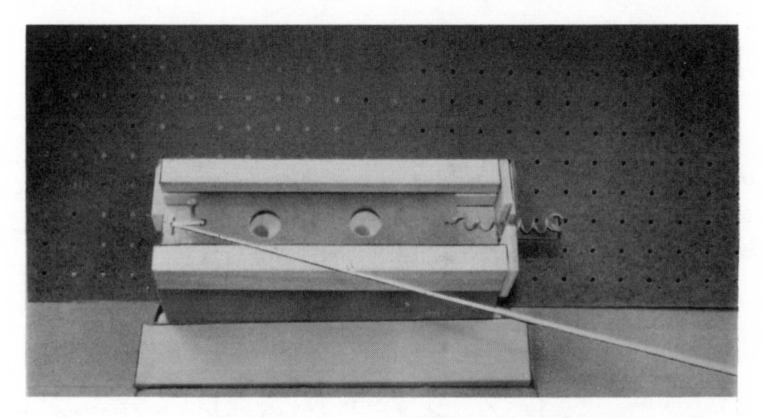

Figure 11-13. Bottom-view sketch of jig.

gravity-feed bins to segregate the parts and deliver them to the operator. Figure 11-14 shows the bins and illustrates a workplace arrangement for the improved method of assembling ball-point pens. The parts to be assembled are located in a semicircular pattern so that the operator can use opposite and symmetrical motions simultaneously.

The right- and left-hand chart, filming specification sheet, cycle-selection chart, micromotion worksheet, and micromotion simo chart for the new method are shown in Figures 11-15 through 11-19.

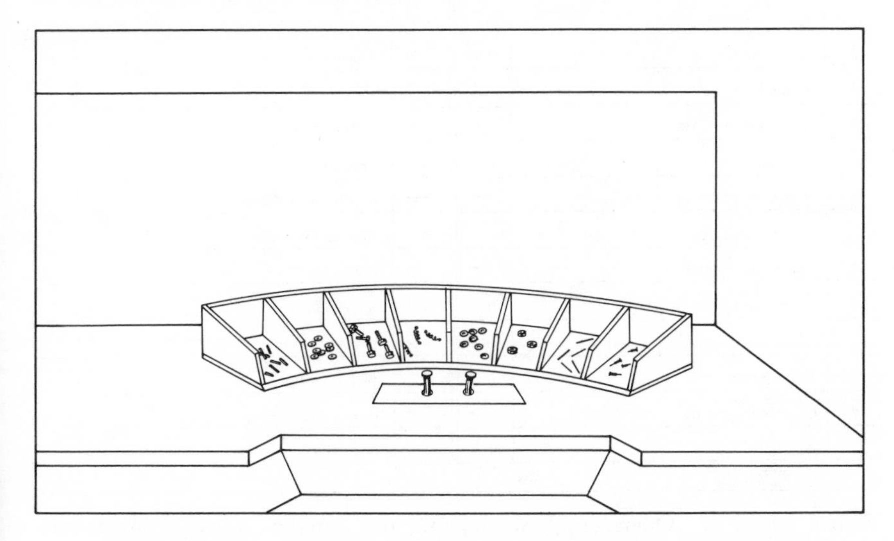

Figure 11-14. Workplace arrangement for the improved method.

RIGHT- AND LEFT-HAND CHART

OPERATION Assembling ball-point pens DATE 10-12-XX

OPERATOR Ray O. Dawn ANALYST U. James Roberts

DEPARTMENT 1410 DIVISION Manufacturing APPROVED

METHOD: ☐ Old ☒ Proposed

Sketch of workplace and indicated motion paths

A - Barrels
B - Springs
C - Cartridges
D - Buttons

E - Barrels
F - Springs
G - Cartridges
H - Buttons

Table
Left-hand paths
Right-hand paths
Holding fixture (jig)
Operator

LEFT-HAND ELEMENTS			RIGHT-HAND ELEMENTS		
DISTANCE (INCHES)	DESCRIPTION	SYMBOL	SYMBOL	DESCRIPTION	DISTANCE (INCHES)
12	Reach for barrel	→	→	Reach for barrel	12
	Pick up barrel	0	0	Pick up barrel	
12	Carry barrel to jig	→	→	Carry barrel to jig	12
	Position barrel in jig	0	0	Position barrel in jig	
10	Reach for spring	→	→	Reach for spring	10
	Pick up spring	0	0	Pick up spring	
10	Carry spring to jig	→	→	Carry spring to jig	10
	Assemble	0	0	Assemble	
8	Reach for cartridge	→	→	Reach for cartridge	8
	Pick up cartridge	0	0	Pick up cartridge	
8	Carry cartridge to jig	→	→	Carry cartridge to jig	8
	Assemble	0	0	Assemble	
6	Reach for button	→	→	Reach for button	6
	Pick up button	0	0	Pick up button	
6	Carry button to jig	→	→	Carry button to jig	6
	Assemble	0	0	Assemble	

Sheet 1 of 2 Sheets

Figure 11-15 *a*. Operator chart showing new method of assembling ball-point pens.

RIGHT- AND LEFT-HAND CHART
CONTINUATION SHEET

OPERATION Assembling ball-point pens DATE 10-12-XX

LEFT-HAND ELEMENTS			RIGHT-HAND ELEMENTS		
DISTANCE (INCHES)	DESCRIPTION	SYMBOL	SYMBOL	DESCRIPTION	DISTANCE (INCHES)
72	Total distance traveled			Total distance traveled	72

Sheet 2 of 2 Sheets

Figure 11-15 *b*. *Continued*

FILMING SPECIFICATION SHEET

OPERATION	DATE
Assembling ball-point pens	10-12-XX

LOCATION	FILM FILE NO.
Dept. 1410, Manufacturing Division	P - 487-2

METHOD: ☐ OLD ☒ PROPOSED

IDENTIFY ITEMS INVOLVED (OFFICIAL NOMENCLATURE AND DWG NUMBERS SHOULD BE USED WHEN APPLICABLE):

Ball-Point Pen Assembly, Dwg No. 446872. Parts: Barrel - Dwg No. 146324-1; Spring - Dwg. No. 146325-8; Cartridge - Dwg No. 146326-2; Button - Dwg. No. 146327-4.

MACHINERY AND EQUIPMEMT (IDENTIFY AND DESCRIBE):

None.

OPERATOR	NAME: Ray O. Dawn	NO.: 22	EXPERIENCE PERFORMING TASK: Regular operator
CAMERA	NAME: Bolex	SERIAL NO.:	LENS: f. 1.9 — SHUTTER TIME: 16 F.P.S. at 1/30 sec.
FILM	NAME: KIN-O-LUN	FILM SPEED: ASA 160	DISTANCE — CAMERA-TO-SUBJECT: 5½ ft.
FOCUS SETTING	FEET: 6	METER: Weston Master #3	"F" VALUE: f 15
LAMPS	1: 500 Watt	2: 500 Watt	3: — 4:
PHOTOGRAPHER	NAME: U. James Roberts	ASSISTANT(S): Charles Fallon	

TIME	STARTED	FINISHED	ELAPSED	FOOTAGE	FILM PROCESSING NO.
	10:11 A.M.	10:14	3	50	4

Bins for parts

Holding fixture (jig)

Operator

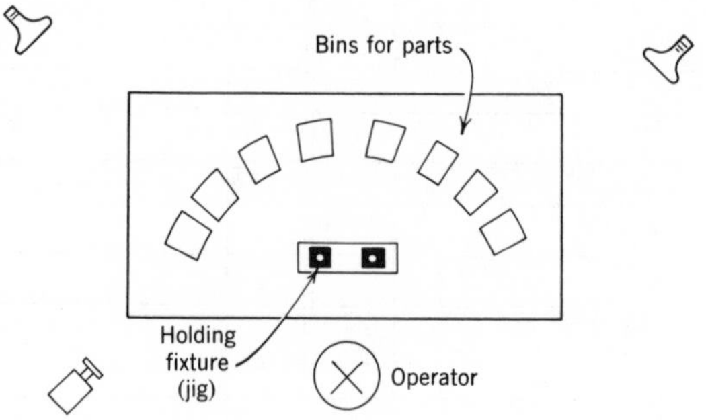

CAMERA

LAMP

NOTES

Figure 11-16. Filming specification sheet—new method.

CYCLE-SELECTION CHART
for
MICROMOTION STUDIES

OPERATION Assembling ball-point pens DATE 10-12-XX

OPERATOR Ray O. Dawn ANALYST U. James Roberts

LOCATION Dept. 1410, Manufacturing Division FILM FILE NO. P-487-2

CYCLE NO.	TIME		COMMENTARY
	CLOCK	ELAPSED	
1	03300	185	Good cycle
2	03135	165	Good cycle – selected for analysis
3	02940	195	Good cycle
4	02728	212	Minor fumble occurred during cycle
5	02485	243	Two fumbles
6	02294	191	Good cycle
7	02084	210	Minor fumble

Figure 11-17. Cycle-selection chart—new method.

MICROMOTION WORKSHEET

OPERATION Assembling ball-point pens FILM FILE NO. P - 487-2

OPERATOR Ray O. Dawn ANALYST U. James Roberts

LOCATION Dept. 1410, Manufacturing Division DATE THIS SHEET IS COMPLETED 10-12-XX

METHOD: ☐ OLD ☒ PROPOSED

Sheet 1 of 1 Sheets

CLOCK READING	SUBTRACTED TIME	THERBLIG SYMBOL	DESCRIPTION LEFT HAND	CLOCK READING	SUBTRACTED TIME	THERBLIG SYMBOL	DESCRIPTION RIGHT HAND
03287	13	.TE	Reaches for barrel	03287	13	TE	Reaches for barrel
03284	3	G	Grasps barrel	03284	3	G	Grasps barrel
03270	14	TL	Carries barrel to jig	03268	16	TL	Carries barrel to jig
03251	19	P&RL	Positions barrel and releases	03250	18	P&RL	Positions barrel and releases
03240	11	TE	Reaches for spring	03237	13	TE	Reaches for spring
03232	8	G	Grasps spring	03233	4	G	Grasps spring
03220	12	TL	Carries spring to jig	03221	12	TL	Carries spring to jig
03213	7	A&RL	Assembles spring and releases	03211	10	A&RL	Assembles spring and releases
03203	10	TE	Reaches for cartridge	03202	9	TE	Reaches for cartridge
03197	6	G	Grasps cartridge	03195	7	G	Grasps cartridge
03187	10	TL	Carries cartridge to jig	03188	7	TL	Carries cartridge to jig
03180	7	A&RL	Assembles cartridge and releases	03179	9	A&RL	Assembles cartridge and releases
03171	9	TE	Reaches for button	03168	11	TE	Reaches for button
03161	10	G	Grasps button	03163	5	G	Grasps button
03152	9	TL	Carries button to jig	03151	12	TL	Carries button to jig
03141	11	P	Positions button	03141	10	P	Positions button
03135	6	A&RL	Assembles button and releases	03135	6	A&RL	Assembles button and releases

Figure 11-18. Micromotion work sheet—new method.

MICROMOTION STUDY
SIMO CHART

OPERATION Assembling ball-point pens FILM FILE NO. P-487-2

OPERATOR Ray O. Dawn ANALYST U. James Roberts

LOCATION Dept. 1410, Manufacturing Division DATE THIS SHEET IS COMPLETED 10-12-XX

METHOD: ☐ Old ☒ Proposed Sheet 1 of 2 Sheets

DESCRIPTION LEFT HAND	THERBLIG SYMBOL	TIME	CLOCK OR METER READING	TIME	THERBLIG SYMBOL	DESCRIPTION RIGHT HAND
Reaches for barrel	TE	13		13	TE	Reaches for barrel
Grasps barrel	G	3		3	G	Grasps barrel
Carries barrel to jig	TL	14		16	TL	Carries barrel to jig
Positions barrel in holding device and releases	P&RL	19		18	P&RL	Positions barrel in holding device and releases
Reaches for spring	TE	11		13	TE	Reaches for spring
Grasps spring	G	8		4	G	Grasps spring
Carries spring to jig	TL	12		12	TL	Carries spring to jig
Places spring in barrel and releases	A&RL	7		10	A&RL	Places spring in barrel and releases
Reaches for cartridge	TE	10		9	TE	Reaches for cartridge
Grasps cartridge	G	6		7	G	Grasps cartridge
Carries cartridge to jig	TL	10		7	TL	Carries cartridge to jig
Places cartridge in barrel and releases	A&RL	7		9	A&RL	Places cartridge in barrel and releases
Reaches for button	TE	9		11	TE	Reaches for button
Grasps button	G	10		5	G	Grasps button
Carries button to jig	TL	9		12	TL	Carries button to jig

Figure 11-19 *a*. Simultaneous motion-cycle chart showing new method of assembling ball-point pens.

MICROMOTION STUDY
SIMO CHART

OPERATION Assembling ball-point pens

FILM FILE NO. P - 487-2

Sheet 2 of 2

DESCRIPTION LEFT HAND	THERBLIG SYMBOL	TIME	CLOCK OR METER READING	TIME	THERBLIG SYMBOL	DESCRIPTION RIGHT HAND
Positions button	P	11		10	P	Positions button
Presses button into barrel and releases	A&RL	6		6	A&RL	Presses button into barrel and releases

Figure 11-19 *b*. *Continued*

Measure of Effect: Estimates of Time Savings and Production Increases

It is important to know time savings and production increases in quantitative terms, as shown in Table 11-1.

Table 11-1. COMPARISON OF ASSEMBLY TIME AND PRODUCTION

	Assembly Time (minutes)		Number of Assemblies per Minute
	1 Pen	2 Pens	
Old method	0.129 [a]	0.258 [b]	7.75 [c]
New method	——	0.165 [d]	12.12 [e]
Difference	——	0.093	4.45

[a] 258 winks (total for the fifth cycle, see Figure 11-10) ÷ 2000 winks (total number of winks to a minute) = 0.129 minutes per cycle to complete one assembly.
[b] 0.129 minutes per cycle to complete one assembly × 2 pens = 0.258 minutes to complete two assemblies.
[c] 1.0 minute ÷ 0.129 minutes per cycle to complete one assembly = 7.75 number of assemblies per minute.
[d] 165 winks (total for the second cycle, see Figure 11-17) ÷ 1000 winks = 0.165 minutes per cycle to complete two assemblies.
[e] 1.0 minute ÷ 0.165/2 = 12.12 number of assemblies per minute.

The percentage-estimate of the reduction in time may be shown by the following formula:

$$\frac{\begin{pmatrix}\text{time/pen}\\\text{old method}\end{pmatrix}-\begin{pmatrix}\text{time/pen}\\\text{new method}\end{pmatrix}}{\begin{pmatrix}\text{time/pen}\\\text{old method}\end{pmatrix}}\times 100=$$

Therefore

$$\frac{0.129-0.082}{0.129}\times 100=36.4\%.$$

In a similar way the percentage-estimate of the increase in output may be calculated as follows:

$$\frac{\begin{pmatrix}\text{pens/minute}\\\text{new method}\end{pmatrix}-\begin{pmatrix}\text{pens/minute}\\\text{old method}\end{pmatrix}}{\begin{pmatrix}\text{pens/minute}\\\text{old method}\end{pmatrix}}\times 100=$$

Therefore

$$\left(\frac{12.12-7.75}{7.75}\right)\times 100=56.4\%.$$

Summary: Benefits To Be Derived from Installing New Method

The facts and data derived from the study of the existing method, contrasted with the facts and data obtained through the practical application of the proposed method, indicate that the innovation will accomplish the following:

1. A substantial reduction in assembly time per unit; namely, 36.4 percent.
2. Greater productivity; namely, 54.4 percent increase per man per minute.
3. A marked decrease in operator fatigue.

As indicated, the economies of the new method will have the effect of reducing the overall cost per unit while at the same time permitting an increase in the total output. Ultimately this means more finished goods for less money.

12

Forms—Kinds and Design

By form we shall mean printed paper or card stock, the dimensions, printed matter, and general arrangement of which have been designed in a standard pattern to accommodate the insertion of recurring but variable information. Simply put, this means that forms are pieces of paper containing preprinted constant data and blank spaces specified for outlining the information to be recorded. In business and industry they are vehicles for conveying facts and figures from one place to another and from one time to another. Their use is constantly necessary because nearly all the important activities of the systems of an enterprise entail the utilization of one or more forms. Unless they are properly designed and constructed, the execution of the affected operations will be greatly impaired.

It is difficult to overestimate the importance of forms; they are a primary essential of most systems. To an undeterminable extent the design of the form is an expression of system efficiency.

Good forms design requires a broad knowledge of the system of which it is part. In designing a form a large number of considerations are brought into play; that is, the place and purpose of the form in the system, its handling and filing requirements, and its economy of use.

A form generally must be designed for a particular purpose; it needs to be tailored for the requirements of a specific system in a definite application. Therefore its design should be considered in terms of its contribution to the optimum fulfillment of the system's objectives. Some of the advantages of a well-designed form are the following:

1. It prescribes the points of information to be inserted and thus serves as a reminder.
2. It permits work to be performed concurrently by providing all pertinent individuals and organizational units with copies of the form.
3. It facilitates the recording and location of information by designating the exact spaces for its placement.

4. It eliminates the necessity for transcribing much information.
5. It makes filing and sorting operations less complex, less difficult, and less time consuming.

Business and industry make extensive use of forms of various shapes, sizes, and colors. Common types include checks, invoices, sales orders, production orders, time cards, and inventory records. Every necessary form has its own particular contribution to make to the welfare and success of an enterprise.

Nearly everyone within a company works with forms, including managers and top executives. These forms provide the framework, or shell, in which information is to be recorded. Often they authorize a particular action, as for example, to disburse money (by voucher and check) to meet a financial obligation (an invoice).

In companies of considerable size the preparing, processing, and filing of forms are routine and enormous tasks. Since the processing requirements of any form give the pattern of the work to be performed, a form may be described as a prescription for recording, communicating, and storing information.

Basic Kinds of Forms. One of the important things to know in systems-and-procedures work is the different forms that are available for use. Current practice in forms design and construction comprehends the following varieties for the purposes indicated:

1. Single-unit (loose) forms
2. Unit sets
3. Book (padded) forms
4. Continuous forms
5. Pegboard forms
6. Stock forms
7. Accounting forms
8. Duplicating masters
9. Tags, tickets, and labels
10. Safety-paper forms
11. Envelopes
12. Punched-card forms
13. Magnetic-strip cards

SINGLE-UNIT (LOOSE) FORMS

Singly imprinted sheets are the most basic and undoubtably the most widely used of all forms. The sheets may be used individually or as a set with carbon paper, depending on the particular requirements.

Recent technological advancements within the reproduction-equipment field have lent substantial impetus to the employment of single-unit forms. New and improved facilities such as duplicating, photocopying, and composing machines have made it both practical and expedient to produce these forms on company premises. As a result the internal duplicating facilities of enterprises are now being extensively employed to make simple, short-run forms.

UNIT SETS

Unit sets are complete sets of sheets with carbons and copies held together as single units. Variations in forms of this kind can be devised in great numbers according to the nature and features they may be desired to possess. Among the many possibilities are (*a*) various quantities of copies (up to 15 and sometimes as many as 20 per set); (*b*) different colors of paper and/or print for each copy; (*c*) combinations of sheets and carbons of assorted widths and lengths, with varying weights of paper stock and quality; and (*d*) provisioning for the removal of self-contained parts of sets at certain stages of processing. Moreover, the sheets of a set can be made to accommodate diversified requirements with respect to perforations, "block out" patterns, spot carbons, strip carbons, and cut-out carbons.

So-called snap sets, or snap-outs, compose a popular and widely used class of unit sets. Printing-supply houses offer them under a variety of trade names. The distinguishing feature of these sets is that by gripping a holding area on the form, usually in the form of a perforated stub, the carbon and sheets may be separated with one coordinated movement, as shown in Figure 12-1.

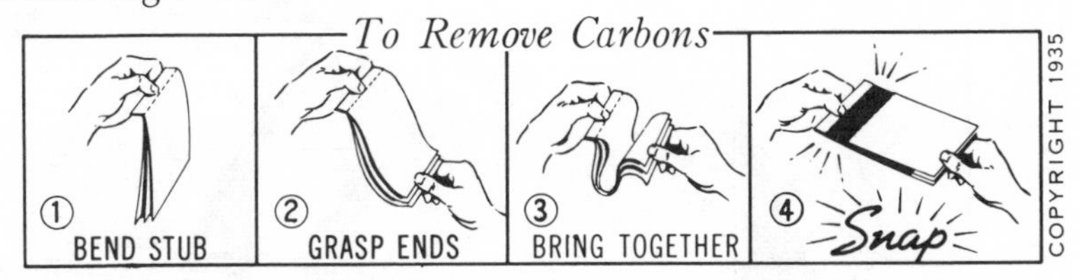

Figure 12-1. Removing carbons from a snap set. Courtesy: Moore Business Forms, Inc.

Snap sets may be constructed with stubs at the top or bottom, left or right side, or on two opposite sides; as a rule, they contain one-time carbons. In the case of *single-stub sets* the carbons are usually attached to the stub and are cut somewhat shorter than the sheets. By holding the perforated stub with the left hand and pulling on the edge of the sheets with

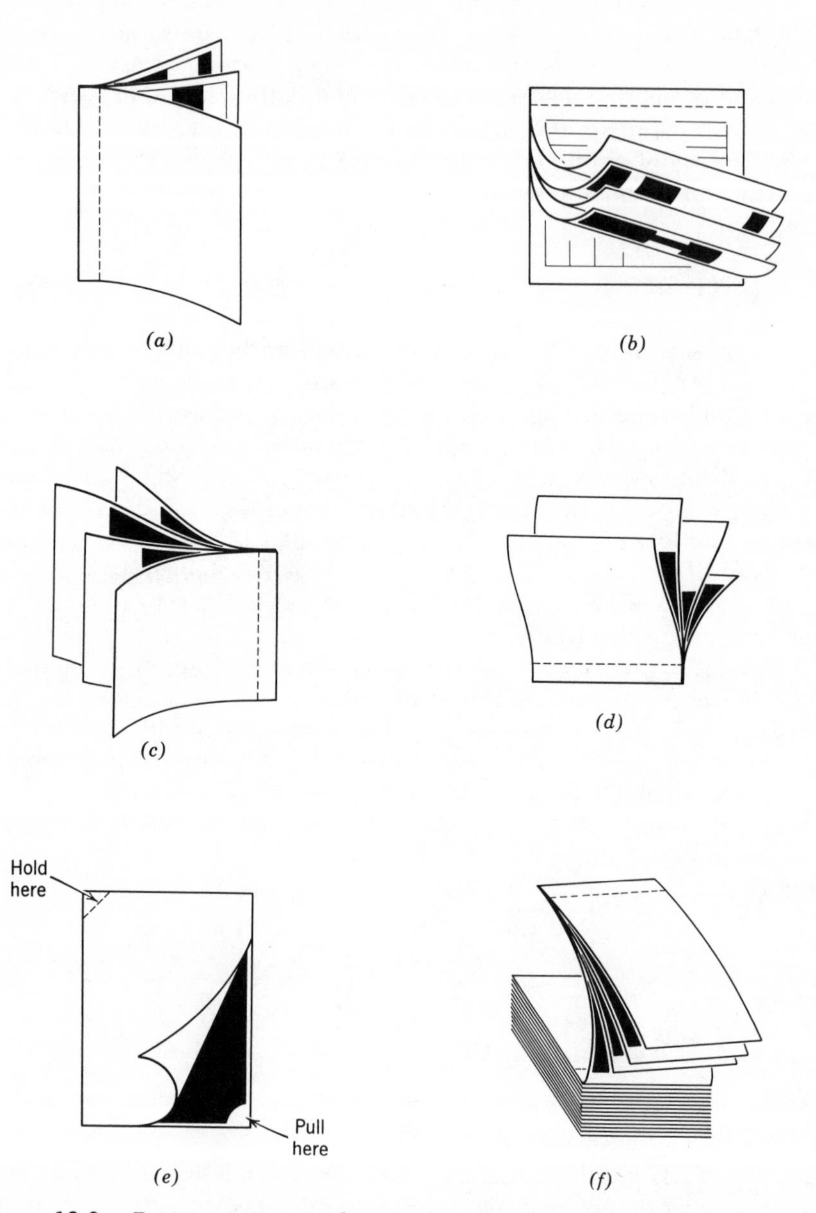

Figure 12-2. Basic variations in the construction of snap sets: (*a*) left-side stub with strip-coated carbons; (*b*) top stub with carbon spots; (*c*) right-side stub with different widths of sheets and carbons; (*d*) bottom stub with different lengths of sheets and carbons; (*e*) corner snap set; (*f*) continuous snap-set forms.

the right hand, all the carbons can be *snapped* from the set. Also, this construction permits the removal of any sheet in the set while leaving the others intact. In the case of *double-stub sets,* the form is separable into two units; thus each stub is detachable independently.

Figure 12-2 shows the basic variations in the construction of snap sets.

Blocked sets are another category of unit sets. Like snap sets they may be fastened at the top or bottom; left or right side. The difference is that the sheets are held together by an application of glue or rubber cement on the outer edges of the paper.

BOOK (PADDED) FORMS

These are single or multiple-copy forms bound in books, the classical example of which is the salesbook. Book forms are commonly made up of three parts: forms, carbons or their counterparts, and bindings. The bindings may or may not include a cover.

Many construction and design variations are possible with respect to book forms. In general, however, the typical book of forms is cemented or stapled on one of its edges and the sheets are extractable as necessary. A strawboard backing is commonly used with each book, and the forms are cemented or stapled to the backing. Course-woven cloth is often cemented over the stub binding to give the book additional strength.

CONTINUOUS FORMS

Continuous forms are an unbroken series of single or multiple-part forms with each set made separable from the adjoining units by means of perforations. Suppliers deliver the printed forms in a flat-pack arrangement or in a roll for easy storing and handling. The use of continuous forms reduces the cost of handling. As shown in Figure 12-3, these forms eliminate nonproductive motions in placing separate forms in a typewriter or other forms-writing machine.

Besides the typewriter, machines designed to handle continuous forms include tabulators, high-speed printers, the Flexowriter, the Teletypewriter, billing machines, and autographic registers for making handwritten copies.

Various mechanical attachments or built-in units are employed on typewriters and other forms-writing machines to automatically actuate, guide, and regulate the imprinting of continuous forms. Among mechanisms that are commonly used are sprocket feeders, vertical and linefinder devices, and carbon-paper holders.

1. Take hold of sheets
2. Take hold of carbon paper
3. Insert carbons between sheets
4. Align carbons and sheets
5. Place each set in writting machine
6. Bring to first imprinting position
7. Imprint
8. Remove from machine
9. Separate from subsequent set*
10. Extract carbons

*Additional step resulting from the use of continuous forms

Figure 12-3. Elimination of repetitive steps by using continuous forms.

A sprocket feeder is a cylindrical element with cogs or toothlike projections that function to engage the holes in marginally punched continuous forms to feed and repeatedly align all sets and copies into correct position for imprinting. The marginal holes for this purpose are usually small, about ⅛ inch in diameter, and placed ½ inch apart from center to center.

The vertical spacer and linefinder is a device for automatically bringing forms to the next typing line; it eliminates the need for manually rolling the platen to succeeding typing lines.

Carbon-paper holders come in many different styles. Common to all is the feature of holding the carbon in place while the form is prepared. Some are self-acting so that in removing one set of forms from the machine the following set is brought into position with the carbon inserted and situated to transfer impressions.

On the Use of Carbon Paper

Carbon-copy requirements may be handled in one of four ways: (*a*) use of one-time carbons, (*b*) use of reusable "floating" carbons, (*c*) use of carbon packs, and (*d*) use of roll carbons.

With the first method it is possible to have any or all parts of continuous forms interleaved with carbon paper. One-time carbons are used where there will be subsequent entries on the form with consequent savings in time and carbon cost. Generally all that is necessary to make good copies is to select suitable grades and weights of paper, use good-quality carbon for transference, and make certain that a firm, even impression is applied.

Reusable *floating* carbons are used on certain kinds of forms-writing machines—for example, cylindrical platen billing machines. In this method carbon paper, wound on rollers, is fed between the folds of a multiple-part form for the typing of successive sets of continuous forms until it is worn out. The general idea is illustrated in Figure 12-4. Observe

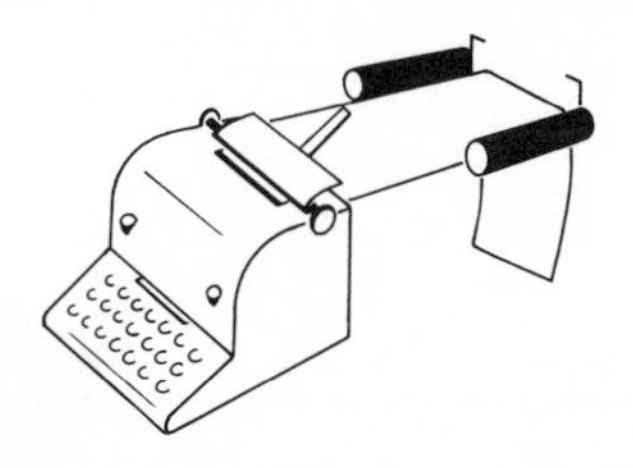

Figure 12-4. Cylindrical platen billing machine utilizing floating carbon.

that there are two rolls: one is for feeding carbon strips between the upper copies of a continuous forms set, and the other is for feeding between the lower copies. Mechanical devices (not shown in the figure) guide the carbon into a right-angle turn as it comes off the roller and "float" it in the same direction as the forms. After the information is typed, the form set is removed from the machine, but the carbon remains in position for reuse. Thus many sets of forms may be prepared with the same section of carbon. When the carbon is worn out, new paper can be fed into place from the rollers, and the worn-out carbon paper can be torn off.

Carbon packs are a particular kind of floating carbon. For typewriters and billing machines a pack is placed in specially designed holding devices where sheets of carbon paper move forward with the forms as they are fed for typing. By means of an aligning-bar mechanism the carbon paper remains in the machine when the forms are removed. The same carbon pack is used repeatedly until it no longer makes good copies.

Roll carbons are features of certain flat-bed writing machines and forms registers. The carbon is reeled from a roll as needed; in one method it is fed traversely over and between the parts of a continuous form. The carbon is reused until it is no longer serviceable.

Types of Continuous Forms

Based on composition, there are two primary types of continuous forms: continuous strip (sides open), and fanfold (alternate sides hinged).

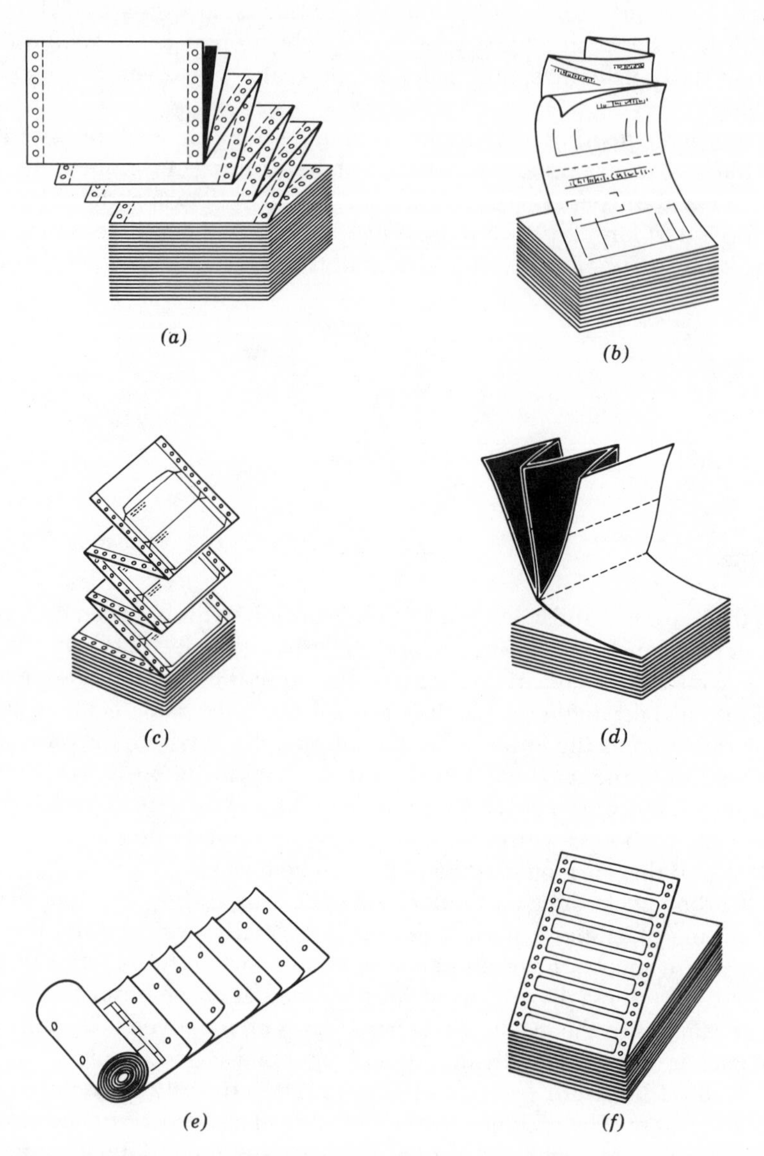

Figure 12-5. Assortment of continuous forms: (*a*) continuous-strip forms, marginally punched on both sides and interleaved with carbon paper; (*b*) fanfold forms with both narrow and wide parts; (*c*) continuous envelopes; (*d*) fanfold forms composed of parts of equal widths, interleaved with carbons; (*e*) rolls of continuous forms used on flat-bed writing machines built to carry carbon paper, on rolls, crosswise; (*f*) continuous labels.

Both are alike. With each type optional features include the use of one-time carbon between sheets and marginal punching. Either type is thus suitable for a variety of the same applications. Nevertheless, the difference in basic construction is significant. An assortment of both types of continuous forms is shown in Figure 12-5.

Continuous-strip forms are printed on as many separate rolls of paper as there are sheets making up the set. (See Figure 12-6.) A set of this type is open at the sides; that is, it does not have every other side edge connected, perforated, and folded over.

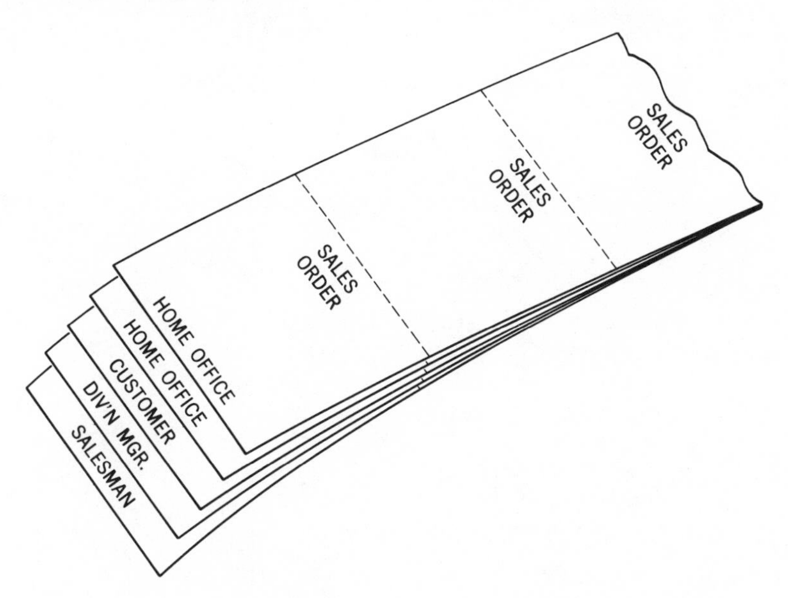

Figure 12-6. Continuous-strip forms.

Some of the construction possibilities with continuous-strip forms are as follows:

1. Forms may have sheets and carbons of different sizes in the same set;
2. Varying colors of paper and/or print for the sheets making up the set;
3. Stripes of different colors and widths for different sheets of the same set;
4. Marginal punching on either or both sides of the forms;
5. Different grades and weights of paper in the same set;
6. Inclusion of a duplicating master as part of the set; and
7. Sheets of the same set may have diverse applications of such fea-

tures as bodily perforations, binding holes, block-out patterns, and spot carbons.

As mentioned, most of the important properties of continuous-strip forms are shared in common with the fanfold-type form. Mutual construction possibilities include all those listed above except for limitations with respect to items 5 and 6.

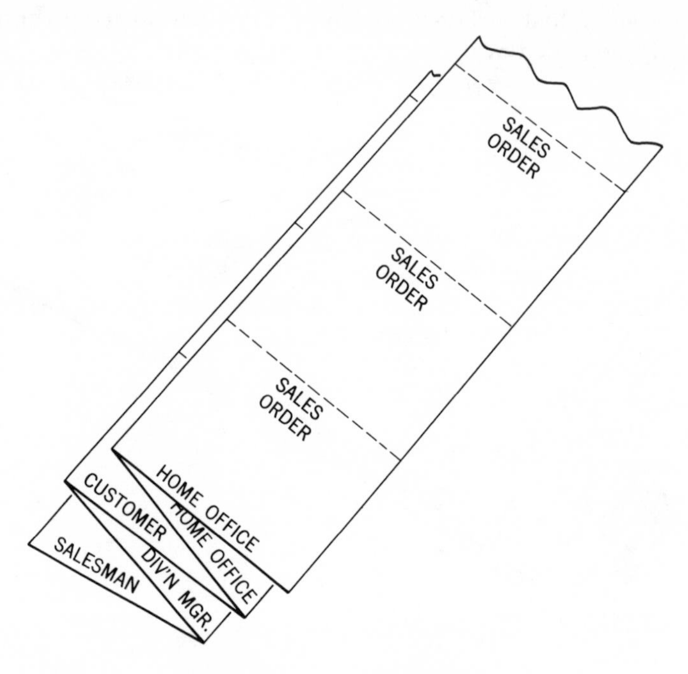

Figure 12-7. Fanfold forms.

Fanfold forms are fashioned in the manner of a fan, as shown in Figure 12-7. The alternate side edges of this type of form are a continuation of the same paper—perforated and creased. The forms are produced on a rotary press from a roll of paper. While traveling through the press the paper is printed on both sides and in such a way that the sheets of the set alternately appear, in strips, on the top and underside of the paper as shown in Figure 12-8. Modern rotary presses are wholly automatic; in addition to the actual printing they dry the ink and perform perforating, cutting, and folding operations as well. Accordingly, in the production of fanfold forms the printed sheets are folded into sets to make the desired combination of parts.

When the sheets of a set are printed from the same roll of paper they

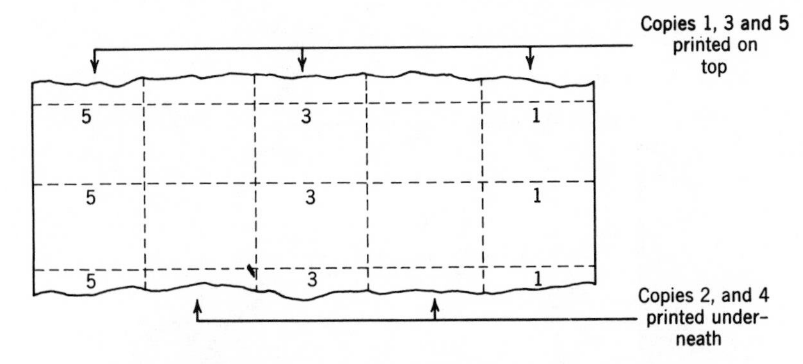

Figure 12-8. Fanfold form prior to creasing.

must be of the same weight and grade of paper. However, fanfold sets can be made up of different grades and weights of paper by collating sheets from one roll with those from another. Tinting can be used to obtain different-color sheets from the same roll of paper. Any number of sheets of the same set may be fully or partially tinted. A common application is to have colored stripes along the borders of one or more parts of a set for the purpose of ready identification.

Mainly because of the construction features, the standard fanfold form is generally considered an economical form to manufacture and buy. However, this type of form is not economical to purchase in quantities less than 15,000 sets of from 4 to 5 sheets.

PEGBOARD FORMS

"One-write systems" and "pegboard systems" are synonymous terms for a number of bookkeeping, statistical, and other record-keeping operations that utilize a pegboard or its counterpart—and generally a guideline bar or unit—as part of the arrangement. Some years ago the method was used almost exclusively in connection with the preparation of payrolls for small- and medium-size businesses. Later on it was adapted for accounts-receivable and accounts-payable operations. At present the basic technique is applied over a wide range of record-keeping functions, both within and without the accounting field, in concerns of all sizes. Large firms use the method in preparing confidential payrolls. A pegboard payroll arrangement is shown in Figure 12-9.

The prime motive behind the use of one-write systems is the reduction or elimination of rewriting. One-write systems have the identifiable feature of aligning a number of forms so that pertinent information may be

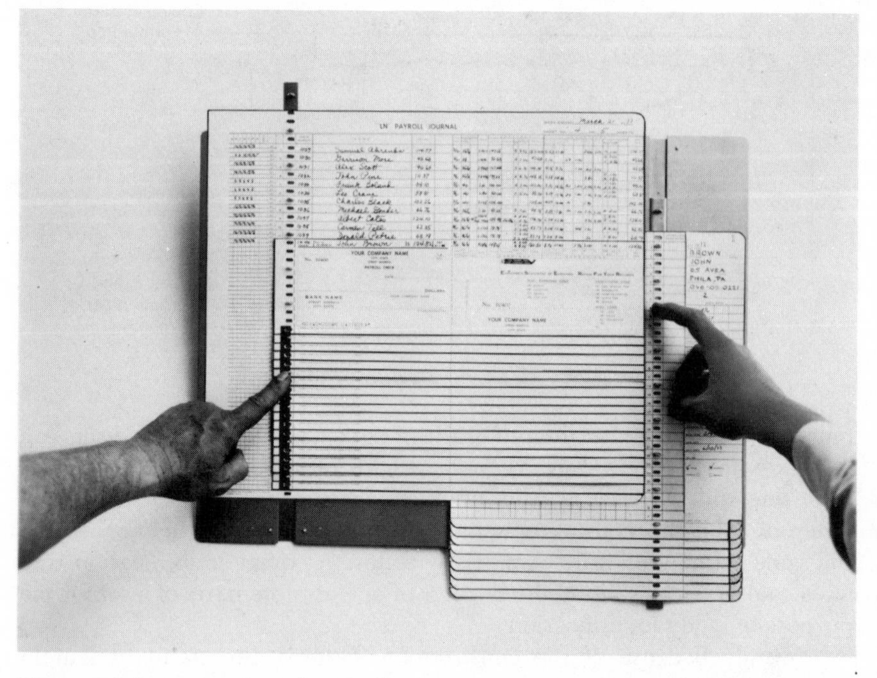

Figure 12-9. A pegboard payroll arrangement. Courtesy: Colonial Pay Systems.

simultaneously recorded on two or more of them. This is usually made possible by a metal strip containing a series of rounded posts or pegs accurately and evenly spaced, and attached to a composition board. Forms are punched to correspond to the spacing of the pegs. The peg-strip is used for mounting the forms, overlapping them somewhat in the manner of shingles, and keeping them in proper alignment. A holding bar may be inserted over the pegs to secure the forms.

Pegboards come in various sizes, a 30-inch width and 18-inch depth being common. They may have the pegs on the top or on either side. The forms may be of various shapes and sizes and may be placed in various relationships to one another for the purpose of obtaining desired registrations. Some pegboards have two or three rows of pegs to increase the number of different forms-alignment possibilities.

Many pegboard systems feature a lineguide to facilitate accuracy and speed in making entries. The typical lineguide is constructed to move up and down or back and forth across the board.

For obvious reasons exactness and uniformity are of the utmost importance in the printing and punching of pegboard forms. The actual make-up of the forms varies. They may be individually constructed as units, or as multiple-part sets. Some pegboard systems use carbonless paper; others use full or strip carbons. And, of course, different-color sheets or card stock may be used for the different forms.

STOCK FORMS

Stock forms differ from custom-made ones in that they are purchased in ready-made form. Known also as "canned forms," they lend themselves admirably to a variety of applications in nearly every business. In a number of situations it is wise to find out if there is a ready-made form on the market before going to the trouble and expense of acquiring a specially designed one. Also, it is sometimes practical to modify stock forms—for example, by stamping or printing additional information on them—and thereby make them suitable for a given purpose.

There is seemingly an endless array of stock forms. They are made in countless variations that serve a multitude of purposes. To mention a few basic types, there are graph sheets, accounting forms, columnar sheets, unit-set forms, manifold books (including sales books), and record-keeping sheets. Moreover, there is an endless number of preprinted tags, tickets, and labels available.

ACCOUNTING FORMS

Because of the number of entries normally made and the handling involved, the forms in this category need definite sustaining qualities. It is necessary for them to be durable and lasting. For this reason most accounting forms are printed on a good grade of medium to heavy paper; card stock and ledger paper are ordinarily used.

Among the more common accounting forms are payroll records, labor and material records, general ledger records, accounts-receivable records, and accounts-payable records. When such forms are to be used on bookkeeping and accounting machines great care must be exercised in their design and printing to assure that they meet the conditions imposed by the equipment. Ill-adapted and inappropriate forms can prove disastrous. The forms need to be uniform in size and accurate in the registration of printed matter.

DUPLICATING MASTERS

In most business firms carbon paper lends itself to an extraordinary number of duplicating applications and probably has produced more copies in the aggregate than any other means. Still, there is a limit to the number of satisfactory copies that can be obtained with carbon paper, and a need for duplicating masters arises. The principal kinds of duplicating masters may be divided into the following classes:

1. Stencil masters
2. Spirit hectograph masters
3. Azograph masters
4. Offset masters
5. Contact masters

Each is indicative of its method of reproduction. Their application depends on the situation and is ordinarily governed by a number of considerations, such as the quantity and quality of the copies desired, degree of legibility needed, handling and retention requirements, possible occasions for rerunning the master, duplicating equipment on hand, cost factors, and the urgency of the situation.

Stencil Masters

This type of master is used in conjunction with "flow through" copying equipment such as the mimeograph. The stencil master, a wax-covered sheet, is cut by means of a writing machine such as the typewriter or Varityper, and/or by hand with the aid of a stylus. As a result ink is able to bleed through the master where the wax has been removed. After preparation the stencil is placed over the cylinder of the machine, which is covered with an ink pad. Orifices in the cylinder allow ink, contained in a tank on the inside, to flow through and soak the pad. A turning of the cylinder causes a roller to be pressed against the stencil. In turn this causes the ink on the pad to penetrate the impressions on the stencil and thus make the desired reproduction on paper.

Satisfactory reproductions are the result of carefully prepared stencils, use of the right paper, and skillful operation of the duplicator. Drawings, lines, special effects, and selective lettering can be done with styli, curve-and-angle guides, and plates or screens for shading. Variations of shading are possible by using different plates or screens. Plates are fabricated from plastic material; screens are made from metal. The master is prepared for shading as follows: the plate or screen with the chosen design is placed under the stencil sheet and rubbed with a special stylus tool. The rubbing perforates the master with small holes that correspond to the pattern on the plate or screen.

Paper for stencil duplicating should be lintless and of a type that is slightly absorbent. Paper that is somewhat blotterlike makes good copies; it accelerates the setting of the ink and makes rapid operation of the duplicator possible. Hard-surface, nonabsorbent paper can be used for delicate work providing each copy is blotted and protected by a technique known as slip sheeting. Slip sheets may be blotting paper, but normally they are extra-long sheets of regular absorbent paper used in stencil duplicating. As the copies are produced, they are interleaved with slip sheets. The insertions may be made by hand or by a mechanical device. After a drying period the long slip sheets are separated from the stack of duplicated forms; they may then be used over and over again on different runs.

Well-made stencils will produce up to, and more than, 5000 copies. When a stencil is not depleted in one run, it may be cleaned and filed for reuse.

Sheets of every conventional color are available for processing on stencil duplicators. Ordinarily black ink is used to transfer the image, but color combinations are possible. There are two alternative methods for printing in two or more colors. One is to have a separate stencil and separate machine run for each color. The other is accomplished by spot inking —a process suited to short runs where the colored areas are not close together. In color work done by this method the different colors are run at the same time, and only one stencil is used. This is made possible by applying the colored inks directly to the appropriate spots on the surface of the cloth pad.

Color work by means of stencil duplicating can be very fussy and is not generally recommended.

Spirit Hectograph Masters

Three sheets of different kinds of paper go into the preparation of this type of master; namely, a sheet of master paper, a sheet of onionskin or other protective paper, and a sheet of hectograph carbon. The arrangement (without the protective sheet) is shown in Figure 12-10.

Unit carbon sets, in which the master sheet and hectograph carbon are attached, are very popular. An unattached tissue divider separates these parts. The tissue remains in place until the actual recording of the copy, at which time it must be removed. After the recording is made the tissue may be reinserted to prevent smudging, until the time when the

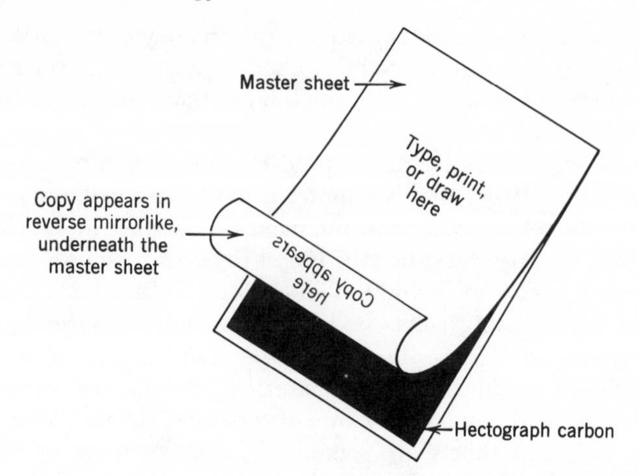

Figure 12-10. Hectograph master.

master itself is run off. It should be noted that with the tissue in place it is possible to construct guidelines and preliminary sketches on the master before making the real and lasting impression.

Hectograph masters may be prepared in several ways. They may be imprinted with a typewriter, drawn on with a pencil or ball-point pen, or printed by the letterpress or relief process. Whatever the method—and combinations may be used—there must be sufficient pressure exerted to obtain the necessary deposit of carbon on the back of the master sheet to assure good reproductions.

Several colors can be combined on a single hectograph sheet and reproduced in one operation. This is made possible through the use of hectograph carbons as well as special inks and pencils that produce different colors. Purple is the most widely used color because it can produce the most copies. Other colors that may be used are red, blue, green, black, brown, and yellow.

Preprinted unit carbon sets may be purchased for systems-and-procedures applications. Such forms fall into two general classifications: printed-through forms and kiss-printed forms. Entries made on the master of printed-through forms are duplicated when the form itself is duplicated. Kiss-printed forms are printed so that no carbon is transferred to the reverse side of the master. Hence only the information subsequently recorded on the form will be reproducible.

Kiss-printed forms cost less than printed-through forms, but the difference in cost is generally not great. One reason for the lower price of the former is that the printing house leaves the tissues in place and thus prints on the face of the forms only.

Problems of registration are largely surmounted with the use of preprinted hectograph forms. Information that is properly recorded on the master must of necessity appear in the right spaces on the duplicated copies.

Hectograph masters can be created by means of hectograph carbon ribbons on standard and electric typewriters equipped for the purpose. Single-color ribbon is available for producing one of several different colors: purple, red, blue, and green. Ribbons in various combinations of two colors are also marketed. The principle underlying the preparation and use of ribbon-made masters is essentially the same as that for the preparation of carbon masters. The hectograph carbon ribbon passes between the master sheet and the platen, so that the imprinted matter appears in reverse on the back of the sheet.

When the hectograph master is ready to be reproduced it is latched to the surface of the cylinder of the duplicator, the carbon facing upward. As the paper for the copies is fed through the machine each sheet is moistened with an alcohol-based fluid prior to its being pressed against the master sheet by a roller. The contact transfers an infinitesimal quantity of the carbon into a positive image on the sheet.

Purple hectograph masters of a good grade will produce 300 to 400 copies. Red, blue, and green masters are good for about 50 copies. Although the copies may be produced in several colors simultaneously, they will be only as serviceable as the least discernible color permits. Copies may be made on various grades, weights, and colors of paper and card stock. Hectograph masters are also available in continuous sets (Figure 12-11). The greatest disadvantage of the method is that during the reproduction process the operator's hands become stained with ink.

Azograph Masters

Much of what has been said about the preparation, construction, and processing of hectograph masters applies also to Azograph masters. Both kinds of masters have approximately the same construction possibilities, are prepared in the same ways, and the method of duplication is essentially the same for each.

The basic difference between the two is in the transfer sheet. Whereas the hectograph carbon possesses an aniline dye that stains the hands of the person coming in contact with it, the Azograph counterpart —a wax-coated sheet containing color-forming compounds—does not soil the hands. Similar but different duplicating equipment is used to process the master.

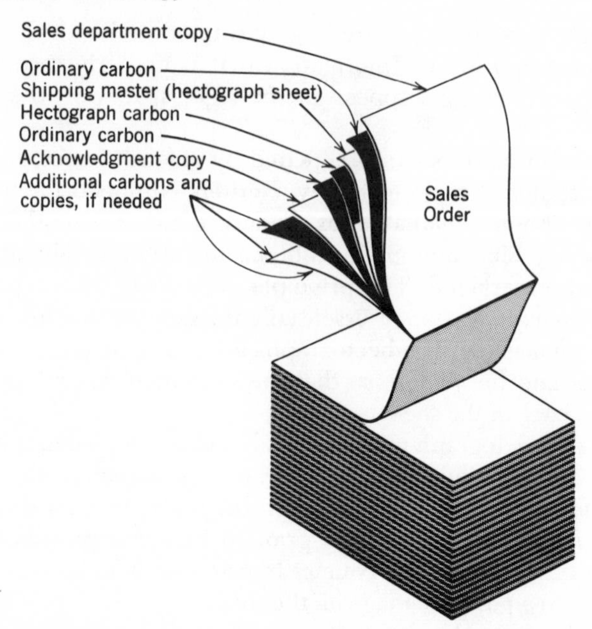

Figure 12-11. Hectograph master as part of a continuous set.

The Azograph method is a short-run duplicating process; it is efficient for producing between 50 to 75 copies.

Offset Masters

Offset masters, also called plates, are of two principal types: direct-image masters and indirect-image masters. The former are ordinarily made from paper or plastic, and may be printed, typed, or handwritten. Regardless of the medium of preparation, the image itself must be greasy. Direct-image masters can be prepared on a writing machine with a grease-base ribbon and/or with a special pencil that has a grease-base lead. Typewriters equipped with such a ribbon may also be used to prepare ordinary correspondence without changing the ribbon. The opposite is not true: a regular typewriter ribbon will not serve to make a direct-image master.

The grease-base feature of the special ribbon and pencil lies at the heart of the offset technique. The method is based on the principle that water and grease repel one another. Fundamentally, this is how it works: A grease-base image is placed on the master by any of the previously stated means. In making the copies a rotary (offset) press is employed. Such presses normally have three cylinders, one above the other. The top one, known as the plate cylinder, holds the master attached to its surface. At every revolution rollers apply water and a grease-base ink to the master. The water-moistened master repels the ink except where the grease-base image appears. As the cylinder turns it comes in contact with the middle cylinder, identified as the rubber-blanket cylinder, and imprints the image, in reverse, on its surface. The negative impression on the roller is then transferred as a positive impression on paper fed by the bottom cylinder. The bottom cylinder, called the ink-impression cylinder, has grippers that hold the paper in place while it is pressed against the rubber blanket.

Reproductions from indirect-image masters are made in exactly the same way. The manner of preparing an indirect-image master is different, however, and entails a photomechanical process such as xerography as an additional step.[1] In this method the image is applied by photographic means to a master made of metal, paper, or plastic. Accordingly, the material to be duplicated—printed matter, drawings, photographs, etc.—can be prepared with conventional instruments: regular typewriter ribbon, ink, and pencil.

The original does not have to be the exact size of the copies desired, since by photographic means it can be reduced or enlarged. By making a large original and photographically reducing the image, it is possible to minimize artistic flaws in the work and come up with more perfect reproductions.

An offset master will produce copies until it is worn out. Contingent on the type of master, what it is made of, and how it is handled, a continuous press run will produce anywhere from several thousand to more than 20,000 copies. For practical purposes we might say that the potential quantity is limitless, because the original or a master can be used to make other masters; and copies can be made therefrom, ad infinitum.

It is the nature of the offset process to produce good copies. Typewritten matter can be made to resemble fine printing, and photographic duplications can be made lifelike in appearance. The method is economical for long press runs.

Color reproductions may be made by the offset process. Some presses will print as many as four colors at one time.

Additional advantages of the offset process are that masters can be stored and used again, information may be added to a master at any time by means of a special typewriter ribbon or pencil, and masters can be made part of unit sets and continuous forms. Paper and card stock of various weights, grades, sizes, and colors may be printed by the offset method.

[1] The xerography process (pronounced ze-rog ra-fe) is also widely employed in self-contained copying machines, the mechanics of which are illustrated in Figure 12-13.

Contact Masters

In general contact copies are produced by exposing a light- or heat-sensitive paper to an image on a master sheet. The sensitized paper is especially manufactured for this particular process. Light-sensitive paper is made by impregnating any suitable paper with a chemical solution; heat-sensitive paper is made by applying a wax coating to any suitable white paper.

As shown in Figures 12-12 and 12-13, four means of "direct contact" reproduction are used; namely, dye transfer, heat rays, photo transfer, and diazo paper. Three of these use the blueprint (or diazo) principle, discussed below, whereas the other accomplishes the duplication process by the medium of heat.

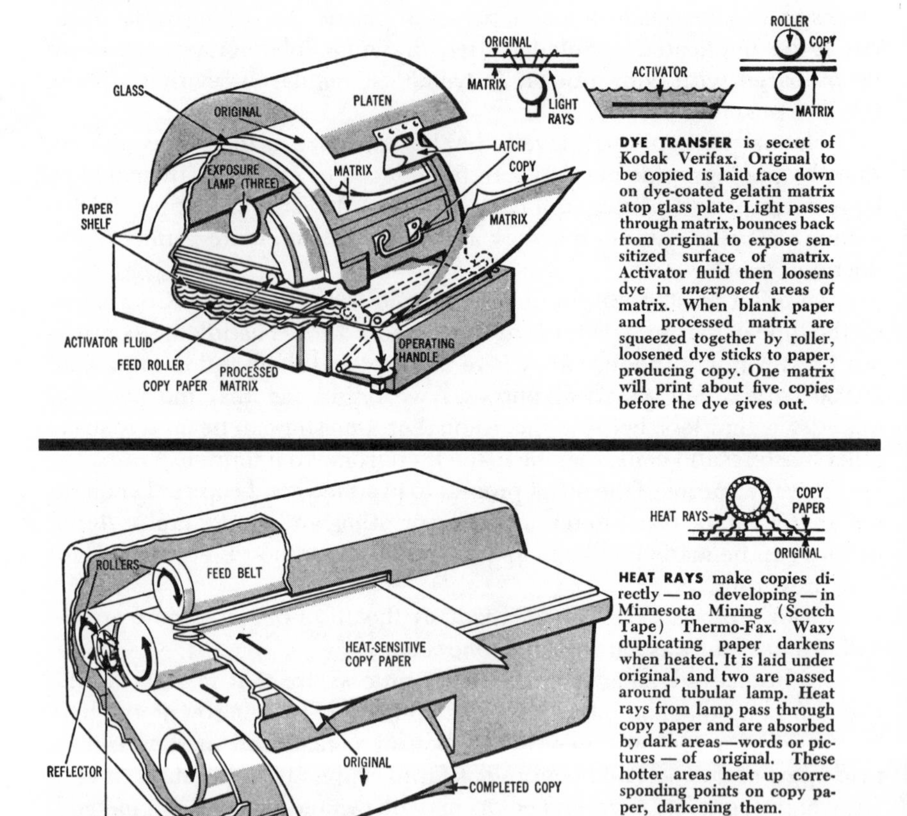

Figure 12-12. Direct-contact copying machines utilizing dye transfer (top) and heat rays (bottom). Reprinted by courtesy of *Popular Science*. (Copyright 1960 by Popular Science Publishing Co., Inc.)

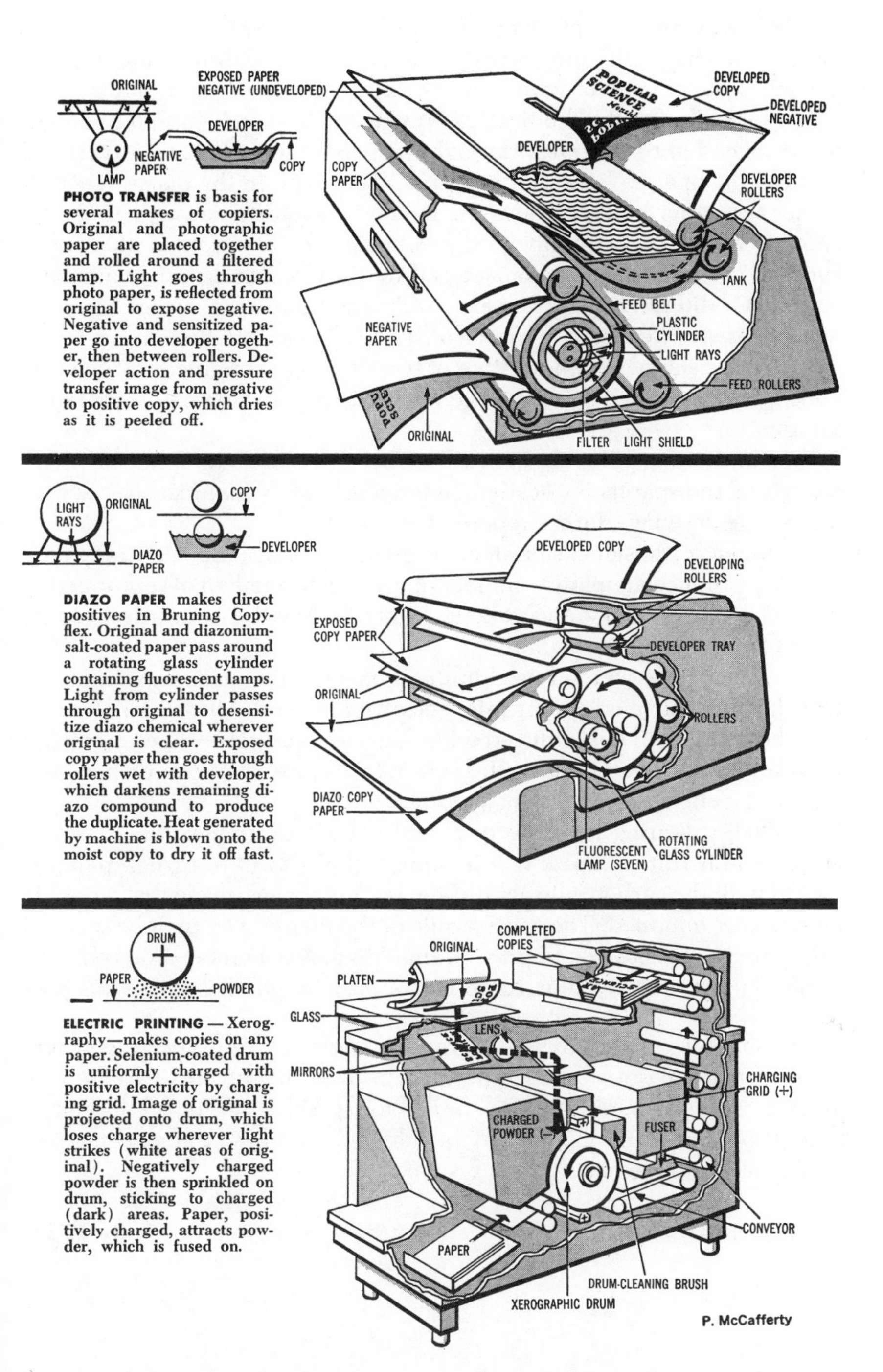

Figure 12-13. Direct-contact copying machines utilizing photo transfer (top), diazo paper (center), and xerography (bottom). Reprinted by courtesy of *Popular Science*. (Copyright 1960 by Popular Science Publishing Co., Inc.)

Briefly, copies are produced by the diazo process as follows. A translucent master—containing typed, printed, or hand-drawn material—is placed in direct contact with a light-sensitive sheet of paper, and as they are passed through the duplicating equipment they are exposed to a light source. The light penetrates through the master sheet but does not shine through the opaque lines that comprise the image. In the places where it strikes the paper light activates the imbued chemicals; subsequent development within the equipment causes a change in color. With certain equipment, such as that manufactured by the Charles Bruning Company and the Ozalid Division of General Aniline & Film Corporation, the copy paper presents the image in brown on a white background.

Diazo machines employ either of two development processes. In some the copy is developed by controlled ammonia vapors; in others, by a liquid solution.

An interesting point about the diazo process is that it permits the preparation of transparencies—called "intermediates"—which can be used to regenerate drawings, forms, reports, and the like.

Although translucent masters, originals, or transparencies, may conceivably produce an infinite number of copies, the method of reproduction itself is relatively slow and practical only for small quantities; 20 to 25 copies is considered maximum.

Whatever the method used in the preparation of masters, it is important that the image on the translucent paper be made sharp and clear, so that the copies will be produced with sharpness and clarity. Lettering and illustrations are often done with India ink because of its opaqueness and general excellence as a drawing medium.

When a writing machine or pencil is used to prepare the master a much clearer run of copies will be obtained if a sheet of carbon paper is placed with the carbon side against the back of the master so that a double recording is obtained. The reverse side of the master may then be sprayed with a protective coating to prevent smudging. Any number of acrylic and similar sprays with various trade marks may be purchased for this purpose.

Deserving of special note is that, for systems-and-procedures purposes, contact masters of the type used in the diazo process may be preprinted and incorporated into a unit set, a padded set, or a continuous form. Processing information may be added to the master at any stage, in conformity with the needs of the system.

A comparison of basic duplicating processes is given in Figure 12-14.

Process	Copies Ordinarily Constituting an Economical Quantity	Speed in Copies per Minute (up to)	Relative Cost per Copy	Maximum Size of Copy (inches)	Types of Material for which Especially Suited	Quality Characteristics
Carbon	12 on a standard typewriter 25 on an electric typewriter	Depending on typing speed	Low	$8\frac{1}{2} \times 14$	Typewritten, Handwritten	Smudges easily
Stencil	5,000	250	Low	$8\frac{1}{2} \times 14$	Typewritten, handwritten, drawings	Durable
Spirithectograph	300	150	Medium	11×17	Typewritten, handwritten, drawings	Serviceable
Azograph	50	150	Medium	11×17	Typewritten, handwritten, drawings	Serviceable
Offset	10,000	150	Low	11×17	All kinds	Clarity exceptionally good but slightly inferior to type printing
Contact	10	4	High	Certain machines will handle very large sheets	All kinds	Properties of copies of some processes adversely affected by long exposure to light

Figure 12-14. Comparison of basic duplicating processes.

TAGS, TICKETS, AND LABELS

For many purposes the tag, ticket, and label are unequalled as a category of forms. All three have this in common: they are used in tremendous quantities to identify objects by means of direct attachment. There is scarcely a business concern that does not make use of at least one of them.

Tags

Small quantities of standard-size tags are commonly printed on a letterpress from ready-make stock. For long-run jobs, however, big tag manufacturers have equipment that prints and fabricates such tags in one series of processes. Figure 12-15 shows standard sizes for tags.

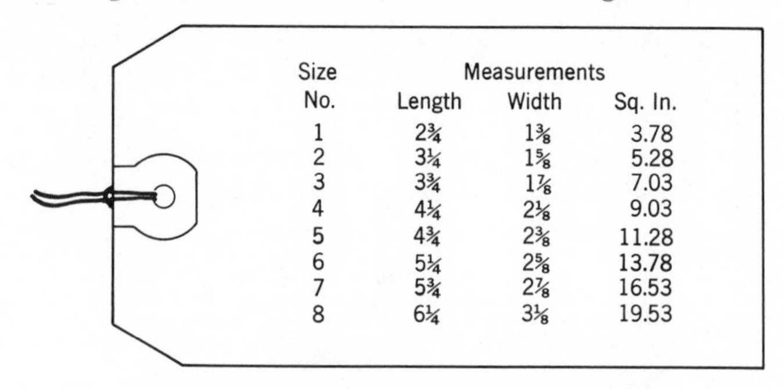

Size	Measurements		
No.	Length	Width	Sq. In.
1	2¾	1⅜	3.78
2	3¼	1⅝	5.28
3	3¾	1⅞	7.03
4	4¼	2⅛	9.03
5	4¾	2⅜	11.28
6	5¼	2⅝	13.78
7	5¾	2⅞	16.53
8	6¼	3⅛	19.53

Figure 12-15. Standard sizes for tags.

Tags can be obtained in larger standard sizes than those shown. Odd sizes may also be purchased, but it is generally necessary to pay a premium when purchasing other than standard sizes.

Tag-Stock Material. The material from which tags are made must suit the conditions of use. Tags used in a machine shop must be able to withstand oil and grease—and remain readable. Usage varies, and from the standpoint of material there is no need to select a better or more expensive tag than the circumstances warrant. Fortunately, there are the following four basic tag-stock materials from which to make a selection:

1. *Sulfite pulp.* Most tags are made from sulfite pulp. They come in a large variety of weights and grades, with a correspondingly diverse range of strength and wear characteristics.
2. *Jute.* This stock is made from plant fibers that are largely reclaimed from used burlap. It possesses substantial strength and can withstand much wear and tear.

3. *Rope.* Tough, long-fibered hemp is the substance of this stock, obtained from used rope. As with jute stock, tags made from hemp are strong and durable.

4. *Cloth.* Tags made from cloth are especially suited to applications where the combined attributes of strength, durability, and pliancy are of major importance.

It is likely that at least one of the foregoing types of material will be well suited for most situations. However, tags are also made of metal and, for special uses, even of glass. Plastics are also used in making tags for certain applications. Metal, glass, and plastic tags are given coarse surface areas so that they may be written upon.

Colors. Tags may be purchased in just about every color of the rainbow. To assure legibility it is advisable to refrain from using dark shades of intense colors such as brown, green, blue, and red.

Tags may also be acquired with the printed matter in more than one color. Sometimes colored stripes are used as an aid to identification.

On Fastening Tags. Devices for fastening tags include strings, wires, and metal clips. At the time of purchase these may or may not be joined to the tags, and should be thus specified when placing the order.

Reinforcements for Tags. The area around tag holes is commonly strengthened by the addition of a paper patch or metal eyelet designed for the purpose.

Continuous Tags. Tags may be acquired in continuous rolls or folds for efficient feeding and imprinting on forms-writing machines such as typewriters and addressing machines. Specially designed tag-imprinting machines—for example, the Dial Set Tag Printer—also utilize continuous tags. This particular machine is for recording repetitive information and consecutive numbers on tags.

Variations. Most tags are fabricated as simple, one-part forms. Many other types of construction are possible, however, and should not be overlooked in designing a system. Tags may be manufactured with perforated detachable stubs, in unit-set style, and in the form of envelopes. Figure 12-16 shows a number of variations.

Tickets

Tickets are used extensively on clothing and other soft-goods merchandise that is sold to the ultimate consumer. Both manufacturers and retailers use them to designate lot numbers, prices, sizes, and inventory codes. There is a large variety of tickets in use, with selection as to size and shape depending largely on the nature of the exact application and the

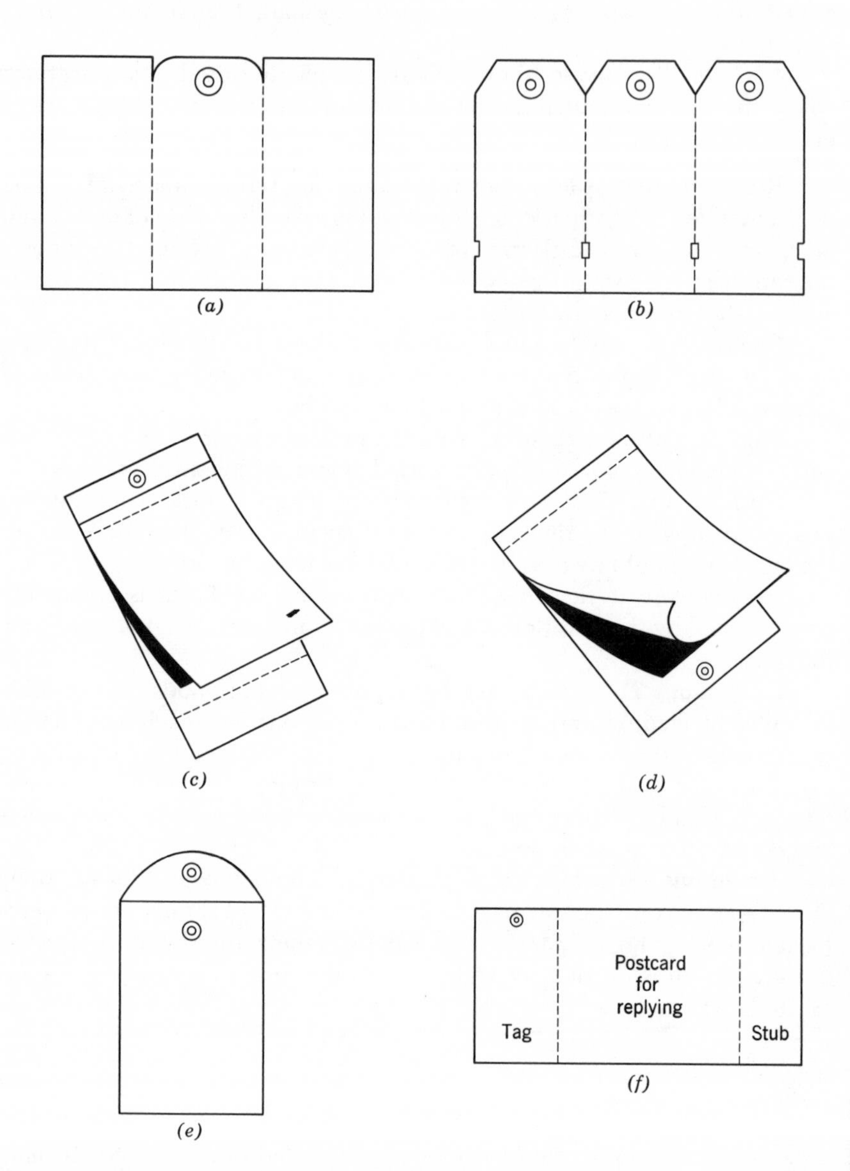

Figure 12-16. Tags and tag combinations: (*a*) tag with stubs, one or more of which may be attached to the tag on either or both sides; (*b*) continuous tags; (*c*) manifold tag with stub in unit-set style; (*d*) manifold tag with carbon-backed front copy; (*e*) tag envelope; (*f*) and tag, postcard, and stub.

amount of information to be recorded on the ticket. Machines are available for marking tickets as they are drawn from a roll, counting, and separating them. Depending on the type, a ticket may be tied, pinned, stapled, or pasted to an item.

Labels

The word "label" as we use it here means a slip of paper with adhesive backing, which enables it to be affixed to something. A typical and widespread use of labels is to facilitate the addressing of mail—letters, packages, cartons, and magazines. When only a few addresses are involved the three or four lines of information for each label may be made by an ordinary typewriter. For more extensive mailings the imprinting may be done by one of a number of automatic or semiautomatic machines. Special labeling equipment is available that will prepare names and addresses on continuous label-stock forms and separate them just before direct application to pieces of mail.

Stock for both labels and tickets can be had in a wide assortment of colors.

SAFETY-PAPER FORMS

Checks, drafts, stocks, and bonds are forms that are included in this classification. These forms are normally printed on safety paper as a precaution against alteration and forgery. No foolproof method of guarding against these old bogies exist, but the use of safety paper does raise barriers against deceptive practices of this kind.

The ingredients that go into the making of safety paper include wood pulp or rags—or a combination of the two. The basic product becomes a sulfite-grade paper, rag-content bond, or ledger paper. In the manufacturing process this paper is treated so that it will have a smooth surface, definite characteristics with respect to ink penetration and drying, and certain firmly infixed protective features that are inherent in the substance or design of the paper.

One method of making safety paper is by means of watermarking. This is an identifying process wherein a translucent design is pressed into the paper as it is being fabricated by a papermaking machine. The impregnation of the watermark pattern makes it all but impossible to satisfactorily restore the design once an alteration has been made.

The most common method of making safety paper is by surface-imprinting it with chemicals or dyes in a two-tone color pattern. The pattern is usually one of a fairly intricate structure, and it is often printed so as to be in equal registration on both sides of the sheet. Since the design is repetitive and pervasive, the paper is referred to as *pantographic safety paper*. The evenness of the design and the two shades of color make undetectable alterations difficult, if not impossible. Because the surface printing penetrates only the top fibers, any erasures by rubber, penknife, or similar instruments will bring a telltale area to the surface in the form of a white spot. Furthermore, paper of this type may be treated to show a brown or white stain if ink eradicator or other chemical is used to effect a change. Where the volume of usage warrants it such papers are purchaseable to individual specifications, and the design may embrace the company name, emblem, or a trademark as the basis of the motif.

An improved version of the pantograph type is a check paper that reveals the word "void" in illuminating letters of dark brown upon application of an ink eradicator. When the eradicator fluid contacts the paper a developing process takes place between it and the chemicals impregnated in the paper. The method affords greater protection because where it might be possible to offer a plausible reason for a brown stain, it would be practically impossible to explain that which has been made void. Safety paper with this feature is generally not for sale as blank sheets but must be purchased from the printing house as a form.

Checks

Although reasonably standardized, the elements of a check can be classified into a relatively small number of categories, as follows:

1. Company identification
2. Serial number
3. Routing symbol
4. Date
5. Payee
6. Amount in numbers
7. Amount in words
8. Drawee (identification of bank)
9. Account number
10. Signature(s)
11. Numbers for magnetic ink character recognition
12. Endorsement Clause

Company Identification. Customarily, the name of the company is centered at the top of the check or placed slightly to the left, at the top.

The name may be accompanied by an address. Often the name will appear with a trademark, pictorial symbol, or scenic illustration.

Serial Number. A serial number facilitates control. For checks that are handwritten or sorted by hand it is usual to place the number toward the upper right-hand corner of the check, since this makes sorting easier.

Serial numbers may be preprinted, and/or prepunched when punched-card checks are used (see Figures 12-17, 12-18, and 12-19), or they may be filled in at the time the check is made out.

Routing Symbol. The routing symbol resembles a fraction, for example:

$$\frac{44\text{–}87}{425}$$

The development of this symbol and its code was the coordinated work of the American Bankers Association and the Federal Reserve System. Both of these organizations recommend that this symbol always be placed in the upper right-hand corner of checks.

The purpose of the routing symbol is to shorten and simplify the process of clearing and collecting checks. In the example shown above the upper left-hand number, 44, designates a large city in which the particular bank is situated. In this location numbers 1 through 49 are used to denote certain large cities, and numbers 50 through 100 denote specific states. Banks within large cities have an assigned city number, and those located elsewhere possess a state number. To the right of the city/state number is the figure 87, which identifies a particular bank. For banks in some states this number may consist of as many as four digits.

Beneath the line is the three-digit number 425. The first digit, 4, signifies a Federal Reserve District. Since the number of districts exceed nine, the complete bottom number may consist of four digits, in which case the first two signify the district. To state it another way: if the number below the line is made up of three digits, then the first one indicates the district; if it is a four-digit number, then the first two signify the district.

The next digit over from the district-identification number, the 2, signifies a branch within the district. One exception to this is the figure 1, which is used to indicate the main office.

The last digit on the bottom half of the routing symbol, the 5 in our example, signifies the number of days of deferred availability of funds.

Date. The custom has been to write the date on the right side of the check, above the figure amount. Although this practice is still widely followed, especially with respect to checks written by hand, the date may be placed in any convenient location on the upper or middle portion of the check. In checks that are prepared by a writing machine, such as the type-

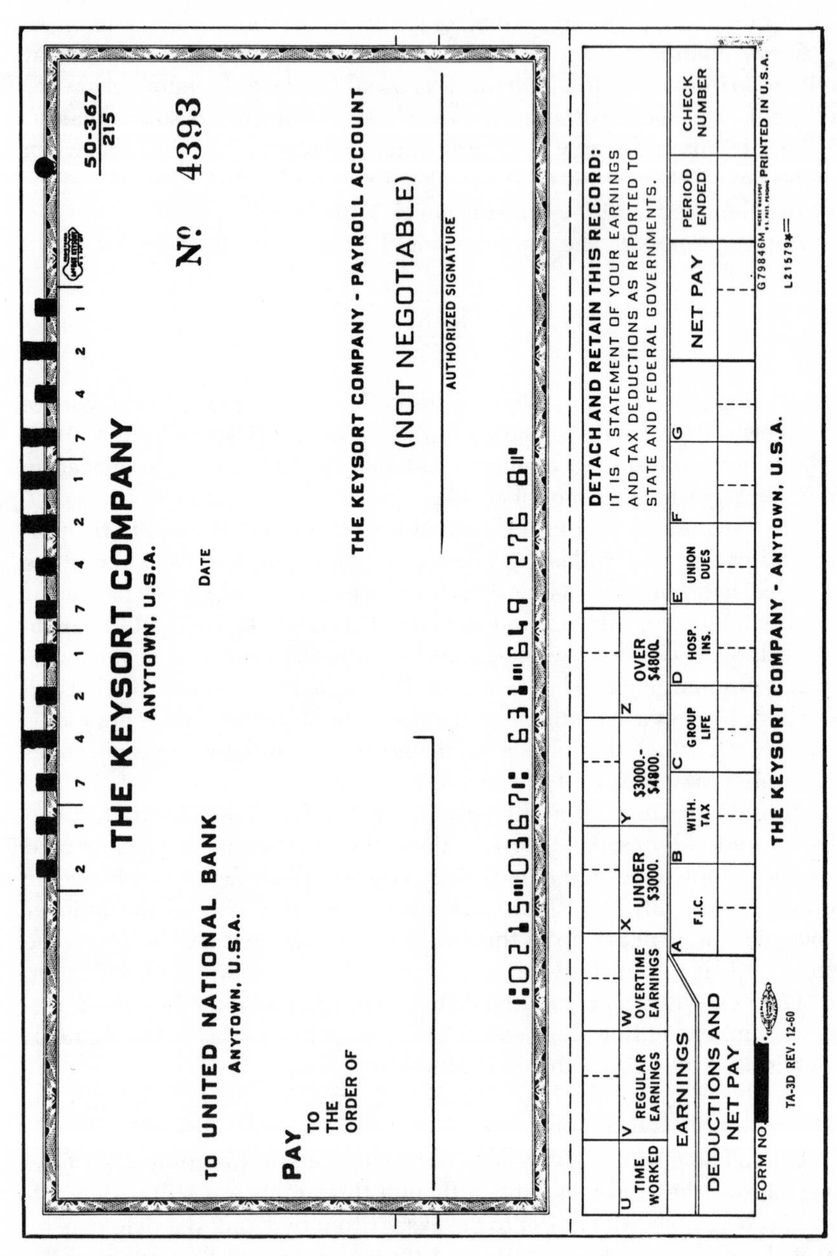

Figure 12-17. Keysort check with earnings statement. Courtesy of Royal McBee Corp.

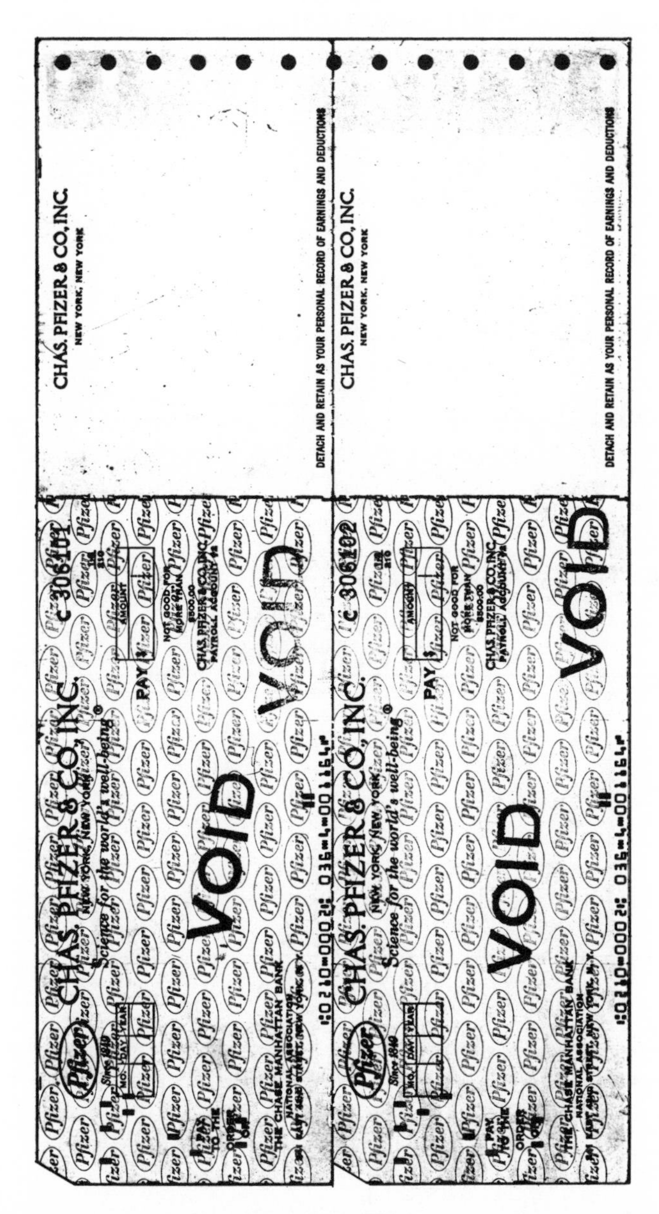

Figure 12-18. "IBM" continuous punched card checks for a payroll application. Courtesy of Chas. Pfizer & Co., Inc.

CHAS. PFIZER & CO., INC.
235 EAST 42ND ST., N.Y., N.Y. 10017
D 072428

VENDOR NO.	INVOICE	INVOICE AMOUNT	DISCOUNT	NET AMOUNT

THIS IS YOUR COPY

Pfizer Since 1849
CHAS. PFIZER & CO., INC.
NEW YORK, N.Y. 10017
Science for the world's well-being®
D 072428
1-2
210
MO. DAY YEAR
EXACTLY DOLLARS AND CENTS
AMOUNT
PAY TO THE ORDER OF
VOID
CHAS. PFIZER & CO., INC.
GENERAL ACCOUNT NO. 1
THE CHASE MANHATTAN BANK
SECOND AVE. & 42ND STREET NEW YORK, N.Y. 10017
AUTHORIZED SIGNATURE
FORMSCARD® WILLOW GROVE, PA. F-5131
0210 0002 036 4 010041

CHAS. PFIZER & CO., INC.
235 EAST 42ND ST., N.Y., N.Y. 10017
D 072429

VENDOR NO.	INVOICE	INVOICE AMOUNT	DISCOUNT	NET AMOUNT

THIS IS YOUR COPY

Pfizer Since 1849
CHAS. PFIZER & CO., INC.
NEW YORK, N.Y. 10017
Science for the world's well-being®
D 072429
1-2
210
MO. DAY YEAR
EXACTLY DOLLARS AND CENTS
AMOUNT
PAY TO THE ORDER OF
VOID
CHAS. PFIZER & CO., INC.
GENERAL ACCOUNT NO. 1
THE CHASE MANHATTAN BANK
SECOND AVE. & 42ND STREET NEW YORK, N.Y. 10017
AUTHORIZED SIGNATURE
FORMSCARD® WILLOW GROVE, PA. F-5131
0210 0002 036 4 010041

Figure 12-19. "IBM" continuous punched card checks for a purchase and disbursement system. Courtesy of Chas. Pfizer & Co., Inc.

writer, the position for the date may be located in front of the name of the payee. Locating the date of the check, name of the payee, and the figure amount on the same line facilitates machine preparation of checks.

Payee. The name of the payee, the one in whose favor the check is drawn, is traditionally placed before the figure amount. Under the name may appear the address of the payee, which may be conveniently used for a window envelope.

Amount in Numbers. The figure amount is best located on the right side of the check, above or near the horizontal center line. Sufficient space should be provided so that the amount can be inserted clearly and accurately.

Checks prepared on forms-writing machines—automatic-data-processing equipment—may have the figure amount printed in two places on the check. The additional recording is intended to take the place of restating the amount in words.

Amount in Words. A line for inserting the amount of the check in words is usually situated below the name of the payee. A check-writing machine is sometimes used for recording this amount. A machine of this kind stamps the amount above the line, in perforated print, so that there is little likelihood of an alteration being made. Check-writing machines afford substantial protection, but a check that is carefully written in longhand or properly imprinted by automatic-data-processing equipment is also reasonably protected against fraudulent practices.

Drawee (Identification of Bank). An area that is often used for the bank name is the lower left portion of the check. This area is generally otherwise vacant. Wherever the name of the bank is situated, it is ordinarily accompanied by the name of the town or city in which the bank is located.

Account Number. Checks commonly have account numbers that designate the categorical and precise nature of the disbursement. Payroll checks have the employee's number; dividend checks have the stockholder's identification number; and regular accounts-payable checks have a voucher number. The preferred location for the number is in the vicinity of the name of the payee.

Signature(s). One and sometimes two persons are required to sign a check. In the case of joint checking accounts the bank is usually instructed to honor either signature, but in the absence of such notification both parties are required to sign. In corporate practice the board of directors authorizes certain officers to sign checks, and as a safeguard frequently stipulates that two signatures are required to validate each check. The titles of those who sign checks are commonly placed under the location for the signatures. If desired, checks may be "signed" with a mechanical check signer that is capable of imprinting both signatures and titles.

Numbers for Magnetic-Ink-Character Recognition. The system of magnetic-ink-character recognition that has been adopted by banks throughout the country requires the printing of machine-language numerals along the bottom edge of checks. These numerals are of special configuration. (See Figure 12-20).

Figure 12-20. Numbers for magnetic-ink-character recognition.

The style of the symbols was created by a national committee representing the American Bankers Association, the foremost manufacturers of electronic equipment, the check-printing industry, and the Federal Reserve System.

The numerals are designed to serve a twofold purpose: to be "read" and processed electronically when printed on a check, and to be readable with the human eye. It is surprising how quickly the figures can be discerned once you have become familiar with them.

Each odd-shaped figure is printed in ink that contains particles of iron oxide. Due to the difference in the style of the characters, each one has a definite and potential magnetic intensity built into it. It is this variance that allows the reader-sorter—the electronic equipment that machine-reads the numerals—to identify each magnetic ink character as it is being processed.

The numbers are printed in code, along the bottom edge of checks. The reason for printing them in this location is that experience has shown that the bottom edge is less likely to become damaged. Referring to Figure

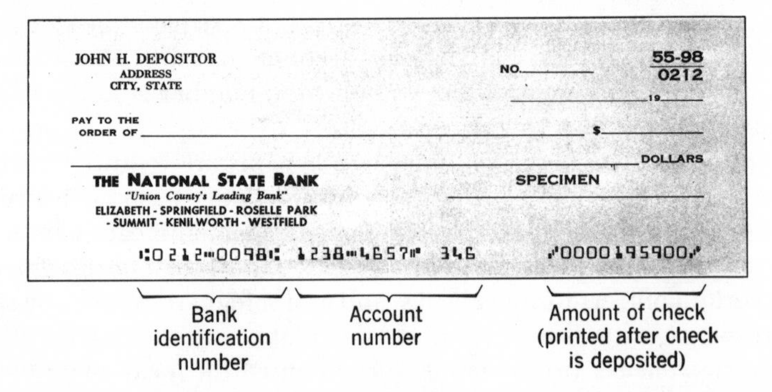

Figure 12-21. Check showing position and coding of magnetic-ink characters.

12-21, it will be observed that three groups of numerals are used. The first gives the bank identification number, which repeats a large portion of the routing symbol; the second, the account number; and the third, the amount of the check—this being printed after the check is deposited.

Although the characters must be printed in magnetic ink in order to be machine-read, the ink is not expensive; the entire check—symbols and all—can be printed during one press run.

Endorsement Clause. An endorsement is usually made on the back of the check, at a right angle to the printed matter on the face. Some companies desire to print an endorsement clause just above where the acceptance words and/or signature will be recorded. One common type that is used a great deal on payroll checks advises the payee to sign his name exactly as it appears on the front of the check. Another type are the clauses that insurance companies have printed on the back of checks to designate the satisfaction of claims.

ENVELOPES

The envelope class of form is capable of making an important contribution to the efficiency and general success of numerous systems. Taken by themselves window envelopes alone constitute a classical example of this. Not only do window envelopes eliminate typing that would otherwise be necessary—they also reduce the probability of errors in addressing and completely eliminate the possibility of putting letters in the wrong envelopes.

In general envelopes serve either a single or a double purpose: they furnish their contents with protection, and they render the necessary information to move the contents from one place to another. Careful selection of the right kind of envelope is important. The various styles that are available are shown in Figure 12-22.

Following is a list and brief discussion of the principal kinds of paper stock used in the manufacturing of envelopes:

Bond with a white wove surface is the commonest type of paper used in making envelopes. This paper is available in various grades, strengths, and finishes. The better bond papers furnish good printing surfaces. Besides white, bond paper may be obtained in various colors, including blue, green, pink, and goldenrod.

Kraft is a strong paper with an irregular structure. It is made from sulfite pulp, possesses good printing qualities, and is usually of a brown color resulting from lack of bleaching. Other colors of kraft paper are

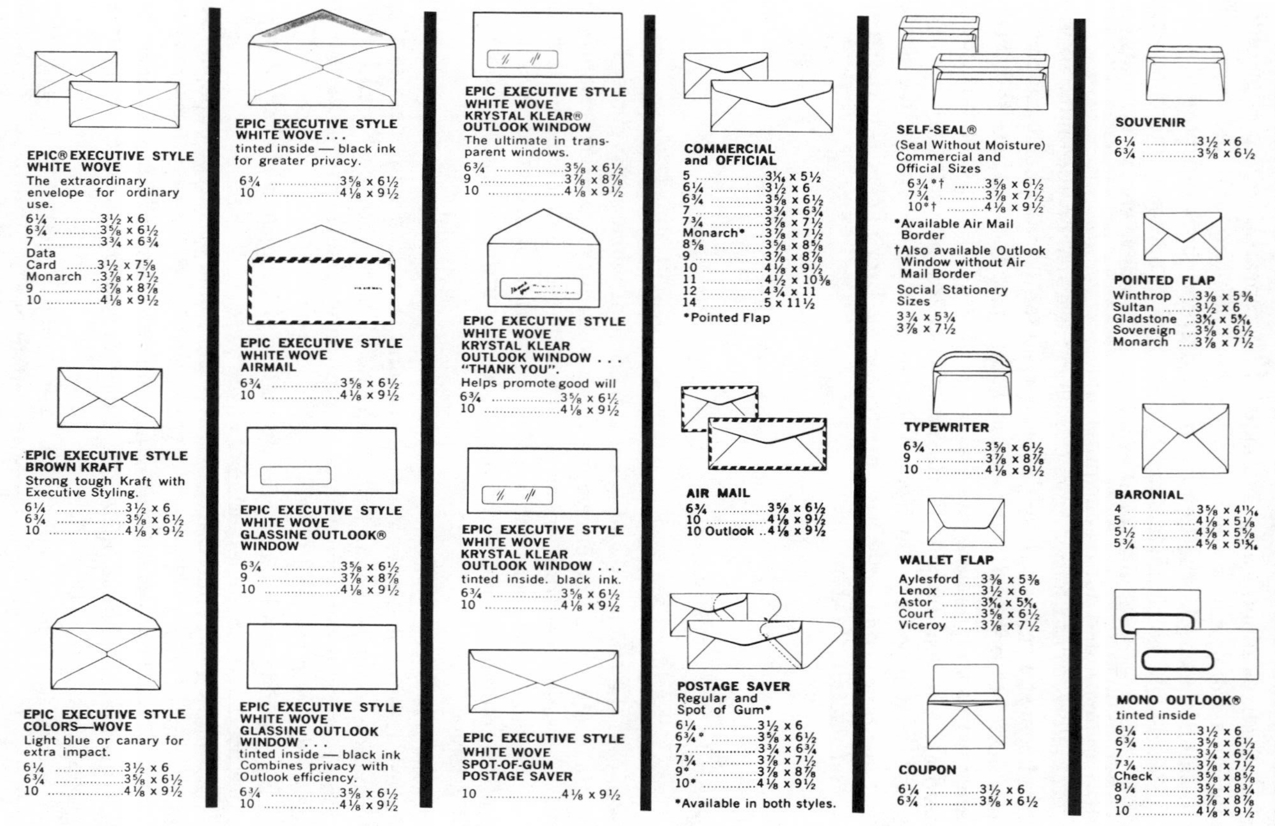

Figure 12-22. Various styles of envelopes. Courtesy of U.S. Envelope Co.

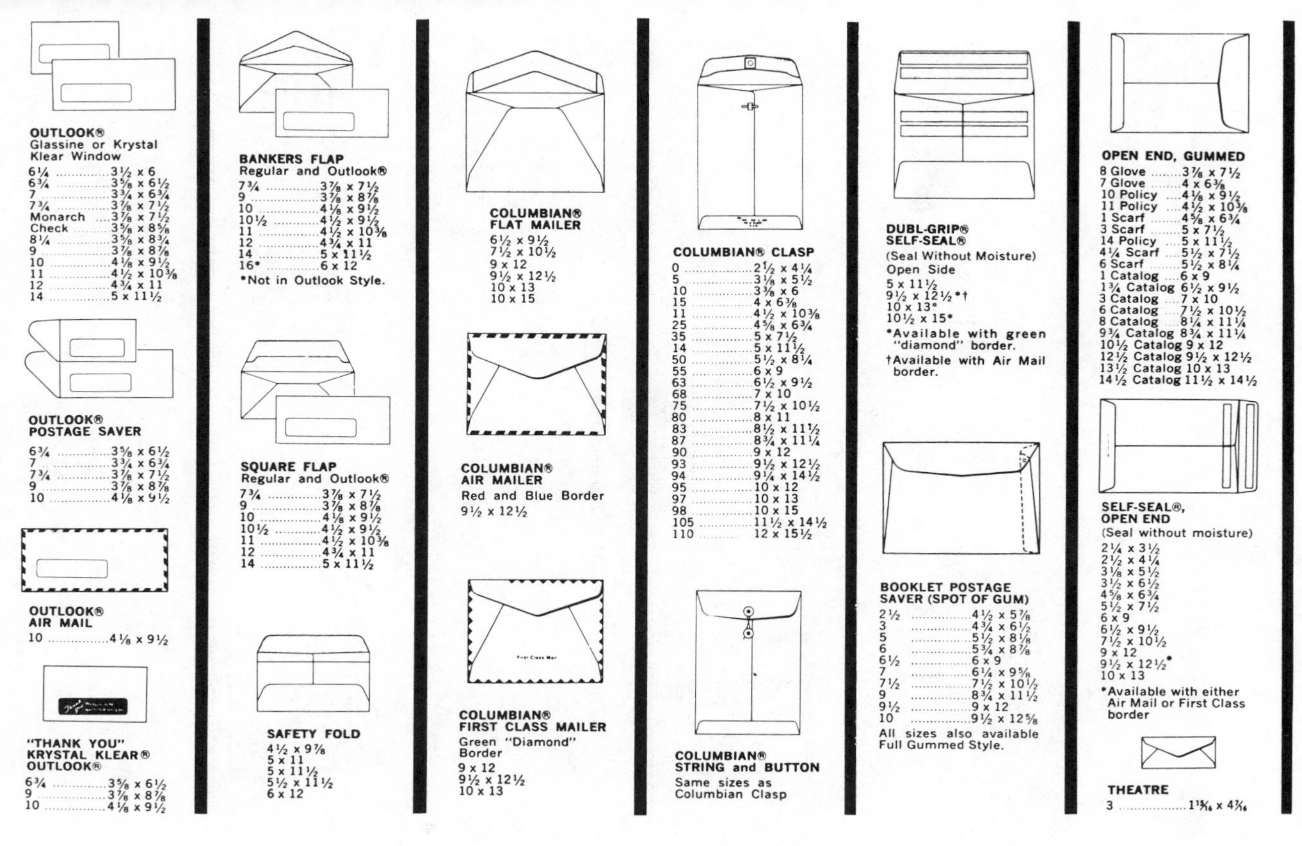

Figure 12-22. *Continued*

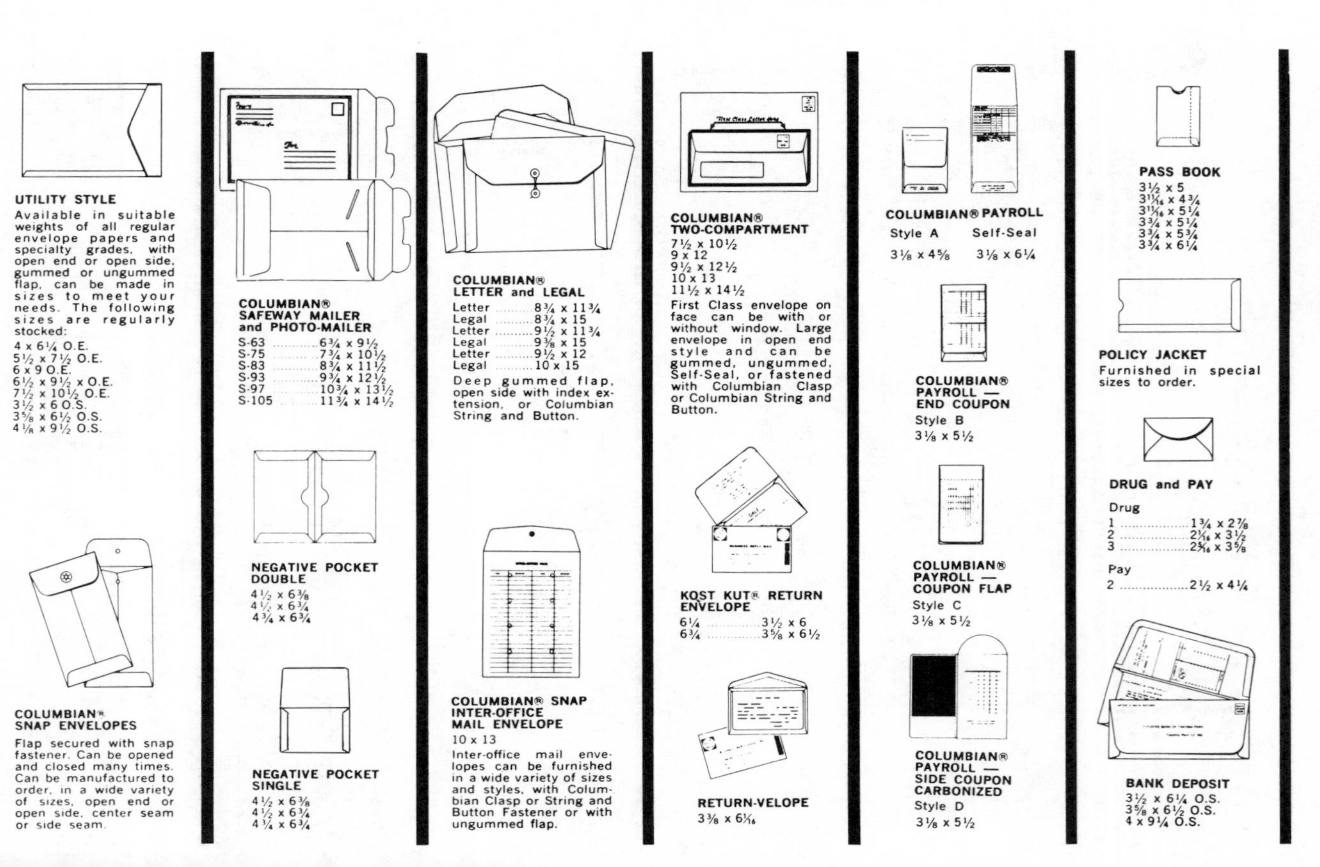

Figure 12-22. *Continued*

COIN

00	$1\frac{11}{16} \times 2\frac{3}{4}$
1-A	$2\frac{1}{4} \times 3\frac{1}{2}$
3-B	$2\frac{1}{2} \times 4\frac{1}{4}$
4-C	$3 \times 4\frac{1}{2}$
4½-D	$3 \times 4\frac{7}{8}$
5-E	$2\frac{7}{8} \times 5\frac{1}{4}$
5½-F	$3\frac{1}{8} \times 5\frac{1}{2}$
6-G	$3\frac{3}{8} \times 6$
7-H	$3\frac{1}{2} \times 6\frac{1}{2}$

POLICY

10	$4\frac{1}{8} \times 9\frac{1}{2}$
11	$4\frac{1}{2} \times 10\frac{3}{8}$
14	$5 \times 11\frac{1}{2}$

BANK COUPON OUTLOOK®
6¼ .... $3\frac{1}{2} \times 6$

BANK COUPON MONO OUTLOOK®
6¼ .... $3\frac{1}{2} \times 6$

COLUMBIAN® PACKING LIST
$4\frac{1}{4} \times 5\frac{3}{4}$

GRIPS-ALL® PACKING SLIP ENVELOPE
(Pressure Sensitive)
$4 \times 6\frac{3}{4}$

GRIPS-ALL® TAPE CARRIER ENVELOPE
(Pressure Sensitive)
$3 \times 7\frac{1}{2}$

COLUMBIAN® SECURITY ORE
$2\frac{1}{2} \times 4\frac{1}{2}$
3×5
$3\frac{1}{2} \times 6$
4×7
$4\frac{1}{2} \times 8$
5×9

TRANSPARENT BAGS AND ENVELOPES

Made from a variety of transparent materials, beautifully lithographed in color. Sizes to order. Choice of sealing methods.

EXPANSION BAG

DIE-CUT ENVELOPE

E-Z PAC BAG with POCKET

VIEW PAC ENVELOPE with FLAP

U.S.E. SPECIAL DUTY ENVELOPES

A Special Duty Envelope is one which has been created to meet a particular need. It may have been specially designed and constructed for this purpose; or it may be a stock envelope adapted to meet the need . . . Over and above the 70 standard styles shown in this folder, U.S.E. makes a wide variety of special styles and sizes. We maintain a file of more than 12,500 envelope dies. With blade cutting and hand-folding facilities also available, there is practically no limit to our ability to meet any envelope requirements.

Your printer or U. S. E. envelope supplier can assist you with samples or other information.

Figure 12-22. *Continued*

available, however, and include gray, green, red, orange, lavender, and white.

Manila is a sturdy buff or light-brown paper made from hemp or similar vegetable fibers. It does not have the strength of kraft, but it does have a superior surface for imprinting purposes.

Cellophane paper is a transparent cellulose material made from the fibers of such plants as cotton, wood, and flax. More than a hundred varieties of this basic paper are manufactured.

Cellophane paper is especially useful where transparency and grease-proof protection are important considerations. Although it is usually colorless, cellophane is available in several tints.

PUNCHED-CARD FORMS

The punched-card forms most frequently employed are the following:

1. IBM cards
2. Remington Rand cards
3. Marginal-punched cards
4. Edge-punched cards
5. Tape-card carriers

The IBM Card

The IBM card is the instrument through which many kinds of financial, statistical, and engineering data are obtained, stored, and processed. Such cards are of a standard size, 3¼ by 7⅜ inches, and measure approximately 150 cards to the inch in thickness. (Many measuring rulers used in designing forms contain a scale that designates the number of cards for each successive inch; for example, a stack of cards up to the 11-inch mark would indicate approximately 1650 cards; up to the 12-inch mark, approximately 1800.)

Chapter 16, entitled "Background and Methods of Data Representation," discusses the format and layout of IBM cards in some detail. Let us note here, however, a few more particulars about their physical attributes.

Manila-paper stock of a natural color is usually used for IBM cards. In addition, solid-color cards are available in yellow, salmon, green, red, blue, brown, and white. Further color contrast can be obtained by ordering striped IBM cards. The stripe may appear in a ¼-inch band in any of the following colors: yellow, salmon, green, red, blue, brown, violet, rose, or gray. Since certain of these colors do not provide a good contrast, the

G = Good contrast F = Fair contrast
O = Obtainable but not recommended

Color of Stripe	Color of Paper Stock							
	Natural	White	Yellow	Salmon	Green	Red	Blue	Brown
Yellow	G	G	O	F	F	F	F	O
Salmon	G	G	F	O	O	F	O	O
Green	G	G	G	F	O	F	F	G
Red	G	G	G	G	G	F	G	G
Blue	G	G	G	F	G	G	O	G
Brown	G	G	F	O	O	O	O	O
Violet	G	G	G	G	G	G	G	G
Rose	G	G	F	G	F	O	F	F
Gray	G	G	G	G	G	G	F	F

Figure 12-23. Striping chart.

chart in Figure 12-23 is presented for guidance. Note that all stripes go well with either natural or white paper stock.

For check, invoice, and similar applications, IBM cards are available in a wide range of colors, with background designs and borders custom-made to meet individual specifications. Background tints of yellow, blue, green, red, brown, and gray are all fairly common.

The special construction features that are available on IBM cards include the following:

1. Cards may be numbered repetitively or consecutively.
2. Cards may be prepunched with repetitive or consecutive numbers.
3. Cards may be printed on the reverse side.
4. Cards may be printed in colored ink.
5. Stubs may be situated at either end of the card, and any desired corner cut may be designated. Stub-cards may be designed so that when the stub is removed at the point of perforation, a regular-size card remains.
6. Cards with stubs may be obtained with holes placed in the stub; and, if desired, with strings or wires looped through the holes.
7. Cards may be purchased in unit-set arrangements, called padded cards.

8. Cards are available in continuous form. They are commonly the standard 3¼- by 7⅜-inch cards, with or without a stub, and joined together at the top and bottom to permit continuous feeding. A 9/16-inch wide punched-carrier strip, on both ends, will permit feeding and alignment on forms-writing machines.

9. Mark-sensed cards are obtainable with either diagonal or horizontal marking spaces.

10. Cards may be designed so that microfilm inserts can be easily and conveniently mounted on them. The microfilm may be fitted into an aperture in any one of a variety of positions that may be designated on the card format.

Remington Rand Cards

Both the International Business Machines Corporation and the Sperry Rand Corporation are major producers of data-processing equipment—in the punched-card (and computer) field—that operates through the use of tabulating cards.[1] Many technical differences exist between the two kinds of equipment, but one of major importance is that in IBM machines electrical contacts are made when fine brass brushes drop through the holes in the card and complete a circuit, which is linked to the printing and adding devices within the machines; in Sperry Rand machines the card-reading and card-printing units are operated mechanically—pins pass through holes in the punched card, actuating the printing and adding mechanism by a linkage of rods and push wires.

Figure 12-24 shows a typical Remington Rand tabulating card.

Remington Rand cards have the same overall dimensions as IBM cards: 3¼ by 7⅜ inches. Both cards, however, have a distinctive format. The Remington Rand card contains 90 vertical columns—45 on the upper tier and the same number on the lower tier.

In the aggregate the Remington Rand card has a total of 540 punching positions (6×90). Through the use of single and multiple punches, as illustrated in Figure 12-24, it is possible to provide for the recording of both numerical and alphabetical information. Of course, only one character may be designated for each column in a tier for any specific application.

Like IBM cards, Remington Rand cards are obtainable in a variety of colors and stripes. They may also be purchased in check-card styles and unit-set arrangements.

[1] The Sperry Rand Corporation was the result of a merger, in 1955, between Remington Rand, Inc. and the Sperry Corporation.

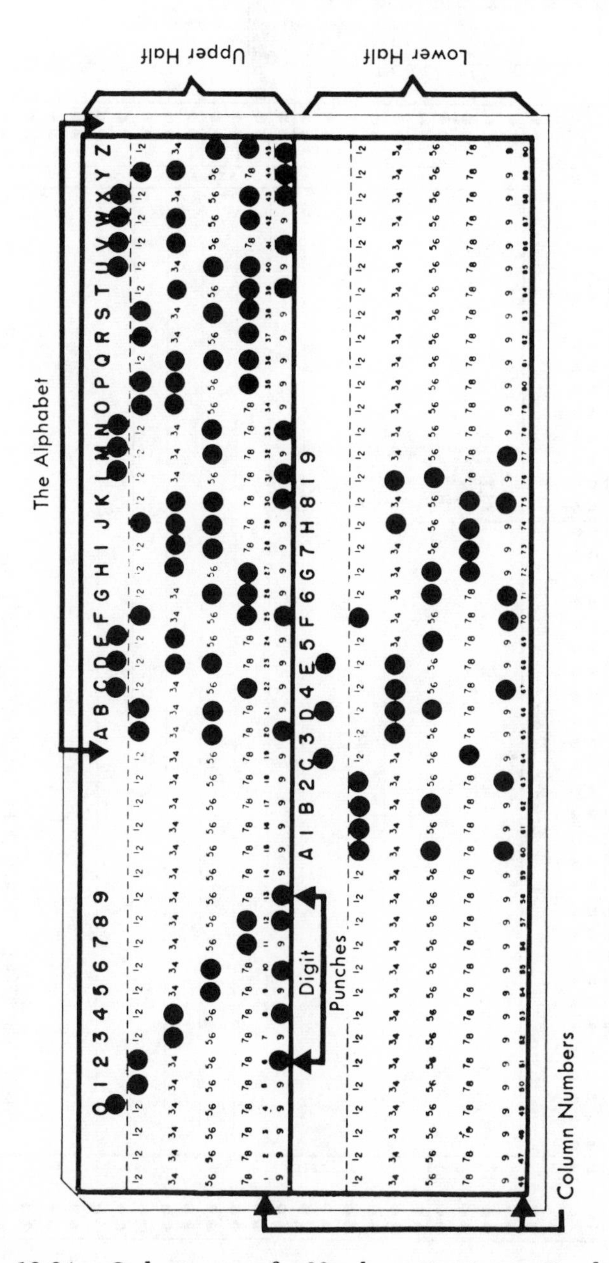

Figure 12-24. Code structure for 90 column Remington-Rand card.

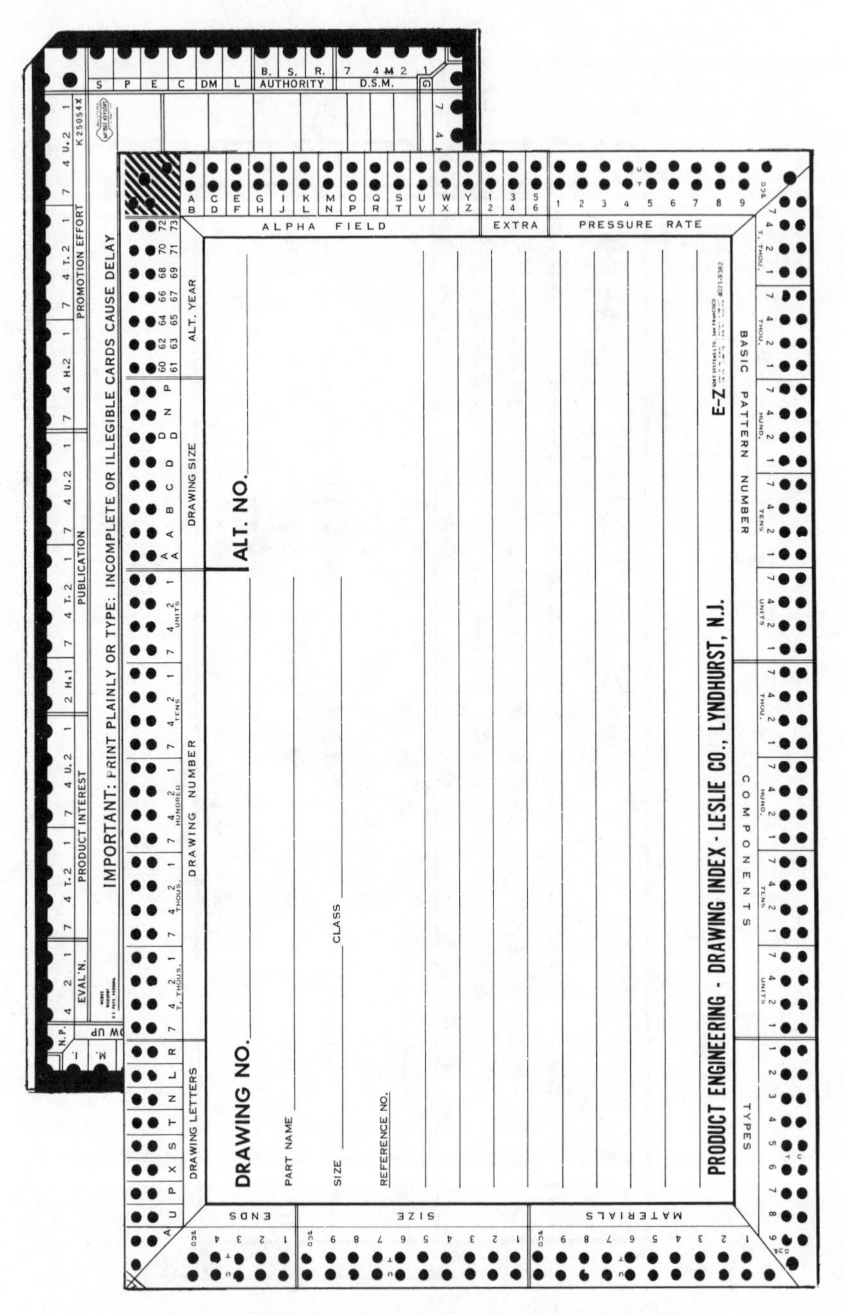

Figure 12-25. Marginal-punched cards.

Marginal-Punched Cards

These cards have coded holes prepunched in certain locations along one or more of their edges, leaving the body of the card free to record or mount information. (See Figure 12-25.) When the desired holes are notched or clipped away to the edge of the card, sorting of the data by key classifications can be accomplished with convenience and ease. As desired, entries on the card proper may be made by hand, or typed, or recorded on a graph. For some applications, paste-up mountings of photographs, clippings, or the like may be employed; for others inserted microfilm may be used. Whatever the technique for storing information or data, the cards must be kept reasonably flat if they are not to become unwieldy.

Hand-sorted punched cards (as they are also sometimes called) are supplied in many different sizes, ranging from 1½ by 2½ up to 8 by 10½ inches. They are obtainable with one corner cut so that it is easy to determine whether all the cards are right side up and facing in the same direction before starting to sort.

When establishing a marginal-punched card file it is necessary to assign meanings to the holes. Normally they are designated to signify a letter, numeral, particular classification, or a code symbol. To indicate the desired holes notches are made on the edge of the card by means of a hand-grooving instrument or special keypunch machine. One notch or a series of notches may be necessary to establish the selected classification.

The actual sorting operation is accomplished by a long, strong needle, similar to a darning needle. As shown in Figure 12-26, when the needle is inserted into the hole representing the selected classification and raised slightly, the cards with holes notched in that position drop from the deck. This technique selects the cards that have been notched to signify a partic-

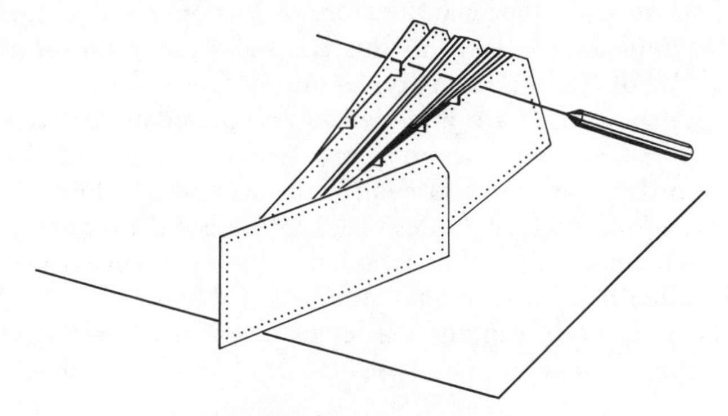

Figure 12-26. Card-sorting operation.

ular designation from those that have not been so indicated. Through a combination of such operations it is possible to accomplish a rather detailed breakdown of a classification.

Special supplies, instruments, and machines are available to implement the carrying out of hand-sorted punched-card filing systems. Among these are manual and electric punches, gummed stickers to correct notching errors, sorting trays, sorting needles, and mechanical sorters for handling a large quantity of cards.

Marginal-punched cards may be purchased as made-to-order forms, coding and all; or with special printing but utilizing a standard card and code.

Two prominent suppliers of marginal-punched-card forms and equipment are E–Z Sort Systems, Ltd. and the Royal McBee Corporation.

Edge-Punched Cards

An important feature of some integrated data-processing systems is the use of punched tape or edge-punched cards as the medium for storing information and activating equipment to print or transmit information. Automatic writing machines such as the Friden Programmatic Flexowriter and the Computyper are examples of equipment that may be operated by either or both media. Such machines can sense holes in a tape or card and cause the automatic typing of a document, and at the same time punch selective coding into an additional tape or card.

Channel codes for paper tape and edge-punched cards are treated in Chapter 18, which is concerned with the full scope of integrated data processing. Let us note here, however, that the same codes that are used for paper tape are used along the edges of cards.

Although the same channel code may be contained in either tapes or cards, and the use of either may be more or less optional for certain systems applications, our interest in this chapter centers on forms. Edge-punched cards fall into this category; lengths of tape do not.

Edge-punched cards are intended to accommodate two types of information: punched and written (pen, typewriter, or the like). The punched recordings are used to control the writing machine—for example, to move a form such as a sales order to the correct typing position—and to type whatever information is coded in the card. The written recordings, on the other hand, are for maintaining a visible record.

For the purpose of explanation let us consider a typical integrated data-processing system for purchase orders. We may assume that the system is operational and that it makes use of two groups of cards. One group, known as vendor cards, contains constant information in the form of vendors' names and addresses; the other group, called item cards, contains constant information on the description of products and related procurement clauses. Thus there is a separate card for each vendor and a separate card for each product. To begin with these cards may have been prepared as a by-product of writing each initial purchase-order document or else they may have resulted from preparation in advance of the first writing. In either case, once properly prepared, they provide a source of error-free input to the automatic writing machine.

When a purchase order is to be prepared the automatic-writing-machine operator pulls the vendor and item cards from separate files. By inserting these prepunched cards into the reading mechanism on the machine, purchase-order documents are prepared with a minimum of manual entries.

The system described above illustrates but one representative application of such automatic writing machines. Actually, these machines may be used for other purposes as well, such as the preparation of sales orders, invoices, and manufacturing orders.

For each application variations of edge-punched cards are possible, as required by the specific system. Figure 12-27 shows examples of both one- and two-card arrangements. It will be observed that each unit or set provides for the recording of certain constant and/or variable information directly on the card.

Tape-Card Carriers

Although it is not an edge-punched card, the tape-card carrier serves essentially the same purpose. A representative one is shown in Figure 12-28; it will be seen that the tape itself is contained within a fold on the edge of the card. As in the case of edge-punched cards, constant and variable information may be recorded on the body of the card.

MAGNETIC-STRIP CARDS

Forms of this classification are designed to be read by machine and by people. They contain *human language* and *machine language* on the same record. The typing of information on the face of the card is translated into computer language on magnetic strips imprinted on the back side of the same card.

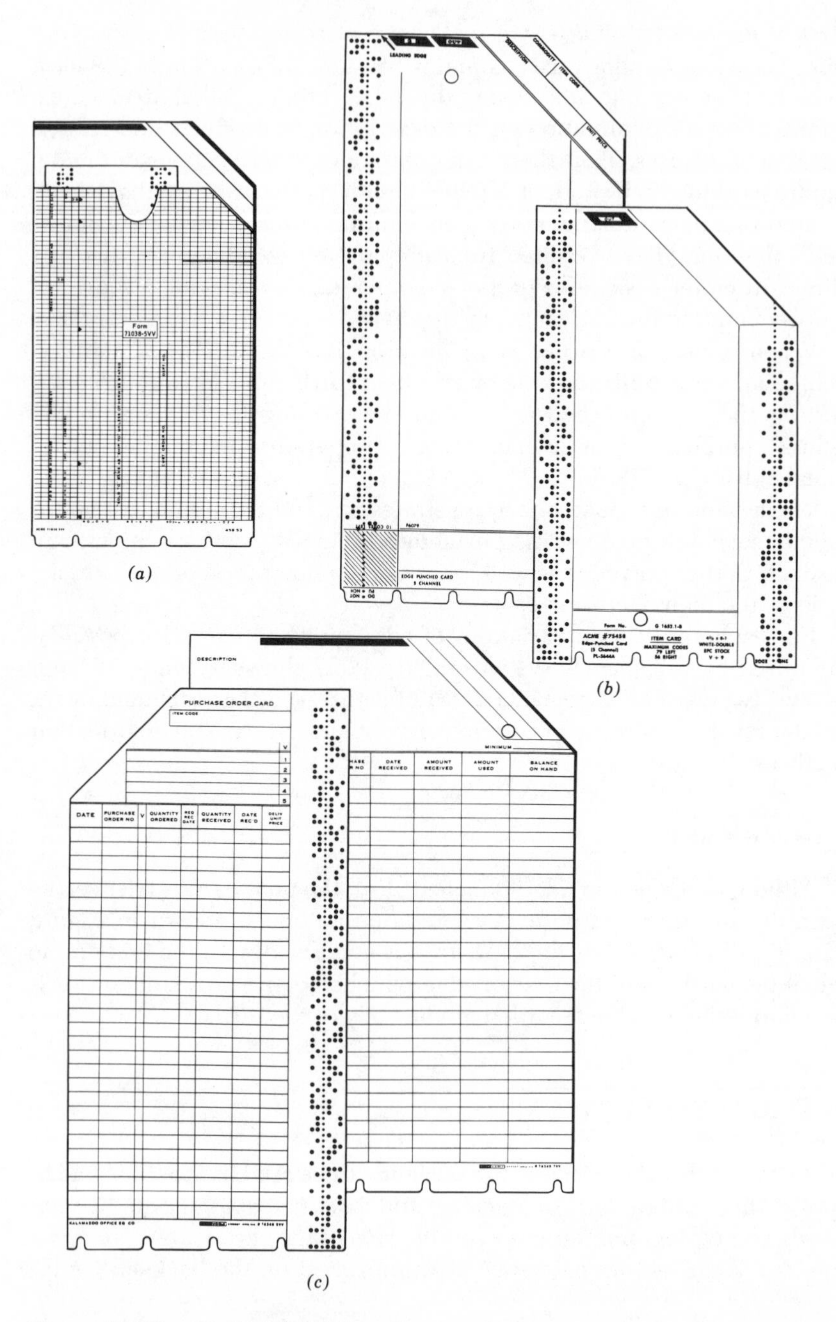

Figure 12-27. Variations of the edge-punched card: (*a*) pocket with customer's record printed on front holds automation edge-punched card; (*b*) single- and double-coded edge-punched cards designed to specifications; (*c*) purchase and inventory record filed as a set—one used as a traveling requisition. Courtesy of Acme Visible Records, Inc.

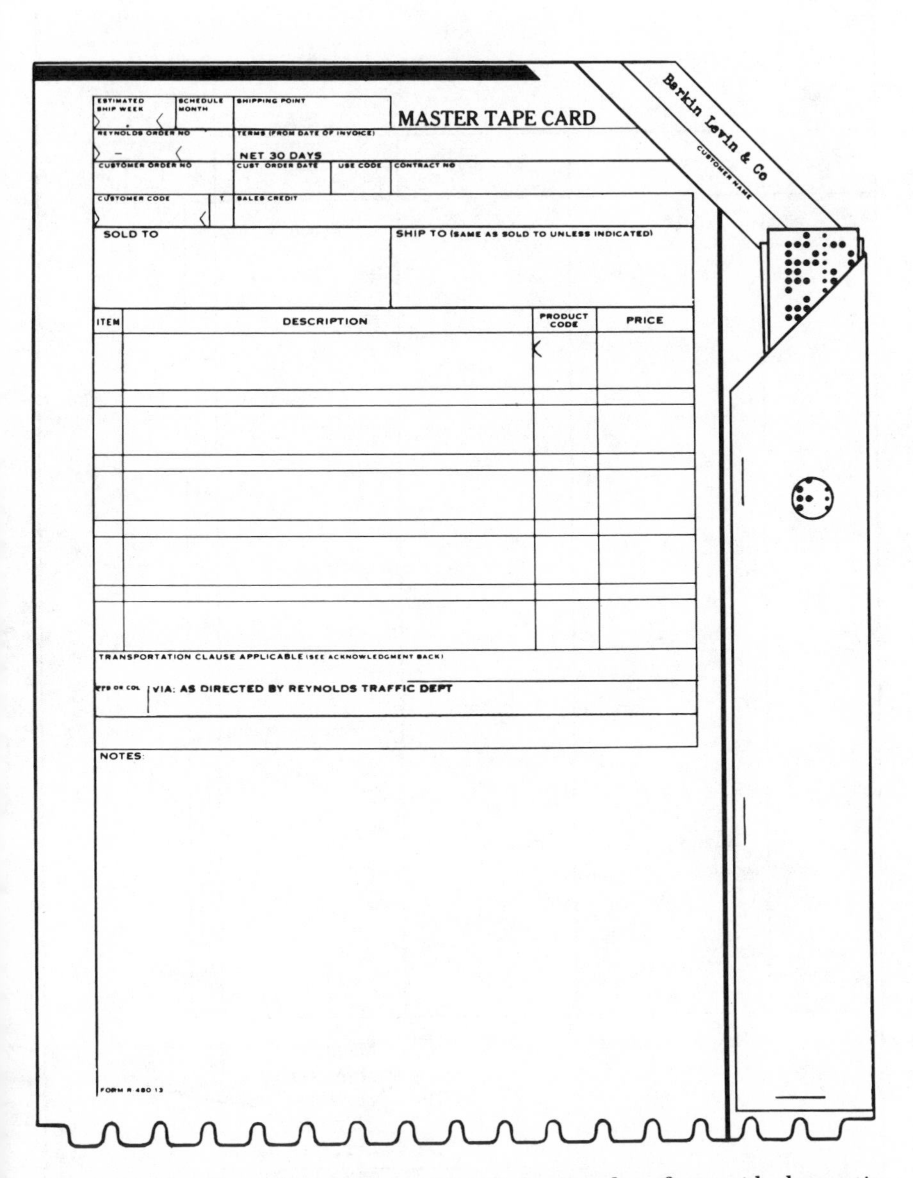

MASTER TAPE CARD

ESTIMATED SHIP WEEK | SCHEDULE MONTH | SHIPPING POINT

REYNOLDS ORDER NO | TERMS (FROM DATE OF INVOICE) NET 30 DAYS

CUSTOMER ORDER NO | CUST ORDER DATE | USE CODE | CONTRACT NO

CUSTOMER CODE | T | SALES CREDIT

SOLD TO | SHIP TO (SAME AS SOLD TO UNLESS INDICATED)

ITEM	DESCRIPTION	PRODUCT CODE	PRICE

TRANSPORTATION CLAUSE APPLICABLE (SEE ACKNOWLEDGMENT BACK)

PPD OR COL | VIA: AS DIRECTED BY REYNOLDS TRAFFIC DEPT

NOTES:

FORM R 480 13

Figure 12-28. Typical tape-card carrier. It is printed on front with the pertinent information contained on the punched tape. Courtesy of Acme Visible Records, Inc.

40

John Jones 2345

TRANSACTION CODE	DATE	AMOUNT	BALANCE
			1,000.00

1 2 3 4 5 6 7 8 9 10 11 12 13 14 15 16 17 18 19 20 21 22 23 24 25 26 27 28 29 30 31 32 33 34 35 36 37 38 39 40

Magnetic strips are located behind this area (see Figure 12–30)

Figure 12-29. Magnetic ledger card. Source: The National Cash Register Co.

The magnetic strips correspond to a section of tape in a reel of magnetic tape in a large computer. Conceptually, the reel of tape may be thought of as being sliced into strips and these strips fastened along the back of cards. An illustration of the magnetic-strip card is the magnetic ledger card shown in Figure 12-29.

The advantages of magnetic ledger cards are many. They include the retention of the hard-copy records of accounting. When a ledger card is selected from a file a visual record of the account (with current totals and balances) is readily apparent. Processing is fast and accurate. The strips on the back of the card store particular information on an account, together with specific instructions on how to process this particular account. Individual account cards may be taken from the file, inserted into a reading device, and the necessary calculations may be carried out by the cen-

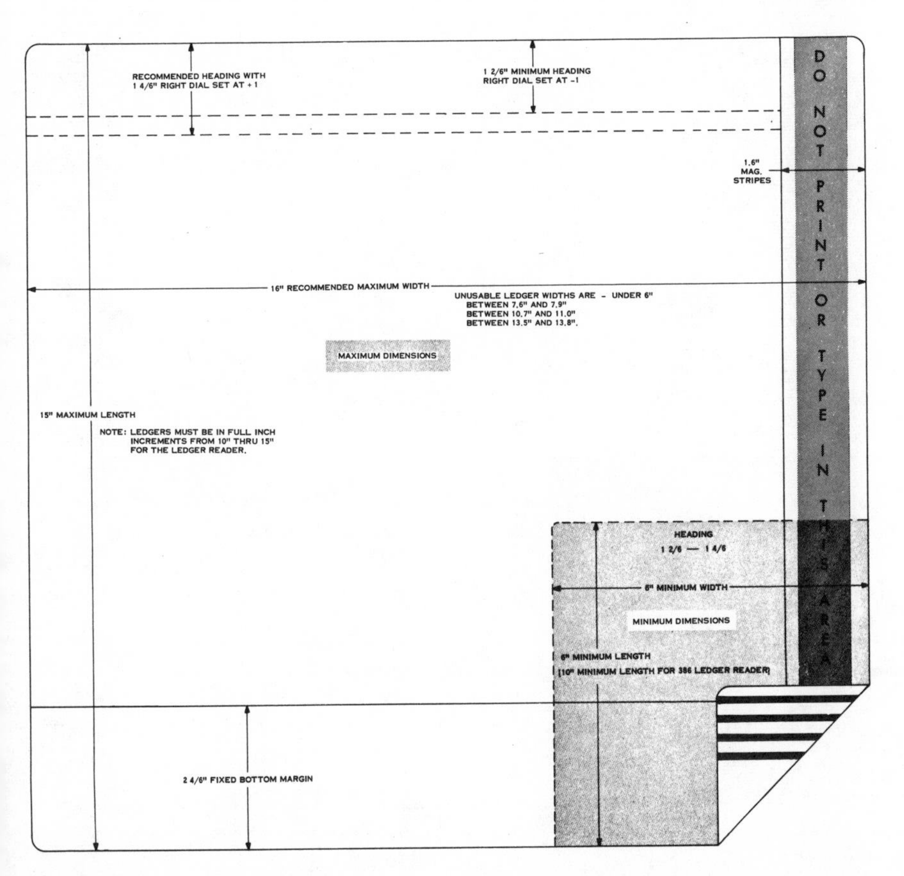

Figure 12-30. Specification for magnetic ledger card. Source: The National Cash Register Co.

tral processor of a computer. Each time an account is processed the totals and balances affected may be updated on the ledger in both printed form and in the language of the computer.

Magnetic ledger cards must be designed to conform to the requirements of the particular equipment with which they will be used. Figure 12-30 presents an example of card-format specifications, and shows where the magnetic strips are located.

13

Forms Creation

DESIGNING THE FORM

Experimental evidence and common sense indicate that a form should be designed to carry out the needs and purposes of the system of which it is a part. The requirements of a particular system are of prime importance in designing forms. The significant relationship is for the form to serve the system, not for the system to serve the form. Still, within the limitations and particular considerations imposed by the system itself, there are generally a number of alternative designs that can be utilized. The objective should be to devise the best possible form for the given system.

The following points of good forms design applied with imagination and judgment should result in forms of the greatest utility.

Form Title

The ideal title is brief, descriptive, and distinctive. Brevity is desirable to conserve space and to have the name of the form easily remembered. A good descriptive title is needed to clearly suggest the purpose of the form; it, too, will help in promoting name retention.

Distinctiveness, the last attribute mentioned, should be emphasized. Too frequently companies have forms with titles that are so nearly alike that they are a constant source of confusion. Seldom is there a need for this. With the application of a little ingenuity it is possible to come up with more than one acceptable and characteristic title.

Placement of the title is another consideration. For forms on 8½- by 11-inch sheets it is common for the title to be printed at the top, where it may be centered or placed to either side. Sometimes titles can be conveniently located at the bottom. This is largely true of certain card forms that are filed in an upright position. On forms used for insertion in books, such as journals and ledgers, the title is suitably placed along the left or right edge.

Form Number

A form number is assigned for identification and general reference purposes; it facilitates the functions of ordering, receiving, storing, and issuing. It is used in written procedures as a handy and ready means of designation.

An easily seen but unobtrusive location should be chosen for the form number. It is ordinarily printed in small type that is clearly discernible.

Rulings

Proper ruling is an important characteristic of good forms design. Lines serve as guides, dividers, or unifiers. As guides they facilitate the recording of information; as dividers they furnish demarcation areas and thus promote accuracy; and as unifiers they help to tie the parts of a form together—for example, by means of a border.

For various purposes such as expediting postings it is advisable to employ different kinds of lines. Striations of heavy and light lines are common for this purpose. Other lines that are frequently used include dotted lines, dash lines, dot-and-dash lines, and double lines.

Caution should be taken to keep the quantity and different kinds of lines to a number appropriate for the purpose. An abundance of lines can severely detract from the appearance and utility of a form.

Control (Serial) Numbers

Forms may be designated with a unique number at the time the form is prepared; or they may be preprinted with consecutive numbers. Checks, purchase orders, invoices, and inspection records are examples of forms that customarily contain a preprinted control number.

Through coding techniques various meanings may be built into control numbers. Such methods may not, however, lend themselves to preprinting the number or may permit only a part of the number to be so recorded. The following are examples of three separate coding schemes for control numbers:

Sales-Order Numbering Scheme

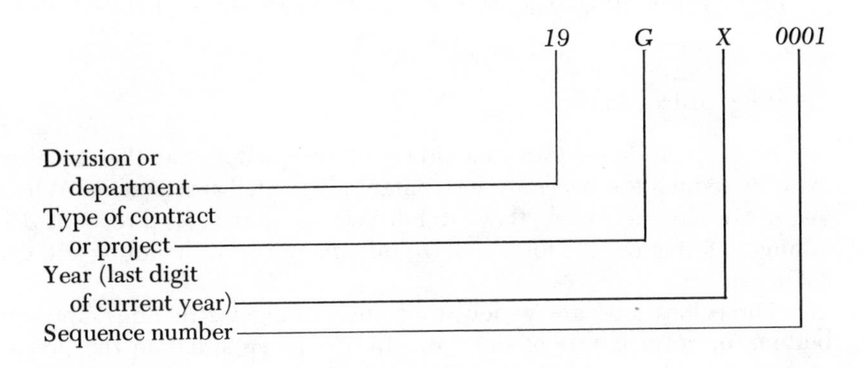

Purchase-Requisition Numbering Scheme

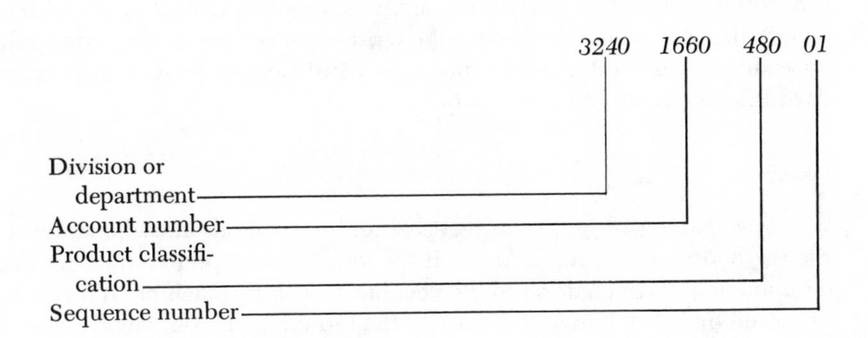

Negotiation Numbering Scheme

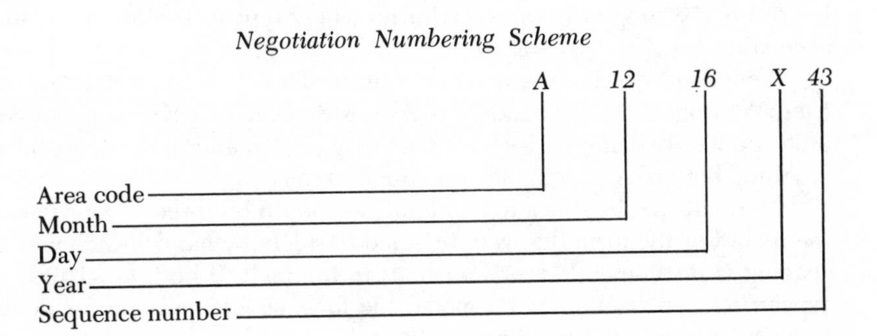

Code construction not only serves for identification but also aids in machine operations that precede and encompass the preparation of reports.

Locating Instructions

Properly designed forms should be self-explanatory and thereby eliminate or reduce the necessity for containing printed instructions. Where instructions are necessary, they fall into two categories: (*a*) directions pertaining to filling out the form, and (*b*) information on its handling and distribution.

Three locations are widely used for placing instructions—the top, bottom, or reverse side of the form. In any given situation the precise placement will depend on overall design considerations. However, when the location is obscure or on the back of the form it is good to signal attention by means of an easily seen notation. Such notes are often printed in a color that attracts attention.

One of the best locations for imparting instructions on forms is in the procedures manual. A correlative act, a practice that is at times recommendable, is to reference the specific written procedure on the form itself; for example: "Complete and route as per instructions contained in Standard Practice Instruction No. 3–14."

Spacing

The spacing on the form, vertical and horizontal, must be suited to the method used in processing it. If the form is to be prepared on writing equipment, it must conform to the peculiarities of the machine. If it is to be prepared by hand, horizontal lines on the form must be sufficiently spaced for making handwritten entries.

Not infrequently forms have to accommodate both machine and longhand entries. This is true of forms that are completed not only in the office but out in the shop or in sales territories where typewriters are not readily accessible.

Longhand entries require more space than do writing-machine entries. Writing machines make impressions that are orderly, compact, and uniform; handwriting varies from person to person and in the aggregate is anything but orderly, compact, and uniform.

Vertical spacing for longhand entries should be three or four lines to the inch. For the form that is to be filled in solely by hand, ¼-inch vertical spacing is recommended, or four lines to the inch. If both longhand and typewritten entries are to be made, the form should provide for double typewriter spacing—three lines to the inch.

In planning horizontal spacing for longhand entries, it is wise to allow ⅛ inch for each character, plus a little extra as a precaution. Horizontal spacing on forms for machine recordings must conform to the style and size of type in the machine to be used. Pica type requires 10 spaces to the inch across the page; elite type requires 12 spaces to the inch. Fourteen spaces to the inch is not an uncommon requirement for small type. Remember that there are many special styles of types on machines, and it is always sound practice to check the number of characters to the inch before laying out a form for use on a particular machine.

Spacing charts such as shown in Figure 13-1 are convenient for designing forms. Generally they are obtainable from printing houses, ruled for any size and style of standard type. In making a form layout dimensions are furnished by the lines on the paper instead of by marking out points with a scale, as required when drawing on blank paper.

For reproduction purposes the color of spacing charts is a light blue. When drawing in black on the sheet, the form may be reproduced and—with certain reproduction processes—the blue guide lines will not appear.

Transparencies offer an alternative method. By overlaying a spacing chart with a sheet of transparent master paper the form may be constructed with the aid of the spacing paper but not directly on it. The lines on the ruled paper underlying the transparency offer guides both for designing the form and for lettering within the prescribed spaces. A tracing board shown in Figure 13-2 is an ideal implement for carrying out the work but is by no means indispensable. It has an inclined working surface of plate glass, the bottom side of which is sandblasted so as to diffuse light. Fluorescent lamps on the inside provide the necessary illumination. A board of this kind can be used on any table or desk.

After the form is drawn copies or dummies can be obtained and then tested on the writing machine for which the form is intended. As suits the occasion, copies may be attached at the top and bottom to simulate continuous forms.

Transparent spacing charts provide for a modification of the above approach. The practice here is to develop the form directly on a regular lined spacing chart. The paper itself is transparent, so that reproductions of the drawing can be readily made by the diazo process.

Composition of the Parts

One essential factor in composition is ease of completion. This demands that the parts of a form fit together to require a minimum of energy, time, and effort in filling it out. Every point of information on the form should be studied to ascertain its best placement not only as a separate

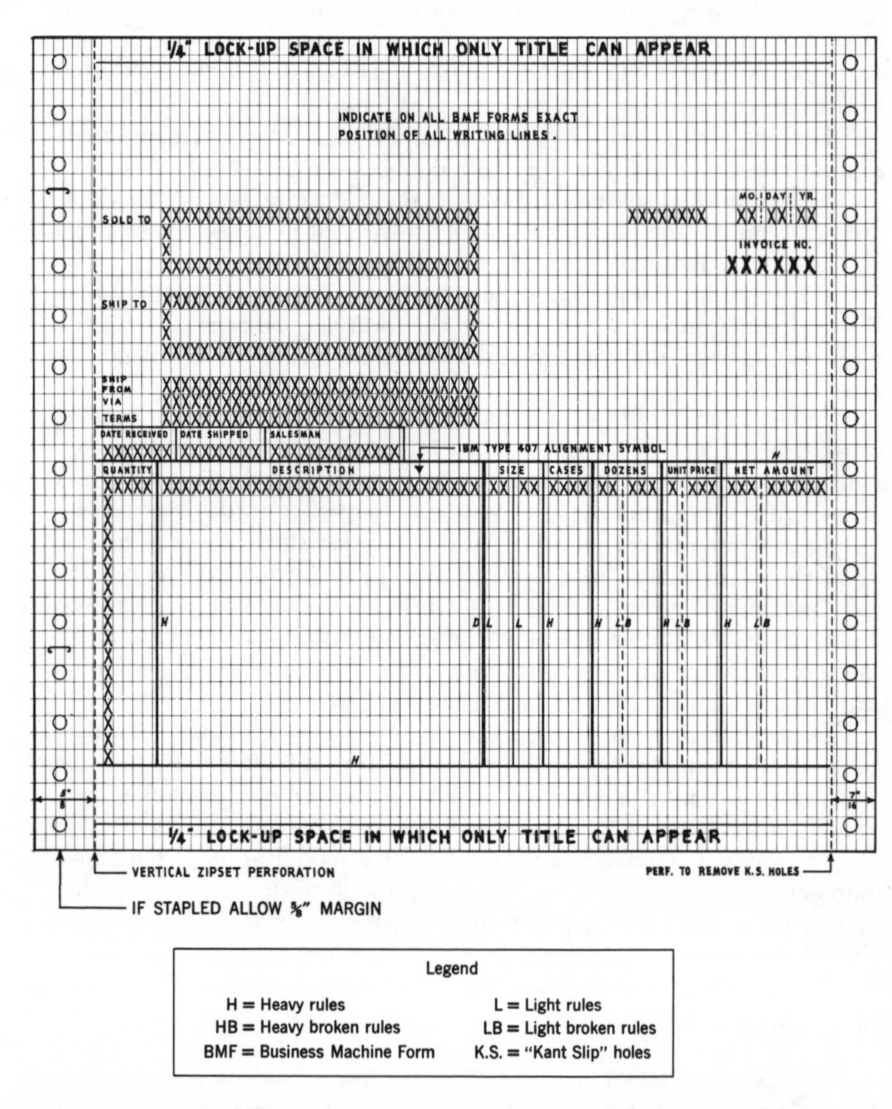

Figure 13-1. Form layout on ten-to-the-inch spacing-chart paper. Courtesy of The Standard Register Co.

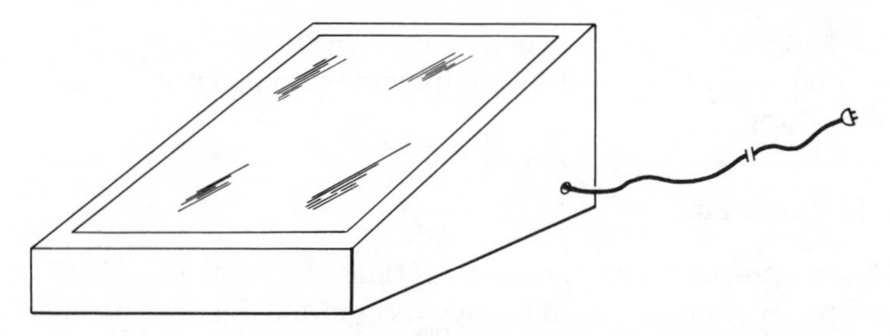

Figure 13-2. Portable tracing board.

unit but also as a factor contributing to the continuity of all the parts. Layout of the elements should be methodical. The well-designed form presents its material in well-organized fashion, with related items logically linked together.

Not to be overlooked is the need for a form to prescribe a sequential pattern of information that is in line with any associated form from which, or to which, information will be transcribed. Appropriate sequence of items on forms that will be used for machine data processing is important. An example is in the punching of tabulating cards from source documents. Unless the information on the source document is laid out according to an arrangement that follows the card format, the key-punching operation will be awkward, slow, and inefficient.

Forms for data processing are not the exception. On all forms that require transcription the proper sequence and grouping of items simplifies operations, cuts down on eye travel, and speeds up performance. Omissions are also avoided because there is less danger of information going unnoticed by its being in an out-of-order place.

Indexing information should be positioned in a convenient spot on all forms. The index relates to the filing categories into which the form will be put. For any particular form or unit copy there are several basic means of indexing—alphabetical by name, subject matter, or geographical area; numerical by date or code. Within these possibilities are many combinations and variations. It is important therefore to consider how the form is going to be referred to after filing so that it can be designed to facilitate referencing. This ties in with the subject material of the next topic, which considers physical filing accommodations.

Filing Considerations

As mentioned, the manner in which the form is to be filed is important. Are copies to be punched and placed in a binder? Are sufficient filing facilities available for use? Will legal-size files be necessary? The answers to these and similar questions must be taken into account. In some cases the size and capacity of existing equipment will influence the size and design of the form.

Copy Identification

Colors, numbers, and printed captions are used to identify copies. Colored paper is a definite aid to copy recognition; for example, white may indicate finance, canary may indicate production, and blue may indicate marketing. Sorting, routing, and filing operations are thus made easier.

In unit-set construction forms may be shingled to further enhance the identification of copies, as shown in Figure 13-3.

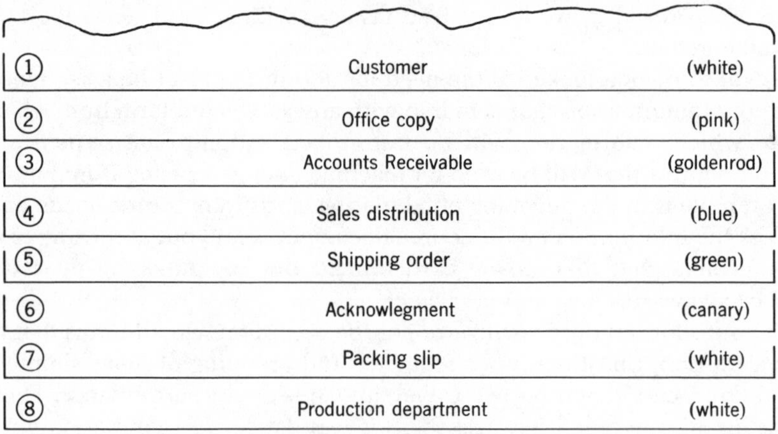

①	Customer	(white)
②	Office copy	(pink)
③	Accounts Receivable	(goldenrod)
④	Sales distribution	(blue)
⑤	Shipping order	(green)
⑥	Acknowlegment	(canary)
⑦	Packing slip	(white)
⑧	Production department	(white)

Figure 13-3. Shingling of an invoice form.

PRINTING THE FORM

Colored Ink

To highlight certain items on a form the printing may be done in more than one color. Aside from the conventional applications of using red for making something stand out, colored print may even be used as a code; for instance, instructions may call for completing items captioned in black or red, black or blue, or any different combination of colors, as suits specific situations.

Two-Side Printing

The reverse side of forms may be used to record instructions, conditions, or to provide for a continuation of the form itself.

Forms that are to be printed on both sides may be printed head to head, head to foot, or head to side. The choice of style should largely depend on how the form is to be used. Forms that will be inserted in a binder and turned from the right side to the left should be printed in the head-to-head style. The head-to-foot style, called "tumble head," is used when forms are to be held together at the top and turned over. Card forms are frequently tumbled so that they can be turned over conveniently.

The head-to-head style means that the printing on the back of the form is at a right angle to that which is on the front. This arrangement is not generally desirable and should be used only when the design on one side requires height and the design on the opposite side requires width.

Paper

Paper may be made from rags, wood, straw, cotton plants, corn stalks, and other fibrous substances. The use of rags as a raw material for papermaking today is used mostly for good-quality paper. Some inferior rag paper is made as a by-product by rag paper manufacturers. Wood is by far the most widely used raw material in the paper manufacturing industry.

Kinds of Paper. Paper that is used for writing or printing with ink is classified and identified below. First, let us clarify the meaning of the word "bond"; it does not, as is popularly supposed, designate a rag-content paper. "Bond," per se, merely indicates a broad category of firm paper used for writing and printing.

Rag Bond. Rag fibers are stronger than wood fibers and result in paper that is exceptional for its strength and beauty. Costly raw material, stringent manufacturing requirements, and a generally superior product are factors that combine to make rag bond a relatively expensive paper.

Sulfite Bond. The vast majority of 8½- by 11-inch forms used in business are in this classification. Sulfite bond is a superb paper for ordinary use; it possesses reasonable strength, can withstand much use (but little abuse), and presents a pleasing appearance. Five grades of this kind of paper are available, the degree of excellence decreasing with the number. Number 1 grade is better than Number 2, etc.

Rag-Content Bond. This paper contains various proportions of rag, usually specified to content as 25, 50, or 75 percent. Ordinarily, the remainder is made up of sulfite pulp. The greater the percentage of rag content, the better the grade of paper.

Ledger Paper. Medium- to heavy-weight bond paper is usually called ledger paper. This paper is used in ledgers, whence it derives its name. It is suitable for applications that involve much handling as well as for permanent records.

Onionskin. The word "onionskin" is picturesque in that it describes a thin, translucent bond paper of light substance and glossy finish. Onionskin contains 25 to 100 percent rag. It is excellent for preparing carbon copies, conserving filing space, and making lightweight mailings.

Manifold. This is a nonrag, lightweight paper that is much like onionskin. Since it has no rag content, it lacks the strength and durability of onionskin. Nevertheless, it serves many of the same purposes equally well. Manifold may be obtained with or without a glazed surface.

Safety Paper. This paper is made from sulfite wood pulp, usually with chemical reagents added to prevent falsification.

Carbon. Tissue or a heavier paper serves as the base for making carbon paper; this is coated with a carbon compound, consisting of color materials ground in an oily or waxy vehicle.

Carbonless Paper (NCR Paper). As used, the letters "NCR" have a double meaning: They stand for *n*o *c*arbon *r*equired and the first three initials of the National Cash Register Company, the distributor of a chemically impregnated paper that produces copies without the use of carbon. On a unit set, for example, the bottom side of the first sheet is coated with a dye emulsion. The top side of the last sheet is coated with another chemical. Any intermediate sheets are correspondingly coated, top and bottom. The arrangement is such that writing on the first sheet causes a chemical reaction where it comes in contact, through pressure, with the one underneath and thus produces an impression. A similar reaction takes place between any additional sheets in the set.

Carbonless paper may be used for various records, such as purchase orders, requisitions, manufacturing orders, invoices, and sales books. Its advantages are that it eliminates the need for disposing of carbon paper, provides copies that are clear and sharp, and it is clean to work with. One disadvantage is its price. On an average forms constructed from NCR paper cost up to 10 percent more than forms interleaved with carbon paper. Another disadvantage lies in the fact that erasures are difficult to make.

Grades of Paper. In descending order of quality grades of paper are customarily identified as follows:

Rag
- 100 percent rag

Rag content
- 75 percent rag; usually 25 percent wood fibers
- 50 percent rag; usually 50 percent wood fibers
- 25 percent rag; usually 75 percent wood fibers

Sulfite bond
- No. 1 Grade
- No. 2 Grade
- No. 3 Grade
- No. 4 Grade
- No. 5 Grade

Attributes to look for in appraising a paper for any particular application include weight, strength, surface quality for the kind of imprinting intended, erasing properties, permanence or aging characteristics, and where applicable transparency factors.

Weights and Grades of Paper for Forms. Important elements to be considered in selecting weights and grades of paper are as follows:

1. The number of copies to be simultaneously produced.
2. The means by which the form is to be filled out; for example, by pen, pencil, or writing machine.
3. The quality and sensitivity of carbon paper.
4. The treatment the form will receive while it is being processed.
5. The length of time the form is to be retained.
6. The plans for filing the form.

The paper industry uses the abbreviation "Sub." (for "substance," meaning weight) to designate the approximate weights of the different kinds of paper as follows: Sub. 7, Sub. 9, Sub. 16, and so on. The suffix number tells the approximate weight of 500 sheets of paper for a definite sheet size. For bond and ledger paper the size 17 by 22 inches is used as a base; 22½ by 28½ inches is used for postcard stock. Accordingly, "Sub. 20" means that 500 sheets of bond paper measuring 17 by 22 inches weighs approximately 20 pounds.

The weight of paper has a direct bearing on the maximum number of satisfactory copies that can be produced at one time. In this respect the information presented in Table 13-1 will be of interest. Remember, however, that factors other than weight must be taken into account (such as the type of forms-writing machine, quality of carbon, and general conditions of use). Different grades and weights of paper may be combined in the same set to yield various results.

Margins

Liberal margins on all four sides add to the attractiveness of forms. Margins may be set equidistant from opposite sides to give the body of the

Table 13-1. SUBSTANCES, THICKNESSES, AND NUMBERS OF COPIES NORMALLY OBTAINABLE WITH SEVERAL DIFFERENT WEIGHTS OF PAPER

Paper	Sub-stance	Number of Sheets per Inch	Point Thick-ness	Approximate Number of Copies per Reproductive Medium: Longhand (Pen or Pencil)	Standard Typewriter or Business Machine	Electric Typewriter
Bond	12			4	6 to 7	7 to 9
Bond	13	440	0.00225	3 to 4	4 to 5	6 to 8
Bond	16	330	0.003	3	3 to 4	5 to 7
Bond	20	285	0.0035	2	2 to 3	4 to 5
Bond and ledger	24	250	0.004	1	1	3
Onionskin and manifold	7	600	0.0015	5 to 6	8 to 10	12 to 14
Onionskin and manifold	9	530		4 to 5	6 to 8	9 to 12

form a balanced appearance. Consideration must be given to such determinants as the method by which the forms will be bound, the quantity to be contained in one binder, and particular requirements for machine processing.

Forms that are to be filed on a ring binder should have a minimum left-hand margin of five-eighths of an inch. For forms that will be placed in a post binder it is advisable to have a generous margin of 1¼ inches or at least a minimum of three-fourths of an inch.

Marginal requirements for forms that are to be processed on a machine depend on the type of equipment and its make. Duplicating equipment such as offset presses and hectograph units usually require a gripper space of three-sixteenths of an inch or more on both sides so that they may be fed through the machine. Forms for use on typewriters need a bottom margin of from one-fourth to three-fourths of an inch in order that they may be held in place while the lowermost portion is typed.

Electronic Tabular Stops

Tab stop positions may be designated on forms by being printed in electrically conductive ink. This feature of design allows typewriters and other forms of writing machines with special electronic attachments to read the form and activate the carriage. The mechanism functions so that the operator can position the carriage from one point to the next by a touch of the tab stop key.

Typography

The appearance and arrangement of the text must be given serious consideration. The ground rules include the following:

1. Unfamiliar typefaces are difficult to read and should generally be avoided.
2. It is usually inadvisable to mix type families on a single form.
3. Select the typeface on the basis of its appearance and legibility. Gothic type is widely used on forms and is highly commendable; it has simple straight lines, presents a trim look, and is easy to read.
4. Type sizes smaller than 10 point are hard to read and should be chosen with care.
5. Long headings printed in capital letters are difficult to read. Where lengthy headings are required readability will be improved by using both capital and lower case letters.
6. Do not permit the text or rulings of the form to overshadow the information that is to be recorded.

APPRAISING THE FORM

When designing a form or redesigning one that is already in existence attention should be given to the principles of forms design that have been presented. As previously stressed, the result must be compatible with the system in question and wholly satisfactory for the purpose intended; otherwise the operations of the entire system can be adversely affected. Elements of the form should be studied with the same degree and care as the operations of the system itself. For the purpose of assuring a comprehensive appraisal, and to make certain that no fundamental design factors are overlooked, a check list such as the one given below is of value.

Forms Appraisal Check List

Mechanics and Overall Considerations

1. Does the form serve the purpose? Is the purpose worthwhile?
2. What is the functional relationship of the form to other forms? What would happen if the form were eliminated?
3. Is the form to be temporary or is it to have an indefinite life?
4. Can the form be advantageously combined with another form? Can an existing form be used instead of this one?

5. Does every copy serve a necessary purpose? Are there a sufficient number of copies?

6. Should one or more copies be extracted and treated as a separate form?

7. Would it be of value to use both sides of the form?

8. Should written procedures or instructions govern the use of the form? If so, should they be printed on the form or published separately?

9. Is this form readily distinguishable from any that it replaces?

10. If the form is to be mailed outside the company, should it be designed for enclosure in a window envelope?

11. Have the proper persons been consulted for opinions, suggestions, and authorizations? Consideration should be given to the people who use the form, other systems analysts, forms salesmen, equipment salesmen, administrators, and supervisors.

Content

1. Is the title descriptive, particular, and meaningful?

2. Are space captions legible, informative, and understandable? Is it perfectly clear as to what information is to be recorded in each space?

3. Would it be advisable to have preprinted serial numbers on the form?

4. Should a space be set aside for inserting an identifying number?

5. Have all necessary items of information been taken into consideration? Are there any requirements for information on the form that are unnecessary and ought to be omitted?

6. Has adequate provision been made for handling all recurring information? Would it be practical to use check boxes, rather than write-in spaces, for denoting certain facts and data?

7. How about placing terms, conditions, and instructions on the form? Could any of these be beneficially placed on the reverse side of the form?

Construction and Composition

1. Is the best possible class of form being utilized for the purpose? (Unit set? Continuous type? Duplicating master?)

2. Does the form look crowded? Out of balance? Is it orderly?

3. Is there sufficient white space?

4. Have the different modes of recording (handwritten and mechanical) been taken into account? Does the spacing suit the mode? (Pen? Pencil? Typewriter? Flexowriter? etc.)

5. Would it be desirable to tailor the spaces so as to accommodate mechanical and handwritten entries?

6. Are the margins on the form adequate? For binding and filing purposes? For catching and holding the form at the top and bottom while being processed on a writing machine?

7. Should holes be prepunched in any of the copies to facilitate filing?

8. Are vertical and horizontal spacings adequate, especially where processing on a writing machine is involved?

9. Are vertical columns arranged to take proper advantage of tabulating stops?

10. Is the order of recording the information efficient? Does it avoid waste of time? Unnecessary hand and/or machine motions?

11. Has the spacing on the form been proved on the writing machine?

12. Should "to" and "from" spaces be provided on the form?

13. Are items that will be constantly referenced (e.g., dates, numbers, and summary descriptions) placed in conspicuous locations?

14. Is the sequence of items satisfactory? With respect to any superseded form? Relative to similar forms?

15. When information is to be reproduced onto or from another form is the sequence of the items in line with such transcribing?

16. Has adequate use been made of codes, keys, and abbreviations for conserving space?

17. Does the size, shape, and tensile strength of the form conform to handling and filing requirements?

18. Has sufficient space been allowed for signatures of initiation and approval?

Paper, Ink, and Printing

1. Would it be desirable to have different colored copies?

2. Should more than one color of ink be used in the printing? If so, which color or colors?

3. Are the weights and grades of paper satisfactory with respect to handling requirements, anticipated life of the copies, and storage accommodations?

4. Is the carbon paper suited to the method of recording (pencil, ball point pen, writing machine)?

5. Does the carbon produce legible copies?

6. Have clear, readable typefaces been selected?

14

Forms Management

THE CONTROL OF FORMS

We have illustrated various kinds of forms and discussed how they are designed. Attention is now directed to the subject of forms control. Its purpose is to prevent waste and increase efficiency with respect to the design, acquisition, and utility of forms. As an additional benefit it also serves as an instrument of control over systems-and-procedures activities, because through the management of forms it is possible to exercise a degree of control over systems.

A well-administered forms-control program should achieve the following:

1. Savings. It is impossible to say how much money can be saved, because this varies according to the business and the circumstances. However, the amount can be substantial and often surprises those who are unfamiliar with the sources. These include the cost of the form itself, the cost of its completion, and the cost of its functioning as part of the system.
2. Elimination of any unnecessary forms.
3. Intelligible and easy to use forms.
4. Consolidation of existing forms wherever practical.
5. Assurance against unwarranted duplication of like forms.
6. Coordination of the forms with related procedures.
7. Regulation over the purchase and reproduction of forms. Make-or-buy policies can be uniformly applied.
8. Maintenance of adequate supplies; less overstocking or understocking.
9. Awareness of procedural alterations. Changes in procedures often affect forms, and requests for new or modified forms are indicative of such changes.

10. Provision for numbering forms so that they might be conveniently identified, and easily referenced in written procedures.

11. Establishment of a central source for indexing forms and furnishing information concerning them.

The above list should be edifying to those who are accustomed to thinking of forms programs as being chiefly for the purpose of eliminating unnecessary forms. Many programs are established with this purpose uppermost in mind, but this phase of the work often becomes secondary to projects of a more positive nature. The design of forms can be highly constructive work.

The necessity for controlling forms has paralleled the rapid rise in the increase of paperwork in business. More forms were the natural outcome of more sophisticated and complex systems. Paradoxically, scientific management as characterized by "the need to know" is largely responsible for the increase. Technological developments of mechanical and electronic equipment have also added to the total number of forms.

FOCAL-POINT REGULATION

The control of forms is best done through a centralized authority. The manpower requirements to carry out the program should be determined by a thorough study of the situation. Factors to be taken into account include the scope of the objectives, anticipated accomplishments, and volume of the work load. In a small company forms management may be the part-time work of one man. For a larger concern one or two persons devoting full time to the tasks may be necessary.

An efficient plan is to have a forms design and control unit as an independent section of the central systems-and-procedures department. One or more forms specialists can carry out the program; and should there be need for intermittent assistance, other systems people may be assigned to help.

One of the chief benefits of a centralized service is that it surmounts departmental considerations; it lessens indifference and promotes cooperation.

ESTABLISHING A FORMS INDEX

A logical approach to the establishment of a complete forms-control program is to catalogue all existing forms. Systems-and-procedures work in general, and effective forms control in particular, is greatly enhanced and facilitated by the ready access to information that such indices afford.

Initially the gathering of the forms may be done all at once, or on a piecemeal basis. In the former case it is customary to request pertinent representatives of the various organizational units to collect and send to a designated person from two to five copies of each of the forms used in their respective departments. It may also be requested that one copy of certain uncommon forms be filled in with typical information to serve as a sample.

After the forms have been collected they must be numbered and filed according to some plan. Forms numbering may be accomplished in one of three ways: (*a*) in consecutive numerical sequence without regard for the function of the form, (*b*) by a functional numbering code, or (*c*) by some hybrid plan that combines categorical and functional codes.

Whatever the method used, its primary purpose is one of identification, and immediate provision should be made to have the assigned number of a form preprinted on all units that are subsequently ordered. Although the placement of the number is not too important, a usual location is the lower left-hand corner of the form; small-size type is generally used.

Consecutive Numbering

In this method the forms are numbered consecutively (1, 2, 3, etc.) without concern for their function. Revision numbers or letters are placed after the sequence number and may be designated Rev. 1, Rev. 2, etc. (or Rev. A, Rev. B, etc.), indicating the number of times the particular form has undergone alterations.

Because of its sequential nature the consecutive-numbering method is the easiest to install and maintain; it also has the advantage of simplifying storage and handling since the forms can be stored in numerical sequence and are easily dispensed when requisitioned.

Two associated drawbacks are that the method does not bring together related forms and could require the establishment of a forms file by function for this purpose.

Functional Numbering

This method actually incorporates consecutive numbering within a functional code. Since the forms are filed by function, the method makes it possible to assemble related forms easily and quickly.

Numbering the forms by function makes it unnecessary to set up a separate functional file. The method allows the grouping of like numbers for stockroom-control purposes and allows related forms to be reviewed when necessary without searching.

Typical categories used as a basis for the classification of forms are as follows:

By department—

0001–0999	Corporate
1000–1999	Accounting
2000–2999	Production
3000–3999	Plant engineering
4000–4999	Sales
5000–5999	Personnel

By subject matter—

1000–1999	Adjustments
2000–2999	Claims
3000–3999	Notifications
4000–4999	Orders
5000–5999	Projects
6000–6999	Records
7000–7999	Reports
8000–8999	Vouchers

As in the case of consecutive numbering, when necessary the functional number should be followed by a revision number or letter.

A potential drawback of this method may be unnecessary work, caused by setting up classifications that are too refined or restrictive. Accordingly the benefits that are derived from the method may not be worth the effort.

Another hazard to guard against is the assignment of inadequate ranges within classifications. If the groupings are not properly established at the beginning, the scheme might require a major revision as the number of forms in each functional category increases.

A good functional plan should be reasonably simple and permit expansion without jeopardizing the basic structure of the scheme.

Hybrid Code

There are many possible variations within this classification. Both letters and numbers are used in assorted combinations. Here again some method of designating revisions is necessary. Such a coding plan may assume the following arrangement:

S–34–P30–3

The letter "S" in this application signifies a standard-size form. In its stead the letter "T" might mean tag; "L," labels; "C," tabulating card; etc. The number that follows signifies what the form is about. Thus the "34" above may indicate capital facilities, finished goods inventory, or personnel records. The letter "P" is the functional designation and tells the purpose of the form—for example, to purchase, to instruct, to move something, to record, etc. The "30" is the serial number, and the "3" is the revision designation.

Some coding plans even have the date and quantity of the last order contained in the coding scheme. An example of such an arrangement is ENG. 42–15M–12XX, where "ENG." is an abbreviation for an administrative engineering form; "42" indicates that this is the 42nd form of the series; the 15M shows that 15,000 copies were printed the last time the form was ordered; and "12XX" designates that December, 19XX, was the date of the printing.

The two examples have been chosen to demonstrate the diversity of coding systems. Plant or geographic location is also commonly built into such codes.

FORMS FOR THE FORMS FILE

As mentioned earlier, two to five copies of the forms used in a company are customarily obtained. Possible uses for the larger number of copies are as follows:

1. *Completed sample copy*. This copy contains representative information to portray the completed form.
2. *Paste-up copy*. Portions of this copy such as lettering and boxes may be cut out and used in designing equivalent or similar forms.
3. *Printer's copy*. This copy may be sent to the printer when the time comes for reordering the form.
4. *Courtesy copy*. From time to time members of management request a particular form. This copy is for the purpose of complying with such requests.
5. *Permanent copy*. This copy is intended as a permanent record and should not ordinarily be removed from the file. It is a good idea to hand stamp the form as follows: "Permanent record—do not remove."

Except for the copies that have representative information recorded on them and those that are marked "permanent," no particular copies are reserved for any explicit purpose. The intention is to have a sufficient number on hand to satisfy anticipated requirements. Copies that will not be returned to the files should be replaced if practical.

Make-up of the Forms File

After the forms have been collected and numbered in accordance with a predetermined plan they must be filed. This is done by preparing a folder for each form and filing by form number. Other items often included in the folder are the following:

1. The request and records that led to the adoption of the form;
2. Sketches and drawings of the original form and each subsequent revision;
3. Records on rejected forms designs and notes as to why they were rejected;
4. Notations with respect to ideas and suggestions for future design changes;
5. Information as to how the form is used in the system. Such information may be in the form of a written procedure; and
6. Letters and memoranda pertaining to the form.

Most filing involves the storage of active and inactive material. When a form becomes obsolete the associated folder and its contents should be removed from the active file, designated "inactive," and retained elsewhere in a special file set up for the purpose.

Cross-Reference Files

One type of cross-reference file that is convenient relates the title of the form to the form number. In addition, if a consecutive-numbering system is used, it will probably be necessary to set up a cross-reference file by function. By referring to the indicated function it will be possible to determine which forms are associated with specific activities. Cross indexes of the kind mentioned are commonly kept on 3- by 5-inch cards.

COMBINING FORMS

Savings are occasionally realized by combining forms. Over the years it is not uncommon for an excessive variety of forms to accumulate within a company, and there is a need to weed out the useless, less desirable, and expendable ones. An examination of related forms frequently brings this to light. One instance took place in a well-known company where, in the course of a few years, 16 variations of essentially the same purchase order were used. With the advent of a forms program all 16 variations were ultimately combined into 4 basic types.

A simple way to compare the features of forms is by means of a recurring-data-analysis chart such as that shown in Figure 14-1. Paper obtained from ordinary analysis pads will serve the same purpose (see Figure 14-2).

Recurring-data-analysis charts are designed so that the left-hand column lists the itemized data that appear on all the forms being studied. Spaces across the top identify the various forms. Check marks and/or comments are placed wherever the item or data apply to a column head-

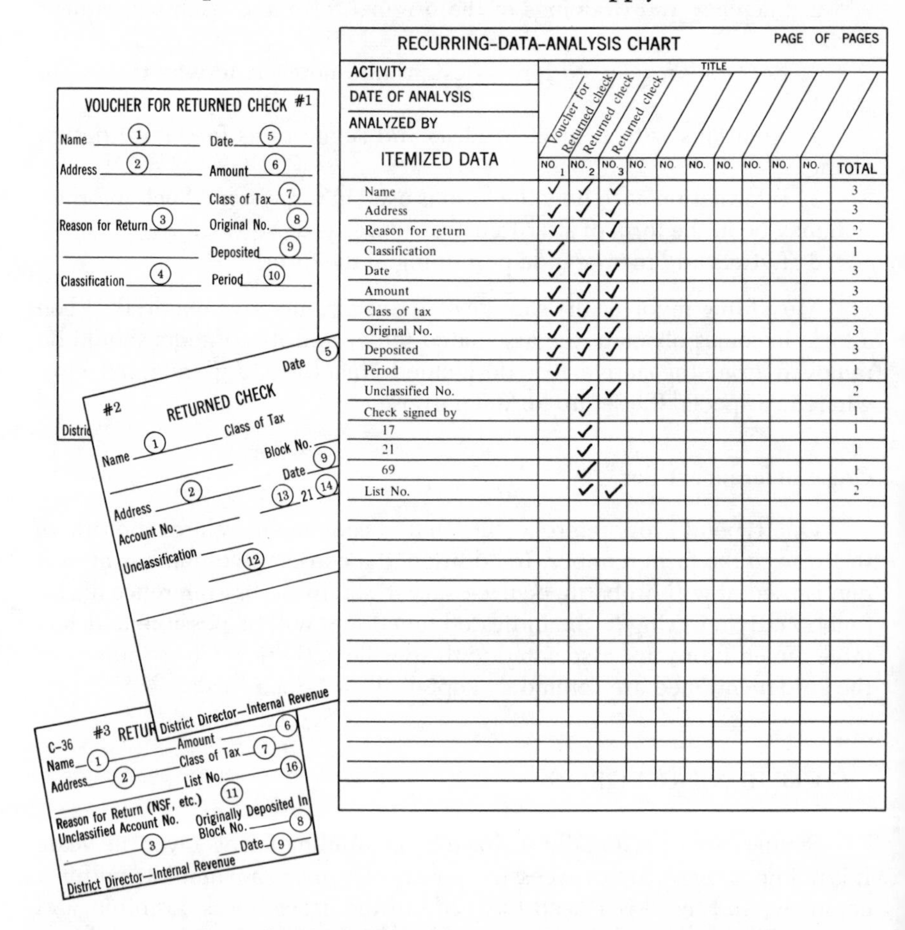

VOUCHER FOR RETURNED CHECK #1

Name (1) Date (5)
Address (2) Amount (6)
Class of Tax (7)
Reason for Return (3) Original No. (8)
Deposited (9)
Classification (4) Period (10)
Distri…

#2 RETURNED CHECK

Date (5)
Class of Tax
Name (1) Block No.
Date (9)
Address (2) (13) 21 (14)
Account No.
Unclassification (12)
District Director—Internal Revenue

C-36 #3 RETUR…

Name (1) Amount (6)
Address (2) Class of Tax (7)
List No. (16)
Reason for Return (NSF, etc.) (11)
Unclassified Account No. Originally Deposited In
(3) Block No. (8)
Date (9)
District Director—Internal Revenue

RECURRING-DATA-ANALYSIS CHART — PAGE OF PAGES

ACTIVITY
DATE OF ANALYSIS
ANALYZED BY
TITLE

ITEMIZED DATA	NO. 1 Voucher for Returned check	NO. 2 Returned check	NO. 3 Returned check	NO.	NO.	NO.	NO.	NO.	NO.	NO.	TOTAL
Name	✓	✓	✓								3
Address	✓	✓	✓								3
Reason for return	✓		✓								2
Classification	✓										1
Date	✓	✓	✓								3
Amount	✓	✓	✓								3
Class of tax	✓	✓	✓								3
Original No.	✓	✓	✓								3
Deposited	✓	✓	✓								3
Period	✓										1
Unclassified No.		✓	✓								2
Check signed by		✓									1
17		✓									1
21		✓									1
69		✓									1
List No.		✓	✓								2

Figure 14-1. Example of information recorded from three forms onto a Recurring Data Analysis Chart. Source: General Services Administration.

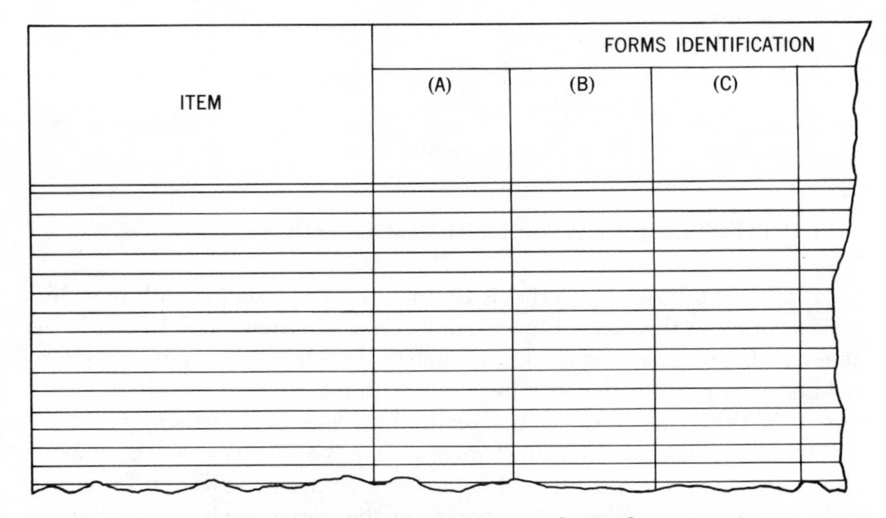

Figure 14-2. Recurring Data Analysis Chart.

ing. For some studies it is advisable to have wide columns, such as in Figure 14-2, to provide plenty of room for showing the precise nature of an item; for example, to picture a segment of the form such as a "yes" or "no" box arrangement. This will permit easy comparison of different presentations of the same data.

When completed the recurring-data-analysis chart features attributes that are common to two or more forms, and thus serves as a means for judging the feasibility of combining or simplifying any number of them.

Even though forms may lend themselves to being combined, such combinations are not always advisable. The mere reduction of the number of forms is not invariably beneficial. It is generally better to have two relatively simple forms than a single form that is highly complex, cumbersome, and difficult to process.

FORMS FOR FORMS CONTROL

That it takes forms to control forms should come as no surprise. Not many are needed, but those that are should be selected with care. Representative of the forms necessary are the following:

1. Request for form
2. Forms registry
3. Printing specifications sheet
4. Stores requisition for forms

5. Perpetual inventory card
6. Purchase requisition for forms
7. Production requisition for forms

Each of these forms will be discussed below.

INSTITUTING NEW AND REVISED FORMS

A widely followed practice is to direct all requests for both new and revised forms to the central forms-control unit. Requests can be made by memoranda or on a preprinted form such as that shown in Figure 14-3. It is customary to require that the petitioner submit a rough sketch of a new form or, in the case of a revision, a marked-up copy of the existing one.

The forms-control unit must review proposed forms as to their need, usage, and adherence to the principles of sound forms design. Provision ought to be made for promptly notifying the persons who made the requests of the acceptance or rejection of their proposals. Once the request for a form is approved the forms-control unit should be obliged to assist the originator in refining the design of the form, if necessary; and may be charged with forms copy preparation if it has the manpower and facilities.

After the design of the form is finalized it is necessary to decide whether the form should be produced on company equipment or purchased from an outside printer. In many cases the structural complexity of the form will dictate the need for a printing house to handle the job. Still, in recent years the technological advancements in the duplicating-equipment field have made it possible for duplicating departments within companies to produce sophisticated reproduction work. They can be especially proficient at turning out simple, standard-size forms.

Allowing the central forms-control unit to render make-or-buy decisions can save time and effect substantial cost savings. Judicial ability and knowledge on the part of the person charged with this responsibility is necessary. He must know the technical capabilities of the internal duplicating department and its cost factors; he also needs to be continually informed of that department's work load.

In ordering noncomplex forms the rule of thumb is to print short-run jobs and to purchase long-run jobs. It is generally agreed that, depending on the capacity and comprehensiveness of the duplicating facilities, quantities under 20,000 constitute short-run jobs. A sample of a production order for the reproduction department is shown in Figure 14-4.

What has already been said about internal duplicating departments leads us naturally into a discussion of their advantages and limitations as producers of printed forms.

REQUEST FOR FORM
(Submit Sketch or Marked-up Copy of Form)

☐ NEW
☐ REVISED

DATE SUBMITTED

DATE REQUIRED

TITLE OF FORM | FORM NUMBER

MANNER OF PREPARATION: ☐ LONGHAND ☐ TABULAR ☐ ELECTRIC TYPEWRITER ☐ OTHER (SPECIFY): ☐ MECHANICAL TYPEWRITER ☐ BOOKKEEPING MACHINE

PURPOSE OF FORM (BE BRIEF BUT INFORMATIVE):

WHEN PREPARED: ☐ DAILY ☐ WEEKLY ☐ MONTHLY ☐ QUARTERLY ☐ AS REQUIRED

FORM WILL BE: ☐ TEMPORARY ☐ PERMANENT

USAGE ESTIMATE: ________ QUARTERLY ________ SEMI-ANNUALLY ________ ANNUALLY

TO SUPERSEDE FORM(S)	QTY ON HAND	RECOMMENDED DISPOSITION
		☐ USE ☐ SCRAP ☐ HOLD (HOW LONG):
		☐ USE ☐ SCRAP ☐ HOLD (HOW LONG):
		☐ USE ☐ SCRAP ☐ HOLD (HOW LONG):

AT PRESENT IS INFORMATION TRANSCRIBED FROM ANOTHER FORM? ☐ NO ☐ YES (SPECIFY):

UNDER PROCEDURES WITH NEW OR REVISED FORM WILL INFORMATION BE TRANSCRIBED FROM ANOTHER FORM? ☐ NO ☐ YES (SPECIFY):

WHAT WRITTEN PROCEDURES ARE AFFECTED?

DISTRIBUTION	COPY 1	COPY 3	COPY 5	COPY 7
	COPY 2	COPY 4	COPY 6	COPY 8

SUGGESTED METHOD OF REPRODUCTION: ☐ OFFSET ☐ CONTINUOUS ☐ LETTERPRESS ☐ FANFOLD ☐ MULTIGRAPH ☐ SNAP OUT ☐ HECTOGRAPH ☐ INTERLEAVED CARBON

SPECIFICATIONS

COPY	PAPER					INK COLOR	NUMBERED SERIALLY	COPY IDENTIFICATION	PERFORATIONS	CARBON		
	SIZE		STOCK									
	WIDTH	LEN.	GRADE	WEIGHT	COLOR					WIDTH	LEN.	KIND
1												
2												
3												
4												
5												
6												
7												
8												

ADDITIONAL SPECIFICATIONS:

REQUESTED BY________________ APPROVED BY________________

Figure 14-3. Form for requesting a new or revised form.

Job No.	PRODUCTION ORDER Reproduction Department		Priority
Quantity	Form No.	Date Wanted	Extra Copies for 1. 2. 3.
Title			
Department	Location	Attention of	

OPERATION

METHOD	✓	OPERATOR	TIME
Phototype			
Vari-typer			
IBM Machine			
Justification			
Proofread by			
Xerox			
Paper mat			
Metal mat			
Press			
Multilith			
Photo plate			
Direct image			
Rerun			
Xerox—paper			
Xerox—metal			
1300 Plate			
Mimeograph			
Bindery			
Punching			
Drill No.			
No. of holes			
Folding			
Collating			
Stapling			
Addressing			
Padding			
Sheets to pad			
Pads to package			
Sheets to package			
Round corner			

DISTRIBUTION

CODE	Direct	Envelope	Other	CODE	Direct	Envelope	Other

Special Instructions:

PRESS

Proofs Required	Pages	
Paper No.	Sheets	
Cover No.	Black Ink	
Size	Other Ink	
	Wrapped and Mailed Date By	

Figure 14-4. Sample of a production order for a reproduction department.

Potential Advantages of Internal Duplicating Departments

In relation to the reproduction of forms the following benefits may be derived from having an internal duplicating department:

1. Fast service is provided for when necessary.
2. The cost for certain kinds of reproductive work—mostly simple, small-size jobs—is lower.
3. Management can conveniently and readily experiment with new forms.
4. Since the printing of forms contributes to the total work load of the reproduction department, there is less idle equipment time. Thus maximum use may be gained from the existing facilities.
5. Centralized control of different types of equipment permits selection of the reproduction method best suited for the job.
6. Total dependence on outside suppliers of forms is eliminated or reduced. This ties in with the first benefit mentioned above. At times convenient outside sources can be nearly impossible to obtain, and, in the event that a form runs out or a temporary form is quickly needed, the internal duplicating department may readily accommodate the situation.

Potential Disadvantages of Internal Duplicating Departments

From the standpoint of producing forms the following disadvantages are commonly associated with internal duplicating departments:

1. The production of forms of relatively simple construction is encouraged even though a form of complex structure would be preferable. This is often done to make greater or sufficient use of existing facilities.
2. In order to cover costs there may be a disposition to do work that is best done on the outside. Commercial printers can perform many jobs better and more economically than company duplicating departments.
3. Saving money by not purchasing a form may prove to be "penny wise and pound foolish." This is true when a self-printed form of inferior quality or construction handicaps the operations of a system.
4. Due to a lack of knowledge of costs there is frequently a tendency to underestimate the cost of producing forms on one's premises and to overestimate savings.

An important point of caution must be made about the propensity to expand central duplicating departments beyond reasonable bounds. This is to be averted; an organization of optimum size should be kept for accomplishing the tasks involved. All the operations pertaining to the processes of forms reproduction should be regulated to secure a high rate of performance at low cost.

Registering the Numbers of Forms

Prior to ordering a form—internally or from an outside source—it is necessary to assign it a permanent number and to record the number in a forms registry. The registry may be maintained in a notebook, loose-leaf binder, or on cards.

Besides providing for the form number and title, space should be provided on the record to designate the initiator's name, his departmental affiliation, the date the form is instituted, and, for recording at some later date, the reason for the form's obsolescence; when applicable, the number of the form to be superseded should also be included.

PROCUREMENT CONTROL

As an aid to purchasing forms the printing specifications sheet is invaluable (Figure 14-5). It serves to make certain that all pertinent factors are taken into consideration and made known to the printer.

When inviting bids from printers it is expedient to forward a copy of the printing specifications sheet along with the request for price and delivery quotations. In many companies the actual contacting of vendors for bids is done by the purchasing department. Nevertheless, the printing specifications sheet, supplied by the forms-control unit, puts the printer on notice of the requirements that must be met.

Failure to obtain satisfactory forms usually does not occur from a lack of "know-how" on the part of the printer but from misconceptions and misundertandings that arise from impersonal dealings. The time to prevent a mistake is before it happens. Good advice for the systems analyst or forms man is to get to know the printers and work with them closely. Seek their opinions. They can be especially helpful when it comes to working out the details of a new or novel form.

Before placing an order for carbon-interleaved forms, it is wise to ask the printer to submit what is known in the printing trade as a paper-and-carbon dummy. The dummy is a model of the colors and grades of paper, carbon and regular, that will be used in the actual forms construction. Its early use in testing under real or simulated conditions allows sufficient time to make whatever changes may be necessary. The dummy should be processed on the machine for which the form is intended to ascertain that

THE ILLUSTRATIVE CORPORATION

Printing Specifications	FORM TITLE	DATE
		FORM NO.

NOTICE

TO VENDOR: Deviations are not to be made in these specifications unless they are authorized in writing.
TO BIDDER: Any proposed deviations from these specifications must be explained in your quotation. Acceptable reasons for same are restricted to those that will benefit your production methods, lower the price, or shorten the period for delivery. The successful bidder will receive another specification sheet attached to a copy of our purchase order, and this sheet will incorporate any changes that are acceptable.

PART NO.	SIZE (WITHOUT STUB)	PAPER STOCK				PRINT						MARGINAL CAPTIONS (TO INDICATE ROUTING, DISPOSITION, ETC.)
		COLOR	WGT.	GRADE	FINISH	INK BLK	INK OTHER	FACE ONLY	HEAD TO HEAD	HEAD TO FOOT	HEAD TO SIDE	
1	X											
2	X											
3	X											
4	X											
5	X											
6	X											
7	X											
8	X											
9	X											
10	X											
11	X											
12	X											

CONSTRUCTION	STUB	PUNCHING	METHOD OF COMPLETION
☐ TAG	☐ NONE	☐ NONE	☐ PEN
☐ ENVELOPE	☐ LEFT SIDE	☐ ROUND HOLE	☐ PENCIL
☐ HECTOGRAPH MASTER	☐ BOTTOM	☐ SLOTTING:	☐ ELECTRIC TYPEWRITER
☐ TAB CARD	☐ TOP	☐ OTHER: ______	☐ MECHANICAL TYPEWRITER
☐ CONTINUOUS	☐ RIGHT SIDE	______	☐ BOOKKEEPING MACHINE
☐ SNAP OUT	ADHESIVE METHOD:	______	☐ TAB EQUIPMENT
☐ SINGLE SHEETS	☐ INSIDE GLUE	NUMBER OF HOLES ______	☐ OTHER: ______
☐ PADDED	☐ OPTIONAL	DIAMETER ______	MACHINE HAVING—
☐ OTHER: ______	WIDTH:	POSITIONING:	☐ ELITE TYPE
______	☐ OPTIONAL	CENTER-TO-CENTER	☐ PICA TYPE
	☐	CENTER TO EDGE	☐ OTHER:

CARBONS: ☐ NONE ☐ ONE-TIME ☐ REGULAR ☐ HECTO ☐ ______ ☐ TYPEWRITER ☐ PENCIL ☐ BLACK ☐ BLUE ☐ ______ ☐ ______

NUMBERING: ☐ NONE COLOR OF PRINT ☐ BLACK ☐ RED ☐ ______ ☐ OPTIONAL RANGE OF SERIAL NUMBERS LOWEST NO. ______ HIGHEST NO. ______

ADDITIONAL PROVISIONS:

DESIGN: ☐ IN ACCORDANCE WITH ATTACHED DRAWING ☐ IN ACCORDANCE WITH ATTACHED SAMPLE FORM ☐ AS PREVIOUSLY PRINTED OR ☐ AS MARKED UP

PACKAGING: WRAP ______ UNITS ☐ PADS ☐ SHEETS ☐ SETS PER PKG MARK THE SMALLEST END OF EACH PACKAGE WITH THE FOLLOWING INFORMATION: FORM NUMBER, QUANTITY, AND WHEN APPLICABLE THE RANGE OF THE SERIAL NUMBERS COVERED BY THE CONTENTS.

CARTONS CONTAINING PACKAGES MUST NOT BE LARGER THAN 13″ x 16″ x 19½″

PROOF: ☐ REQUIRED ☐ NOT REQUIRED PAPER AND CARBON DUMMY: ☐ REQUIRED ☐ NOT REQUIRED WE REQUIRE THREE SAMPLE FORMS DELIVER TO: ______ ______

K-9723F

Figure 14-5. Printing specifications sheet for forms.

it provides legible copies. In soliciting a quotation the request for the dummy can be stipulated on the printing specifications sheet.

So far the printing specifications sheet has been mentioned only with respect to events leading up to the placement of an order. When an actual order is placed this document may serve another very useful purpose; it can then be referenced on the purchase order so that the printer is obligated to produce the form in accordance with its requirements.

As part of the contractual requirements the printer may be instructed to submit a proof of the form before making the run. Time permitting, it is wise to request proofs, especially where a new form is concerned. The proof is a trial impression of the form and should be a good replica of what the printed form will look like in regard to shape, size, typefaces, and rules.

Upon receipt the proof should be verified by comparing it, item for item, with a copy of the model form sent to the printer. Should changes or corrections be in order, they may be marked directly on the proof itself or indicated on a separate sheet of paper and attached to the proof.

After the proof has been checked it should be returned to the printer promptly. Any changes affecting the proof will delay the delivery of the forms. It is therefore advisable, in the case of a reorder, to check the quantity of forms on hand to determine if the supply is adequate for interim requirements. In instances where the supply will run short or where a much needed new form will be delayed, it is often necessary to provide temporary forms for the intervening period.

STOCKROOM-CONTROL PROGRAM

For receiving, storing, and distributing forms and stationery supplies it is customary to have a centralized stockroom that services all organizational units. An adequately supplied stockroom in a medium or large company may have several hundred different forms in stock. Of major concern is to have a sufficient supply of each type on hand at all times without overstocking forms that may soon become outmoded.

The responsibilities of the stationery storekeeper normally include (*a*) checking and putting away forms received from both vendors and the internal duplicating department, (*b*) giving notice to appropriate departments and persons of the arrival of ordered forms, (*c*) systematic storage of forms with provision for their safekeeping, and (*d*) the issuance of forms on the basis of authorized requisitions.

Forms should be wrapped, labeled, and stored to provide protection, ready identification, and easy access. A good stockroom-control program begins with the placement of the order. Provision can be made at that time to have the forms wrapped in convenient quantities with packages appropriately marked as to content. Reference to the printing specifications sheet shown in Figure 14-5 reveals a "packaging" section for making such requirements known to vendors.

Routine for Drawing and Replenishing Purchased Forms

The procedures for drawing and replenishing forms revolve around three basic records; that is, the stores requisition, the perpetual inventory card, and the purchase requisition.

Stores requisitions differ in details but basically they show the name of the person and/or department needing the form, date of preparation, description of the form, quantity desired, and the signature of the individual authorizing withdrawal from stores. As an additional feature they commonly have preprinted consecutive numbers for control purposes.

In general practice the stores requisition is prepared in duplicate. The originator retains one copy, and the other is forwarded to the stockroom.

Four classes of information are found on inventory cards: (*a*) the name and number of the form (one card for each form having a separate base number is usually best), its description, and location; (*b*) rate of usage (this is ordinarily given on either a monthly or quarterly basis); (*c*) cost data; and (*d*) inventory figures. The inventory account shows a continuous record on the receipt of forms, quantities placed on order, disbursements made, and the balance on hand. Each entry should be recorded with the date and the control number of the document causing the entry to be made.

The well-designed inventory card has a space for inserting a quantity known as the "reorder point." When the level of forms reaches this point it is a signal to process a requisition. Arrangements may be made to route purchase requisitions via the forms-control unit, the purpose being to permit the forms-control unit to approve or disapprove a repetitive order.

In establishing the reorder point it is necessary to consider the rate of usage, time required for making the purchase, delivery schedules, and quantities that are purchasable in economical lots. To these considerations we need to add another: Sufficient lead time should be allowed to permit the forms-control unit to notify users of the form that it is about to be reordered, so that those concerned have an opportunity to make recommendations for its improvement, as necessary.

Routine for Drawing and Replenishing Internally Produced Forms

Forms produced on duplicating equipment in the company's own printing shop are also stored for subsequent release.

The paper-work routine for handling the drawing and replenishing of internally produced forms is much like that for purchased forms. The same stores requisition may be used, and the perpetual-inventory cards for both may have the same general format. If desired, two different card colors may be used to differentiate between forms printed in the company's own shop and those that are purchased.

What has been said about the reorder point applies basically to internally produced forms as well. It is customary for the storekeeper to send a production requisition to the internal duplicating department. This requisition designates the printing specifications and authorizes the duplicating department to produce the quantity noted. Here again the production requisitions may be routed to the duplicating department via the forms-control unit. Requiring the forms-control unit to check the merits of requisitions helps establish a tight and purposeful control.

Variations in Routines

Not all stockroom-control programs are carried out as explained. There are various modifications due to organizational patterns, preferences as to controls, or any number of individual factors. The forms-control plan outlined in Figure 14-6 presents some procedures that are slightly different and are adapted to suit a particular enterprise.

OBSOLESCENCE OF FORMS

The disposal of forms is necessary because of deterioration, obsolescence through revisions, and outright elimination without the need for replacement.

Obsolete forms, like many other things of the past, are easily forgotten, put aside, and allowed to accumulate. This is why centralized stockrooms and departmental cabinets become filled with forms of so many yesteryears. The situation is not uncommon and is fairly prevalent throughout industry. Procedures need to be instituted for ferreting out, handling, and discarding obsolete forms.

Two approaches are desirable: First, provision should be made for supervisors and other management personnel to advise the forms-control

Standard Practice Instructions

Title: Forms Control S.P.I. No. 42
Rev. A
Issued: 9-7-XX Page 1 of 2 pp.

1. Purpose

1.1 To provide procedures for controlling the design, requisitioning, purchasing, and storing company forms.
1.2 To establish the responsibility of the Systems-and-Procedures Department of the Eastern Division for company forms with respect to
1.2.1 Basic requirement for the form as an essential part of approved operating procedures.
1.2.2 Principles of good forms design.
1.2.3 Economy in method of production and in quantity ordered.
1.3 To establish the responsibility of Stationery Stores for
1.3.1 Maintaining an adequate supply of each form stocked in accordance with the provisions of this instruction.
1.3.2 Ordering replenishments of stock in time to allow Purchasing to obtain favorable prices.
1.4 Forms that are processed entirely within a single department, have a low volume of activity, and are reproduced on company-operated printing facilities are exempted from the provisions of this instruction.

2. Requests for New or Revised Forms

2.1 Requests for new or revised forms will be sent to Systems and Procedures with a sketch or marked-up copy showing constant and variable data, and other information regarding the requirements of the form and the monthly usage.
2.2 To provide sufficient lead time for processing form revisions all proposed changes should be sent in promptly. Do not wait until the form comes up for reorder.

3. Procurement of New or Revised Forms

3.1 Systems and Procedures will investigate the use of the form; if approved, will recommend whether to make or buy and forward approved sketch or marked-up copy to Stationery Stores, with specifications and proposed

Figure 14-6. Written procedures covering forms control.

Standard Practice Instructions

S.P.I. No. 42 Rev. A Page 2 of 2 pp.

order quantity.

3.2 Stationery Stores will place a requisition for quantity required upon Purchasing or company-operated facilities.

3.3 Purchasing and all company-operated printing facilities will not accept requisitions for printed company forms which lack the approval of Stationery Stores.

4. Storage, Requisitioning, and Reorder of Forms

4.1 Forms with relatively heavy usage, or usage by more than one department, will be stored by Stationery Stores. The supervisor of Stationery Stores will determine whether usage is heavy or light.

4.1.1 Requisitioning of all forms from Stationery Stores will be made on Supply Requisition, Form No. 86–B–2, as provided in Standard Practice Instructions (S.P.I.) No. 41.

4.2 Forms with relatively light usage will be stored by the using department, except that a minimum quantity of all outside-purchased forms will be stored in Stationery Stores.

5. Records

5.1 Stationery Stores will maintain a stock record card for all company forms under its control, showing minimum/maximum quantity, as established by Supervisor of Stationery Stores on the basis of usage, with minimums set at a point that will allow sufficient lead time for Purchasing to obtain favorable prices.

5.2 Stationery Stores will maintain a Purchase Requisition Traveler for each company form, to be processed in accordance with S.P.I. No. 87.

5.3 Stationery Stores will maintain a Numerical Forms Register for issuance of form numbers.

5.4 Systems and Procedures will maintain a file of all company forms, arranged by functional classification.

6. Disposition of Obsolete Forms

6.1 When a company form has been revised or superseded Systems and Procedures will amend any S.P.I.'s that are affected and will advise Stationery Stores of the disposition of any obsolete stock.

Figure 14-6. Written procedures covering forms control.—Continued.

unit when a form has become obsolete. This method is helpful but not always reliable. The more reliable approach is to see that appropriate members of management are periodically given a list of existing forms (usually once or twice a year) and asked to indicate their status as "in use" or "obsolete."

Once a form has been declared officially obsolete, the forms-control unit should notify all affected departments and agencies to disregard it. If the form has salvage value, notification may be given to forward all supplies of the form to a central location for special handling. It should be noted that most carbon-interleaved forms do not have any reclamation value; they cannot even be sold as waste paper.

When the occasion is one of an internally produced form having gone out of use, further steps are called for. The duplicating department should be instructed to destroy affected masters—hectograph originals, mimeograph stencils, or paper offset plates. Obsolete zinc offset plates should be marked for reclamation. Photographic negatives used in the offset process may be forwarded to the forms-control unit and filed in the inactive forms file as part of a historical and permanent record.

15

Manuals and Written Procedures

A systems-and-procedures manual may be viewed as a collection of instructions for the purpose of communicating information on policies and practices. Manuals contain facts and data such as ground rules, operational procedures, directives, and, instructions. Members of management on almost every level of performance will find it helpful to have clear-cut policies and procedures written down. If supervisors and staff personnel are to accomplish work efficiently, they must know what is expected of them. Manuals go a long way in this direction.

Company size and complexity will dictate the need for the number of manuals needed. The small-size company may find a single manual sufficient for the purposes intended. Sections of it may be devoted to accounting, production, marketing, and quality control. At the other end of the spectrum is the large firm with at least one manual for each of these fields; a "controlling," or interdepartmental, manual; plus, perhaps, a few additional ones. Separate industrial-relations manuals, specialized engineering manuals, and even legal manuals are not uncommon in giant corporations.

HOW MANUALS CONTRIBUTE TO THE EFFECTIVENESS OF AN ENTERPRISE

The principal benefits to be derived from manuals may be summarized as follows:

1. They function as a work guide that is especially useful for obtaining information. The executive, administrator, and supervisor are freed from having to know countless details.

2. They provide clear expressions of policy, authority, and responsibility in matters relating to operational practices.

3. They disseminate information to all concerned, with certainty.
4. They ensure that like operations will be handled in a uniform manner wherever desired.
5. They make smooth transitions possible when changes in personnel occur.
6. They promote accuracy and understanding. To formulate a written instruction it is first necessary to follow a thinking-writing process that is of itself constructive and clarifying. Putting things in writing helps to generate ideas, eliminate vagueness, and fuse together the chains of thought. It keeps the writer on the path of logical thinking. Also, whatever is put into writing is subject to scrutiny. Any omissions or errors that may be contained in the written procedures will be brought to the surface for correction.
7. They facilitate the orientation and training of personnel. Correlatively, they save the time of supervisors that would otherwise be spent in making explanations.
8. They promote the concept of management by exception, with attention to the exception—or unusual, rather than to routine, matters.
9. They induce critical appraisals that frequently lead to new or improved operating conditions.
10. They serve as a coordinating agency through which activities are integrated into meaningful wholes.

CLASSIFICATION AND TITLES OF MANUALS

In a broad sense company manuals may be classified as interdepartmental, departmental, and special. Within these categories, some common titles of representative handbooks are as follows:

I. Interdepartmental (General) Manuals
   A. Organization Manual
   B. Standard Policies and Procedures Manual
      Alternative titles for the same manual are
      1. Interdepartmental Instructions Manual
      2. Standard Operating Procedures Manual
      3. Standard Practices Instruction Manual
      4. Manual of Standard Practices

II. Departmental Manuals
   A. Engineering Instructions Manual (Administrative)
   B. Accounting Manual
   C. Manual of Quality-Control Orders

- D. Industrial Relations Manual
- E. Production Control Manual
- F. Sales Manual

III. Special Manuals

- A. Parts Manual
- B. Manual of Instructions on Equipment

Our interest will center mainly on manuals containing written procedures, such as standard policies and procedures manuals, and certain of the various departmental manuals. Much of the discussion that follows is of a general nature and applies to manuals of every kind and description.

MANUALS OF WRITTEN PROCEDURES

Physical Features of the Manual

The ideal binder for a manual is attractive, strongly made, and durable. Simple construction is best; it should permit changes in content with ease and convenience. Avoid the type of contrivance that constitutes a challenge when it comes to inserting or removing pages. Such binders are more than a matter of annoyance; they are the cause of written procedures not being placed in them and piling up on desks instead (in general), thus negating the utility of the entire plan.

The physical form and size of the manual is important. A good manual has a heavy stiff cover measuring approximately 9¾ by 11½ inches, an outside surface of durable leather substitute, and carries standard 8½- by 11-

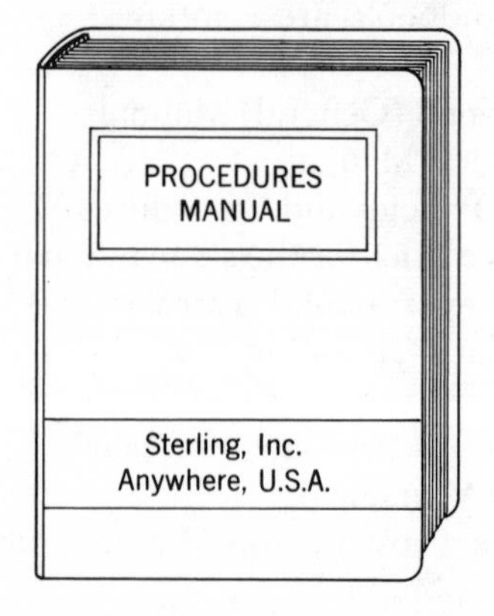

Figure 15-1. Engraved binder.

inch paper on three snap rings. An authoritative-looking book is an added attraction. For this reason some companies go to the trouble and minor expense of having the title of the manual and the name of the concern engraved on the cover. (See Figure 15-1.) It is an effective touch and worth considering. The title should be brief but clear; lettering and any other art work should be correctly designed.

Contents of the Manual

The main contents of a systems-and-procedures manual may be outlined as follows:

1. Preface
2. Numerical table of contents
3. Alphabetical index
4. Written procedures

Preface. The preface introduces the entire manual to the reader. It is customary to stress the importance of abiding by the written procedures, to relate the role of the individual in carrying them out, and to issue an appeal for cooperation and loyalty. All of this should be stated in general terms, kept brief and to the point. Exaggeration is to be avoided; it is essentially a destructive quality, burying the message and tiring the reader.

To be of maximum value a manual must have the full support of management. For that reason the preface usually gives official sanction to the manual, and is commonly and properly presented over the signature of the executive in charge.

Numerical Table of Contents. The contents of the manual is an orderly arrangement of the titles of the procedures in accordance with a numbering scheme. (See Figure 15-2.) The number indicates the general location of a procedure in the manual.

Ordinarily the topics covered develop out of the needs of the enterprise and the situation. Systems-and-procedures manuals represent long-term preparation. When a written procedure is determined to be necessary it is prepared and included in the manual.

Alphabetical Index. Systems-and-procedures manuals contain an alphabetical index to enable the reader to readily locate specific information (Figure 15-3). The prevailing practice is to place the index at the back of nonfiction books, but it is quite normal to place the index at the front in the case of systems-and-procedures manuals. Book construction accounts for the difference. Indices in the back of loose-leaf binders can be difficult to utilize; and, in the process of using them, there is a tendency to cause excessive wear and tear on the contents.

MARK-WELL DIVISION
THE ILLUSTRATIVE CORPORATION

STANDARD POLICIES AND PROCEDURES
CORPORATE OFFICES

SPP NO. 01–001R PAGE 1 OF 3
RELEASED August 1, 19XX
EFFECTIVE Immediately
CANCELS SPP No. 01–001Q

TITLE: Numerical Table of Contents

PURPOSE: To provide a numerical outline for facilitating reference to the titles and dates of issuance of Standard Policies and Procedures (SPP's)

No.	Rev.	Title	Released
01—FRONT MATTER			
01–001	R	Numerical Table of Contents	8–1–XX
01–002	N	Alphabetical Index	6–1–XX
01–005	C	Publication and Distribution of Standard Policies and Procedures	5–8–XX
02—GENERAL			
02–001	A	Retention of Records	9–4–XX
02–002	A	Transactions Between Non-purchasing Department Personnel and Vendors	9–15–XX
02–005	A	Safeguarding "Trade Secrets"	3–15–XX
02–007	A	Forms Control	8–11–XX
02–009	B	Obtaining the Services of Temporary Personnel from Outside Agencies	6–1–XX
02–012	C	Vacation Allowances	7–14–XX
02–015	A	Leaves of Absence	10–29–XX
02–018	A	On the Preparation and Handling of General Correspondence	5–15–XX
02–020	C	Putting Decisions in Writing	8–13–XX
02–022	E	Permanent Records	5–12–XX
02–025	B	Control of Office Supplies	8–20–XX
02–026	D	Suggestions, Assessment, and Treatment of	2–9–XX
02–029	A	Patent Applications	1–16–XX
02–031	A	Subscriptions to Periodicals—Trade Magazines, Newspapers, Professional Publications	4–2–XX
02–034	D	Security Regulations	4–28–XX
02–035	B	Travel Authorizations	6–10–XX
02–038	C	Property Passes	11–5–XX
02–040	F	Reporting Illnesses	5–7–XX

03—ACCOUNTING AND FINANCIAL			
03–001	A	Credit Cards	2–29–XX
03–009	A	Cash Advances	2–20–XX
03–012	B	Expense Reports	5–13–XX
03–019	A	Withholding Deductions	1–17–XX
03–100	B	Stock Purchase Plans	7–19–XX
03–112	B	Payroll Classifications	9–8–XX
03–117	C	Salary Structure	8–15–XX
03–121	A	Control of Capital Equipment and Facilities	8–16–XX
03–123	D	Overtime Payment	4–1–XX
03–127	C	Changes in Salary	4–12–XX
03–128	C	Material Damaged in Transit	9–20–XX
03–140	B	Interplant Transfer of Material	11–23–XX
03–145	B	Recording Scrap and Rework Charges	6–7–XX
04—ENGINEERING			
04–001	B	Project Orders, Engineering	11–20–XX
04–008	D	Review and Approval of Technical Material for Verbal Presentation or Publication	4–21–XX
04–011	A	Reviewing and Processing Ideas for Patents	2–2–XX
04–014	A	Handling Unsolicited Ideas for Inventions	6–1–XX
04–119	C	Progress Reports	7–27–XX
04–121	F	Issuance of Change Orders and Product Modifications	7–8–XX
04–138	B	Origination and Processing of Product Design Changes	3–2–XX
04–165	D	Preparation of the Confidential Payroll	5–1–XX
04–187	B	Annual Physical Inventory	2–19–XX
05—MARKETING			
05–001	H	Product Sales Code	8–1–XX
05–019	C	Sales-Order Processing	3–23–XX
05–042	B	Consignments to Customers	4–8–XX
05–087	D	Samples of Products	6–1–XX
05–110	G	Quotations	9–10–XX
05–115	I	Proposals	8–20–XX
05–121	F	Service Guarantees	11–17–XX
06—PRODUCTION			
06–008	A	Requisitioning Tools, Dies, Jigs, Fixtures, and Gages	5–8–XX
06–020	C	Operation Sheets	6–16–XX
06–027	B	Manufacturing Drawings	8–11–XX
06–062	D	Care and Maintenance of Test Equipment	2–25–XX
06–148	G	Safety and Health	5–28–XX
06–197	I	Training New Operators	10–7–XX
06–208	A	Storage of Tools and Fixtures	6–17–XX
06–214	G	Production Orders	4–4–XX
06–217	A	Trouble Reports	7–18–XX
06–225	D	Work Numbers	9–13–XX
06–234	B	Control of Customer-Owned Property: Tools, Jigs, Fixtures, Gages, and Test Equipment	4–29–XX

Form No. 438–A

Figure 15-2. Table of contents on an interdepartmental manual.

MARK-WELL DIVISION
THE ILLUSTRATIVE CORPORATION

STANDARD POLICIES AND PROCEDURES — SPP NO. 01–001R PAGE 3 OF 3

07—INSPECTION AND QUALITY CONTROL			
07–001	B	Quality-Control Orders	7–12–XX
07–002	I	Incoming Inspection	1–14–XX
07–003	A	Blueprint Interpretations	6–2–XX
07–024	A	Gage Control	7–30–XX
07–034	A	Source Inspection	1–23–XX
07–048	A	Magnetic Particle Inspection	6–1–XX
07–078	B	Quality Standards	8–18–XX
07–080	E	In-Process Inspection	9–13–XX
07–083	B	Control of Inspection Stamps	7–8–XX
07–085	C	Materials Review	8–2–XX
07–091	A	Return-to-Vendor Material	10–1–XX
07–204	C	Scrap Material	11–6–XX
07–205	C	Rework Material	8–1–XX
07–206	D	Final Inspection	3–17–XX
07–207	F	Salt Spray Test	3–21–XX
07–209	A	Stop Orders	2–4–XX
07–210	A	Welder Certification	12–16–XX
07–222	C	Classification of Defects	2–27–XX
07–224	C	Sampling Inspection	4–23–XX
08—TRAFFIC			
08–006	C	Overshipments and Undershipments	9–4–XX
08–008	A	Returning Reusable Containers to Vendors	5–29–XX
08–010	A	Payment of Freight Bills	8–6–XX
08–011	C	Material Damaged in Transit	8–23–XX
09—PRODUCTION CONTROL			
09–014	C	Withdrawal of Production Material from Stores	1–12–XX
09–022	C	Numbering Parts	6–2–XX
09–038	C	Drawing Changes	7–30–XX
09–104	H	New and Changed Parts Notices	1–23–XX
09–116	D	Stores Records	6–1–XX
09–204	K	Bills of Material	2–28–XX
10—PURCHASING			
10–001	D	Preparation of Purchase Requisitions	9–29–XX
10–003	B	Issuance of Purchase Orders	8–8–XX
10–007	A	Renegotiable Contracts	6–29–XX
10–011	A	Purchasing from Other Divisions of the Corporation	5–15–XX
10–028	A	Numbering System for Purchase Orders	9–2–XX
10–040	D	Return of Discrepant Material to Vendors	12–21–XX
10–041	F	Rework of Vendor Material	2–4–XX
10–062	A	Cash Purchases	7–19–XX

Form No. 438–B

Figure 15.2 *Continued*

In preparing the index remember that its purpose is to assist the reader in locating information. Visualize yourself in his position. Ask repeatedly: What topics should be indexed? Under what lead-off word or significant phrase will the topic be sought? As you proceed, shape and evaluate in your own mind the various elements that will go into the make-up of the index.

Written Procedures. The sinew of systems-and-procedures manuals are the written procedures themselves. This leads into a discussion of their shape, size, and general composition.

FORMAT OF WRITTEN PROCEDURES

General Considerations

In the preparation of written procedures the following suggestions are recommended:

1. Use a preprinted form when feasible. Such forms reduce copying, insure uniformity of appearance, and give an impression of authenticity. A good technique is to have the basic composition printed on offset masters such as the Multilith. It is then practical to have the variable information, including the procedures themselves, typed directly on the form-master on a typewriter equipped with an offset (Multilith) ribbon. Copies made in this way have a professional appearance, and if the form itself is well designed, it can be aesthetically satisfying.
2. Use a good grade of paper of suitable weight and strength. Shun the cheaper grades of duplicating paper. A bleached white paper with a regular finish is best. (*Note:* using both sides conserves paper, making manuals less bulky and easier to handle.)
3. Have lettering that is attractive and easy to read. For any one manual it is best to use typewriters with the same typeface and style. A popular face is the elite, which is somewhat small but has the advantage of 12 characters to the inch. Pica rates higher for readability, but it permits only 10 characters to the inch. Whichever face is used, and there are a number of fine new styles, it is wise to employ both upper and lower case letters. Capitals make excellent headings, but when used exclusively they result in hard-to-read copy.
4. Leave a liberal amount of white space. Keep ample margins and provide for double spacing between sections and paragraphs.

Parts of Written Procedures

In describing written procedures there are three major parts to be considered:

MARK-WELL DIVISION
THE ILLUSTRATIVE CORPORATION

STANDARD POLICIES AND PROCEDURES
CORPORATE OFFICES

SPP NO. 01–002N PAGE 1 OF 3
RELEASED June 1, 19XX
EFFECTIVE Immediately
CANCELS SPP No. 01–002M

TITLE: Alphabetical Index

PURPOSE: To provide an alphabetical outline for facilitating reference to numbers and locations of Standard Policies and Procedures (SPP's)

Topic	*No.*
Absence, leaves of	02–015
Applications for patents	02–029
Authorizations, travel	02–035
Bills of material	09–204
Blueprints, interpretation of	07–003
Cash advances	03–009
Cash purchases	10–062
Certification, welder	07–210
Change orders, issuance of	04–121
Changed parts notices	09–104
Changes, drawing	09–038
Codes	
Numbering system for purchase orders	10–028
Product	09–001

Topic	*No.*
Consignments to customers	05–042
Containers, return of	08–008
Contracts, renegotiable	10–007
Control of	
Capital equipment and facilities	03–121
Customer-owned property	06–234
Gages	07–024
Inspection stamps	07–083
Office supplies	02–025
Correspondence, general	
Preparation and handling of	02–018
Credit cards	03–001
Decisions, recording	02–020
Deductions, withholding	03–018
Defects, classification of	07–222

Topic	*No.*
Design changes (See "Product design changes")	
Drawing changes	09–038
Drawings, manufacturing	06–027
Equipment and facilities, capital control of	03–121
Equipment, test	
Care and maintenance of	06–062
Expense reports	03–012
Final inspection	07–206
Forms control	02–007
Freight bills, payment of	08–010

Gage control 07–024
Guarantees, service 05–121

Health, safety and 06–148

Illnesses, reporting of 02–040
Incoming inspection 07–002
In-process material, inspection of 07–080
Inspection
  Final 07–206
  Incoming 07–002
  In-process 07–080
  Magnetic particle 07–048
  Sampling 07–224
  Source 07–034
  Stamps, control of 07–083
Interplant transfer of material 03–140
Inventions, handling unsolicited
  Ideas for 04–014
Inventory, annual physical 04–187
Issuance of
  Change orders and product modifications 04–121
  Purchase orders 10–003

Magnetic particle inspection 07–048
Manufacturing drawings 06–027
Material, interplant transfer of 03–140
Material, vendor
  Damaged in transit 03–128
  08–011
  Return of discrepant 07–091
  10–040
  Rework of 07–205
  10–041
  Scrap 07–204
Material review 07–085

New and changed parts notices 09–104
Numbering parts 09–022
Numbers, work 06–225

Office supplies 02–025
Operation sheets 06–020
Orders
  Production 06–214
  Quality control 07–001
  Stop, quality control 07–209
Overshipments and undershipments 08–006
Overtime payment 03–123

Parts, numbering of 09–022
Passes, property 02–038
Patent applications 02–029
Patents, reviewing and processing ideas for 04–011
Payment of freight bills 08–010
Payment for overtime 03–123
Payroll classifications 03–112
Payroll, confidential, preparation of 04–165
Permanent records 02–022
Personnel, temporary, obtaining the services of 02–009
Processing of sales orders 05–019
Product design changes, origination and processing of 04–138
Product sales code 05–001
Production material, withdrawal of 09–014
Production orders 06–214
Products, samples of 05–087
Project orders, engineering 04–001
Proposals 05–115
Publication of technical material 04–008
Purchase orders, issuance of 10–003
Purchase requisitions, preparation and processing of 10–001
Purchases, cash 10–062
Purchasing from other divisions of the corporation 10–011
Putting decisions in writing 02–020

Quality-control orders 07–001
Quality standards 07–078
Quotations 05–110

Recording scrap and rework charges 03–145
Records

Form No. 439-A

Figure 15-3. Alphabetical index of an interdepartmental manual.

MARK-WELL DIVISION
THE ILLUSTRATIVE CORPORATION

STANDARD POLICIES AND PROCEDURES

Topic	*No.*	*Topic*	*No.*	*Topic*	*No.*
Permanent	02–022	Sales-order processing	05–019	Training new operators	06–197
Retention of	02–001	Salt spray test	07–207	Travel authorizations	02–035
Stores	09–116	Samples of products	05–087	Trouble reports	06–217
Reports		Sampling inspection	07–224		
Expense	03–012	Scrap and rework charges, recording of	03–145	Undershipments, overshipments and	08–006
Progress	04–119	Security regulations	02–034		
Trouble	06–217	Service guarantees	05–121	Vacation allowances	02–012
Requisitioning tools, dies, jigs, fixtures, and gages	06–008	Sheets, operation	06–020	Vendors, transactions with	02–001
Requisitions, purchase, preparation of	10–001	Source inspection	07–048		
Returning reusable containers to vendors	08–008	Standards, quality	07–078	Welder certification	07–210
Review and approval of technical material for verbal presentation or publication	04–008	Stock purchase plans	03–100	Withdrawal of production material from stores	09–014
Rework and scrap charges, recording of	03–145	Stop orders, quality control	07–209	Work numbers	06–225
Rework of vendor material	07–205 10–041	Storage of tools and fixtures	06–208		
Safeguarding "trade secrets"	02–005	Stores records	09–116		
Safety and health	06–148	Subscriptions to periodicals	02–031		
Salary, changes in	03–127	Suggestions, assessment and treatment of	02–026		
Salary structure	03–117	Temporary personnel, obtaining the services of from outside agencies	02–009		
Sales code, product	05–001	Test equipment, care and maintenance of	06–062		
		Tools and fixtures, storage of	06–208		
		"Trade secrets," safeguarding	02–005		

Form No. 439-B

Figure 15-3. *Continued*

1. Heading(s)
2. Body
3. Authorization and approval section

Heading(s). A written procedure may have two headings: a full heading for the first page and an abbreviated one for succeeding pages. Typical of the information found on a complete heading is the following:

1. Company identification. The heading usually gives the full name of the company. Since many business firms have trademarks and insignias these may also be included at the top of the form.
2. Name of the (manual) set of procedures of which it is part. The ideal designation is short and expressive, has a pleasing ring to the ear, and can be conveniently abbreviated. Standard practice instructions, called SPI's, standard policies and procedures, called SPP's, and interdepartmental instructions, called II's, are all titles that meet these specifications.
3. Title of the procedure. The title should be meaningful and distinctive. It should suggest the nature and purpose of the contents.
4. Purpose. This calls for a clear-cut statement of purpose to provide the reader with a summary telling of what the procedure is about. One or two well-put sentences is all that is necessary.
5. Reference. Some procedures carry a "reference line," which directs the reader to one or more other documents. Subjects for possible reference include related procedures, associated documents, and pertinent memoranda such as might have initiated, authorized, or suggested the action.
6. Number. Procedural numbers are commonly expressed in three sections: a category, or group, number; a sequential number; and a revision designation. An example is given below.

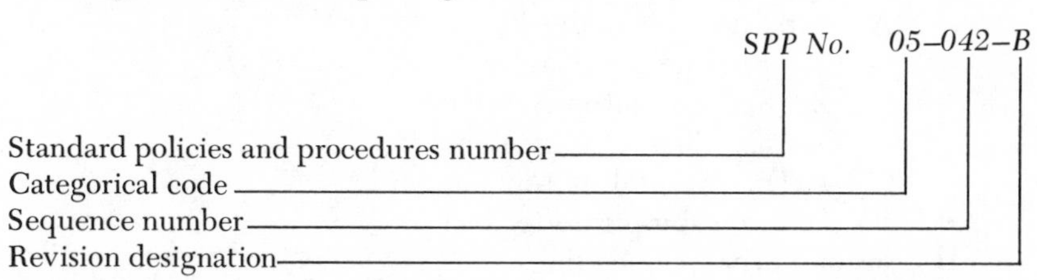

Either letters or numbers may be used to indicate revisals. Single letters, A through Z, will serve for 26 revisions, and thereafter double letters may be used; for example, AA. Numbers, of course, may be used ad infinitum.

7. Page identification. In order that written procedures be easily collated, filed, and consulted it is necessary to number the pages. Ways of

doing this are: "Page______of______pp." and "Page______." The second method is meant to place the number of the page in a more conspicuous location, toward the edge of the sheet.

8. Date of release. This is the date the procedure is issued or distributed.

9. Effective date. This is the date the procedure becomes law. It may or may not be the same as the date of issuance. When the dates coincide it is customary to indicate "immediately" as the date the procedure goes into effect. (*Note:* Once the method of writing dates has been decided on, such as 12–12–XX, 12/12/XX, or Dec. 12, 19XX, be consistent and always use the same method on all procedures of the same set.)

10. Canceling information. So that the recipient will know which procedures are canceled or superseded, space needs to be provided for the necessary entry. Usually a new procedure replaces another in the manner of a regular revision. However, a new procedure may cancel not only a preceding one with the same basic number but also others that have different basic numbers. An entry such as "Cancels SPP Nos. 05–368–D, 07–004–B, and 10–111–A" is not unusual.

There are many plans for arranging the elements of headings. Figures 15-2 and 15-3 show typical preprinted forms. Notice that pages other than the first contain abbreviated but mutually related headings. However, a single well-arranged compact heading may be used for all pages of procedures and is even preferred by some.

Body. The nine major factors to be considered here are as follows:

1. *Methods of outlining.* Two methods are successfully adapted to the outlining of subject matter. The following arrangement, an alphanumeric combination, is the customary one:

I. First principal heading
  A. First subtopic under I
    1. First subtopic under A
    2. Second subtopic under A
      a. First subtopic under 2
      b. Second subtopic under 2
  B. Second subtopic under I
II. Second principal heading
  A. First subtopic under II
  B. Second subtopic under II
III. Etc.

The other method, not nearly so widely used but still popular, employs digits and decimal points. No letters or Roman numerals are used. The pattern is called *decimal-numeric* and is developed along these lines:

1. First principal heading
    1.1 First subtopic under 1
    1.2 Second subtopic under 1
        1.2.1 First subtopic under 1.2
        1.2.2 Second subtopic under 1.2
            1.2.2.1 First subtopic under 1.2.2
            1.2.2.2 Second subtopic under 1.2.2
    1.3 Third subtopic under 1
2. Second principal heading
    2.1 First subtopic under 2
    2.2 Second subtopic under 2
3. Etc.

Observe that the subdivisions undergo a series of indentions in both cases. However, a ramification of the second method eliminates the necessity for beginning subtopics further to the right than main topics; it takes the following form:

1.0.0.0 First principal heading
1.1.0.0 First subtopic under 1.0.0.0
1.2.0.0 Second subtopic under 1.0.0.0
1.2.1.0 First subtopic under 1.2.0.0
1.2.2.0 Second subtopic under 1.2.0.0
1.2.2.1 First subtopic under 1.2.2.0
1.2.2.2 Second subtopic under 1.2.2.0
1.3.0.0 Third subtopic under 1.0.0.0
2.0.0.0 Second principal heading
2.1.0.0 First subtopic under 2.0.0.0
2.2.0.0 Second subtopic under 2.0.0.0
3.0.0.0 Etc.

This type of classification is very familiar to persons who work with technical specifications (such as engineers, technicians, and mechanics). It has the advantage of saving space. However, indentions provide white space, which enhances readability.

Actually, all three of the arrangements shown have commendable separate qualities, and each is suitable for the presentation of procedures in written form. The selection is more a matter of personal preference than anything else. Once a choice has been made it should be used consistently throughout the manual.

Regardless of the outlining method used, topics of the same rank should be listed in the same manner. Subordinations are rarely needed beyond the fourth division.

2. Main headings. Headings of the first magnitude should be de-

signed to attract attention. This is done effectively by making them short, printing the words in capital letters, and underlining them for emphasis. Putting an occasional heading in the form of a question is also effective; it helps break up monotony by calling attention to the subject matter.

3. Additions and changes. Whenever practical, revisions to existing procedures may be designated by one or more brackets in the left-hand margin. Economy of effort is thus made possible: Readers who are familiar with the former instruction can skim over the old material and focus their attention on whatever additions and changes are involved.

4. Abbreviations and condensations. Long names in procedures need be written out only once. Thereafter it is permissible to express them in the form of condensations or abbreviations. Thus, writing, "Duplicating and Photocopying Request (Form 4691–C)" one may simply refer to "Form 4691–C"; similarly, "the National Association of Accountants (NAA)" can become "the NAA."

Many relatively short names may also be abbreviated; for example, "P.O." for "Purchase Order," "A/R" for "Accounts Receivable," and "Q.C." for "Quality Control."

Whatever the length of the full name, the standard rule is to place the abbreviation in parentheses following the first use of the term so that in all subsequent applications its meaning will be clear.

5. Formal definitions. As Homer put it, ". . . words are shifty, multiple in meaning, and dangerous." To prevent misunderstandings of a semantic nature terms whose meanings are in need of clarification should be defined. Three parts must be given: the term itself, the class to which it belongs, and the particular characteristics that give it individuality within the class.

Normally, a satisfactory definition can be set forth in one or two sentences. Bear in mind that it is best presented before entering into a detailed discussion of the subject matter to which it applies.

6. Visual aids. Graphs, flow charts, and specimen copies of forms are visual aids that can be advantageously included in manuals of written procedures. Modern reproduction methods duplicate them easily and clearly.

The introduction of visual aids into procedures depends on the preference of the systems analyst who prepares the instruction. Considerable ingenuity may be exercised by him in this matter. Worthy of special mention is the practice of placing encircled numbers within the spaces on a form, duplicating this image as a part of the procedure, and cross-referencing the numbers to explanatory material in the text. The technique is especially good where complex procedures are involved. An alternative method is to show segments of the form in conjunction with the development of the written material.

7. Paragraphs. Experience shows that short paragraphs make for easier reading than long ones. Procedural paragraphs should therefore be short or at least of moderate length. The *ideal* paragraph may be thought of as containing four or five sentences of no more than 14 to 16 words each.

8. Language. The language of procedures should possess essentially the same attributes as those that are laid down for preparing reports. Written procedures should be readable, realistic, informative, concise, direct, clear, truthful, and timely. (See Chap. 24.)

9. Titles of personnel. A prudent measure is to avoid the use of personal names in written procedures. Titles are used instead. Then, if individuals change jobs, the currency of a procedure is not impaired.

Authorizations and Approvals. Official standing is given to a written procedure when it is signed by the person who produced it and/or the person or persons charged with authorizing its use. Interdepartmental procedures are commonly signed by the systems analyst and countersigned by one or more other persons, such as the systems-and-procedures manager and the comptroller. In some companies such procedures are released over the signature of only one person, usually either the systems-and-procedures manager or a vice-president.

Departmental procedures normally call for the signature of the systems analyst and the department head. Here, also, variations are possible, and one signature may be all that is required.

Some companies make it a policy for both interdepartmental and departmental procedures to require approval signatures from all responsible persons directly concerned with the material covered. In this way relevant personnel are given an important power of veto, and each feels that he has an important voice in the matter. Once these persons have signed the document, it is expected that they will abide by its requirements.

Figure 15-4 is a layout for recording multiple signatures of approval, with corresponding dates. For convenience an outline drawing of this kind is sometimes included on the bottom part of the first page of a preprinted procedural form. As an alternative, the basic information relative to the recording of approvals may be typed in on the last page, in which case the presentation of signatures will resemble that found on joint memoranda.

DISTRIBUTION OF MANUALS AND PROCEDURES

Manuals are ordinarily issued to members of top management, department heads, and various administrators. It is through these people that the contents are made available to all concerned, including operating

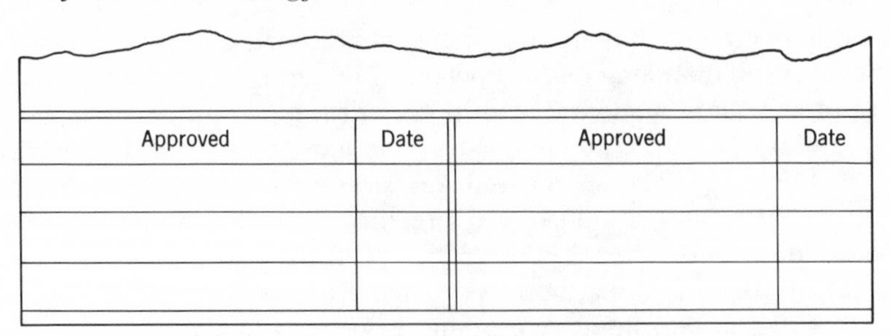

Approved	Date	Approved	Date

Figure 15-4. Form for recording signatures of approval.

personnel. Where manuals pertain to specific operations the necessary information should be readily available to those who need it, and this includes employees on all rungs of the organizational ladder.

Not all persons possessing manuals need to receive all the written procedures. Two extreme positions are taken on their distribution. One is to issue procedures to all persons holding manuals, regardless of their individual concern. The other is to distribute the procedures on a selective basis, using the "need to know" as the criterion.

When the need to know is used as the determinant, it is practical in the case of some companies to issue complete manuals to certain persons and abridged versions to others; for example, division or department managers are provided with all current procedures, whereas section managers receive only those with which they are directly concerned. The plan can work satisfactorily if the person holding the complete manual is situated in a convenient location. Then if the holder of an abridged version needs to refer to a complete manual, he may do so without traveling a great distance. In all events, provision should be made to allow responsible persons to obtain copies of procedures in which they feel interest.

MANUALS CONTROL

An accurate record of the distribution of manuals is essential. It serves two purposes; namely, to show the names of the holders who are to receive new and revised copies of procedures, and to regulate the whereabouts of the manuals. With respect to the latter it must be recognized that manuals contain private information that should be safeguarded. An employee should not be permitted to retain a manual when he leaves the employ of the company.

A satisfactory method of control requires that the persons receiving manuals sign for them under a plan similar to that used by libraries for lending books. Each manual is serially numbered and equipped with a pocket on the inside cover for holding a card designed to contain the name of the manual, its serial number, signatures of *borrowers*, and dates of issue. When a manual is charged out, the card is removed, signed by the assignee, and retained in a central file. When a holder leaves the services of the company, he must return the manual to the manual-control agency as part of the process of termination.

MAINTENANCE OF MANUALS

Manuals must be kept up to date if they are to give maximum service. In the majority of companies changes in manuals are made necessary by events such as improvements in existing operational practices, new and different work, and the problems of growth associated with expansion. Modification of manuals is a recurring activity. Procedures are added, revised, or deleted.

Information on changes must be forwarded to the holders of manuals. Partial changes may be contained on single sheets; complete changes may consist of entire supplements or replacements. The new or revised material is frequently sent out with a cover memorandum that instructs the recipients on how to make the change; for example, in the case of an entire substitution, a memorandum may simply say, "Attached is SPI No. 46–B of the subject manual. You are requested to remove SPI No. 46–A and replace same with SPI No. 46–B".

At periodic intervals it may be advisable to issue a listing of the changes that have affected a manual. This is done either quarterly or on a monthly basis, depending on the circumstances. The purpose is to make certain that all manuals are updated.

WORKING DRAFTS

In certain classes of work, as in the introduction of a highly complex system, it is often desirable to issue working drafts of written procedures. The overall pattern of these tentative instructions follows the same format that will be used for their permanent form. However, they are usually presented not on the preprinted form but on plain sheets of paper (usually hectograph), which clearly identifies their tentative status. They are given a designation such as "working draft"; for additional mark of recognition they may be duplicated on colored paper.

Working drafts are not only useful as instruments of control but also signal the need for continuing the close surveillance of operations. People in charge are thus apt to be more attentive to the requirements of a situation. As the need for changes becomes known, the improvements should be noted for subsequent inclusion in a standard written procedure.

A WORD OF CAUTION ABOUT MANUALS

A major drawback of manuals seems to lie in the danger of overgeneralization. Procedures are often written in such an indefinite manner as to make them vague and almost useless. Two reasons may account for this—the desire to avoid the errors that so often affix themselves to the preparation of precise presentations, and the wish to minimize the number of procedural changes that will be necessary.

Manuals are of little or no consequence unless they tell you something worthwhile. If they do not, they are destined to become dust collectors. We must make our written procedures meaningful and helpful to the persons who will be referring to them. As in preparing well-formulated reports (Chapter 24), emphasis is on simplicity and clarity of presentation.

part 5

Automatic Data Processing—

Electric Accounting Machines

16

Background and Methods of Data Representation

INTRODUCTION

The "auto" in automatic data processing (ADP) is a combining form meaning "self." In conjunction "auto" and "matic" mean "self-acting." Actually, then, in discussing ADP we are talking about self-acting information processing.

More precisely, ADP may be defined as the gathering and handling of information by means of machines and electronic equipment to minimize human participation.

TERMINOLOGY

Automatic-data-processing equipment and activities are spoken of in terms of three categories:

1. Electric accounting machines (EAM)
2. Integrated data processing (IDP)
3. Electronic data processing (EDP)

There is a good deal of overlapping in the meanings of these terms. It may be said that EDP is the more general one; it is often used to encompass the handling and treatment of recorded information by the agency of electronic devices of many diverse kinds. Considered in this context both EAM and IDP fall within the province of EDP.

From the viewpoint of semantics much equipment of the EAM and

IDP categories may rightly be called *electronic;* it is also a fact that EDP installations *integrate* data.

Still, in practice the terms EAM, IDP, and EDP have rather specialized meanings: EAM refers to conventional punched-card equipment, also known as tabulating equipment; IDP refers to various units of equipment that are linked together by common-language media in the form of magnetic or punched tape or cards to constitute a unified operation; and EDP has to do with the processing of information on computers and auxiliary units. This breakdown of ADP functions provides a convenient basis for the treatment of our subject matter. Accordingly, Part V discusses equipment and functions of the EAM classification, Part VI is devoted to an explanation of IDP, and Part VII treats computers and associated equipment.

Before discussing the different kinds of equipment in the EAM category, we should give attention to the tabulating card as a medium for storing and processing information.

THE TABULATING CARD

The tabulating card, generically known as the tab card, is the mainstay of most ADP systems. It is used in connection with EAM, and both IDP and EDP installations.

The basic coding format of the IBM [1] type of tab card is illustrated in Figure 16-1. There are 80 vertical columns. Within each of these columns there are 12 locations for punching a hole or combination of holes to represent a single number, letter, or special character. Only 10 of the 12 locations are typically printed with numerals. The other two, comprising two additional rows on the vertical plane at the top of the card, are left unmarked. Reading down the card, from top to bottom, the horizontal rows are: 12, 11, and 10 (which is indicated by 0), followed by 1, 2, 3, 4, 5, 6, 7, 8, and 9. The twelfth and eleventh rows are not indicated numerically and are also referred to as the R and X rows, respectively.

Two types of punching are used: zone punching and digit punching. Perforations in rows 12, 11, and 10 are called zone punches; those in rows 1 through 9 are known as digit punches.

To record a character in any of the positions within a column, a special machine is used. An electric card punch will pierce the card with the necessary hole or holes by pressing the proper key.

Special characters are formed by three punches: two digit punches and one zone punch, in row 12, 11, or 10; for example, by depressing the "%"

[1] Registered trademark of the International Business Machines Corporation.

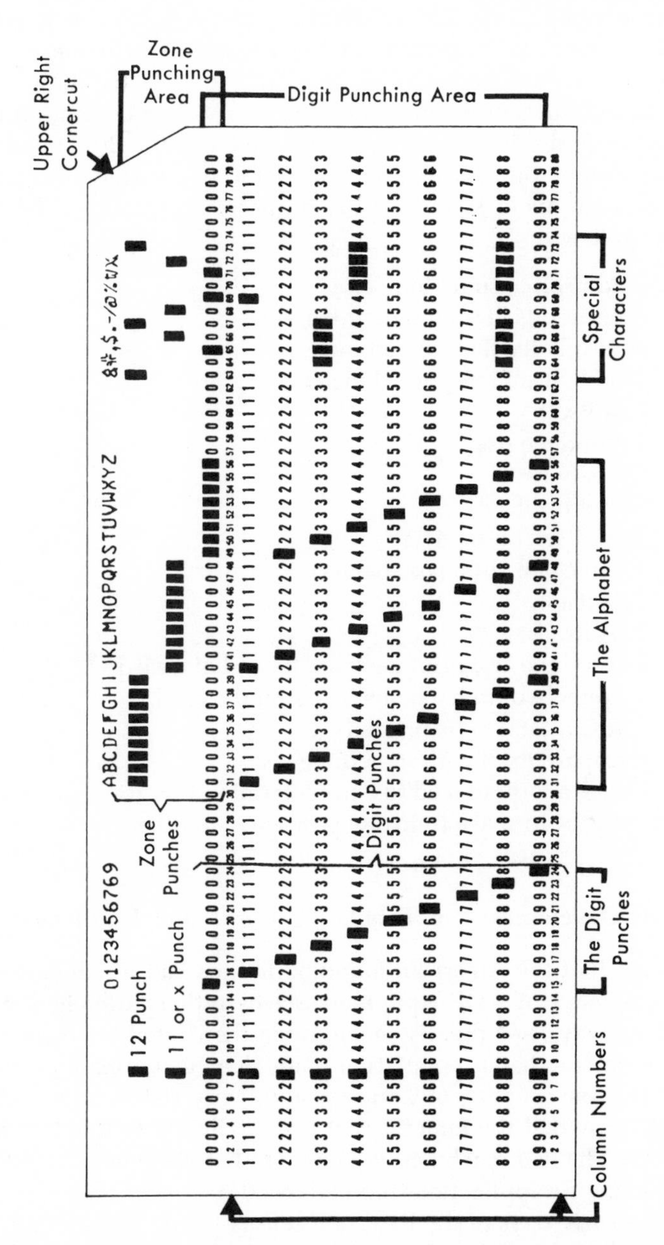

Figure 16-1. An IBM tabulating card showing how punched holes represent specific letters, characters, and figures. The standard card is 7⅜ inches long, and 3¼ inches high.

key, any selected column is automatically and simultaneously punched in the tenth, fourth, and eighth positions. Depressing a numerical key will perforate a column in only one place—that is the selected digit.

Once information is represented on the card, it becomes a record of convenience and utility that may be retained indefinitely. Coded facts and data can then be processed and manipulated in a variety of ways, at high speed, on equipment suited for the purpose. As published by IBM, the *punched hole* will:

1. Add itself to something else.
2. Subtract itself from something else.
3. Multiply itself by something else.
4. Divide itself into something else.
5. List itself.
6. Reproduce itself.
7. Classify itself.
8. Select itself.
9. Print itself on the IBM card.
10. Produce an automatic balance forward.
11. File itself.
12. Post itself.
13. Reproduce and print itself on the end of a card.
14. Be punched from a pencil mark on the card.
15. Cause a total to be printed.
16. Compare itself with something else.
17. Cause a form to feed to a predetermined position, or to be ejected automatically, or to space from one position to another.

Historical Development of Tabulating Cards and Equipment

To say that the American founding fathers had anything to do with the development of ADP equipment seems rather farfetched. Still, in a roundabout way, they did. By including in the United States Constitution the requirement that a population count had to be made within three years of the first meeting of Congress and every 10 years thereafter, they indirectly provided the impetus that eventually led to the design of such equipment. The first machines were used to handle census data. Commercial and scientific applications soon followed.

In the early days of the Republic census-taking was a relatively easy matter. At the time of George Washington the 1790 census revealed a population of 3,929,214. The tasks of compilation and computation were done by hand. By 1880, ninety years later, the population of the country totaled in excess of 50 million persons. Today the population exceeds 200 million.

Because of the rapid increase in population and requirements for information of a more complex nature, the 1880 census took seven years to accomplish. The methods of processing the information were laborious and time consuming. Conditions were so harassing that the Director of the Bureau of the Census announced that, unless better ways were found to do the job, the next decennial census would require 10 years to complete.

Faced with this prospect, the Bureau employed Dr. Herman Hollerith, a statistician and engineer from Buffalo, New York, to devise a mechanical means for processing census data. The challenge was welcomed by him, as he had been experimenting with punched-card mechanisms.

In the mid-1880s Dr. Hollerith went to work for the Bureau. The idea of punched-card mechanisms was not new at the time. As early as 1801 a French inventor by the name of Joseph Marie Jacquard had developed a loom (Figure 16-2) that successfully wove intricate patterns of cloth by the use of cards with holes punched into them. The Jacquard method calls for working the design out on squared paper and then perforating cards to correspond to the design. In appearance the cards faintly resemble Remington Rand cards without the corner cut. After perforation the Jacquard cards are laced together at the edges and placed on a mechanism near the top of the loom. As the cards move they pass under a battery of mounted needles, and the prearranged holes in the cards allow selected needles to drop through and lift certain strings, which in turn lift heddles in the harness of the loom. The pattern that is produced can be changed by using different cards.

Akin to the principle employed by Jacquard is the one used to operate player pianos. These mechanized instruments were popular during the latter part of the 1800s until about 1928. Essentially they were a type of ADP equipment. Instead of "reading" facts and data in the form of coded holes, they converted coded holes into sounds of music. Briefly, here is how they functioned. A pneumatic mechanism was encased in the instrument, which contained about 65 felt-covered "fingers," the exact number depending on the make and model. These fingers were positioned in front of a finger board, so that, with the unwinding of a music roll pierced with holes, air pressure was directed to the desired fingers, which in turn activated the keys. The operator provided the necessary air pressure through the pumping of two pedals.

Of even greater pertinence is the work of the English mathematician and mechanician Charles Babbage. From what is known, it seems that Babbage was more than slightly agitated by the number and magnitude of errors he found in astronomical and other numerical tables he had occa-

Figure 16-2. Jacquard's loom. This was one of the first punched card mechanisms. Notice the portrayal of the "Computer Family Tree" on the plaque in the lower left hand corner of the photograph. The tree itself is pictured in Figure 19-9.

sion to use. He set out to do something about it. During this endeavor he became interested in calculating machines, and in 1822 he announced that he had constructed a small-scale model of a machine that would perform sophisticated computations. The following year he gained the professional support of the Royal Society, and on its recommendation the British government provided him with the necessary financial backing to build a full-scale unit. He proceeded to construct no ordinary calculator but a machine that would compute the square of successive integers. Babbage used the name "difference engine" to identify the machine he had in mind. Destiny ordained that the machine was never to be completed. After working on it for about 10 years, and just as his long-laid plans were nearing fruition, he adopted a principle that was different from the one originally conceived. The new plan of action called for building not a "difference engine" but an "analytical engine" that would utilize cards—an adaptation that was similar in principle to the operation of the Jacquard loom. He deduced that, if punched cards could control the needles on a loom, they could control the mechanism on a calculating machine.

As envisioned by Babbage, the completed engine would possess feedback or self-regulating features that would have truly made it the forerunner of modern-day automated equipment; it would also have a *memory* unit of some one thousand digits and be able to print out the results of computations. However, the British government had already made what amounted to a sizable grant, and in the absence of more tangible results it withdrew its financial support.

Dissatisfied by the event but by no means discouraged, Babbage spent much of his personal fortune and the remainder of his life in attempting to make his engine a workable reality. He died in 1871 without perfecting an operative unit. Had he been successful, Babbage would have been the inventor of the first mechanical computer of its kind. Despite his lack of success, the pioneering work he did is considered to have been instrumental in the subsequent development of automatic-data-processing equipment.[1]

Though Jacquard and Babbage pioneered in the early application of punched-card mechanisms, it remained for Hollerith to develop the first electrically operated tabulating equipment. Following a period of trial and error, he devised a mechanical tabulating machine that processed cards on which data was represented by means of holes. An early Hollerith electric tabulating machine is shown in Figure 16-3. Elements of the Hollerith tabulating scheme are shown in Figure 16-4.

Dr. Hollerith's original patents were not granted until 1889. At about the same time the Department of Health in Baltimore, Maryland, obtained a set of the equipment for the purpose of processing public-health information. The distinction of having the first successful electromechanical data-processing center in the world thus goes to this agency of our Federal Government.

In 1890 the Bureau of the Census was able to process the Tenth Decennial Census on tabulation equipment. The installation was comprised

[1] Other interesting highlights concerning Charles Babbage (1792–1871): he was a graduate of Cambridge University, and from 1828 to 1839 he was Lucasian Professor of Mathematics at that institution; in 1820 he helped found the Astronomical Society; he wrote *Economy of Machinery and Manufactures*, which was published in 1832; and in 1834 he assisted in establishing the Statistical Society.

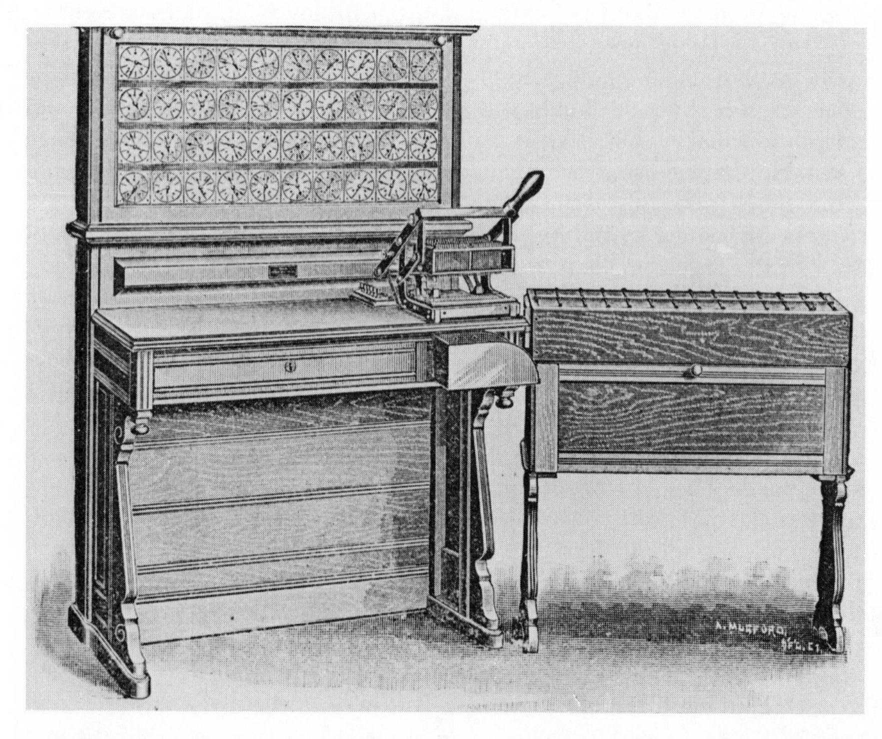

Figure 16-3. An early Hollerith electric tabulating machine, ca. 1900. As each punched card was placed in the press, holes were electrically sensed and totals indicated on dials. Then a magnet raised the lid of the appropriate compartment in the cable-connected sorting box on the right. The totals were transcribed by hand.

of a pantograph punch, a manual gang punch, a hand-fed unit counter, and a sorting box. Though crude in comparison with today's equipment, these units succeeded in substantially reducing the time required for processing census information. Only one-third of the time of the previous census was required to complete the task even though the population had risen by an additional 10 million people.

Hollerith kept working on the development of punched-card equipment. By the turn of the century, while still working at the Bureau, he came up with a card-punch machine, a semiautomatic tabulating machine, and a sorter capable of classifying cards at the rate of 300 per minute.

In the early equipment of Hollerith machine tabulations were made by floating the cards across a pool of mercury and letting telescoping pins in the reading head fall through the hole and contact the mercury. The contact closed an electrical circuit, and the information thus "read" was recorded on dials.

Aside from census work Dr. Hollerith foresaw numerous commercial and scientific applications for the equipment. In 1903 he left the Bureau and established the Tabulating Machine Company, which provided the nucleus for the now well-known International Business Machines Corporation.

Soon after its formation a dispute arose between the Tabulating Machine Company and the Bureau of the Census with respect to patents, rentals, and related matters. The upshot of the situation led the Director of the Census to petition Congress for funds to develop new equipment. Granted the appropriation, the Bureau set out to do what it proposed. About 1905 it established a developmental laboratory. The Bureau acquired the services of James Powers, a noted engineer from New York, to originate equipment of a different design, but of a nature that would still employ punched cards. Powers succeeded, and the equipment that evolved was the principal electromechanical means used in the processing of data in the 1910 census.

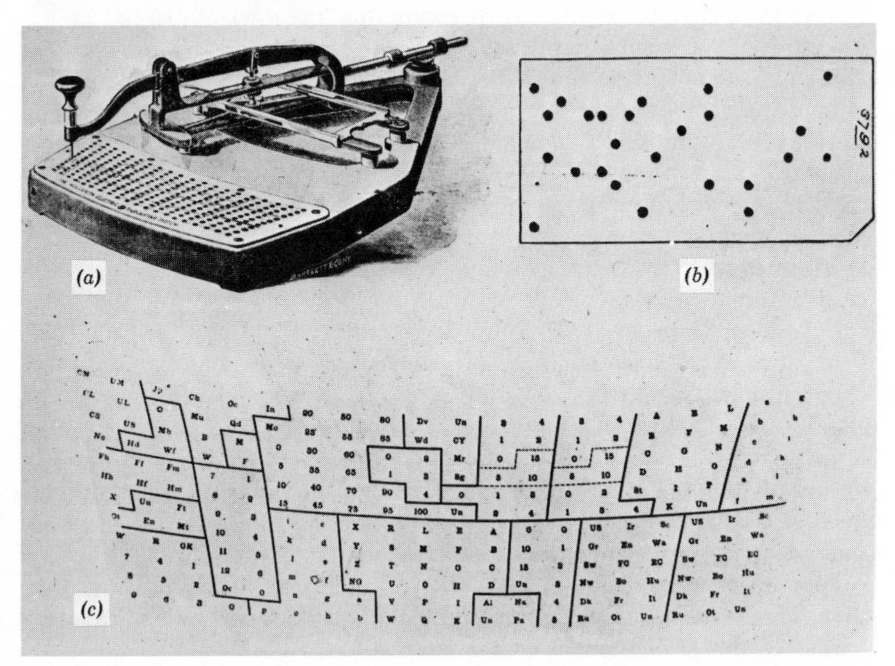

Figure 16-4. Elements of the Hollerith tabulating scheme: (*a*) Hollerith keyboard punch; (*b*) complete card for one person; (*c*) symbols of holes in Hollerith census keyboard.

In 1910 Powers terminated his position with the government and founded the Powers Accounting-Machine Company. At a later date this company was acquired by the Remington Rand Corporation.

Today the IBM alphanumeric card-coding scheme is occasionally referred to as the Hollerith code, after the name of the man who conceived the idea. Similarly, the Remington Rand card format is sometimes called the Powers layout, after the name of its originator.

Layout of the IBM Card

Considerable thought and planning must be given to the laying out of an IBM card. What goes on the card and its precise arrangement will, of course, be governed by the requirements of the documents or reports to be generated. The format must be in accord not only with the needs of the project but with equipment peculiarities and capabilities as well.

Specification sheets of the kind shown in Figures 16-5 and 16-6 are especially handy for designing IBM cards. Ample quantities of each type of layout form are available, in pads, from tab-card vendors—without charge. Actual standard-size cards containing just rows of printed digits, though somewhat small, are useful for experimenting with card designs. Lines may be vertically drawn between the columns of numbers and columnar descriptions placed at the top. (The reason for the different types of card-layout forms will become more apparent as we proceed into the chapter.)

On tab-card formats one or more columns containing an item of information—such as a name, date, or any series of characters that are treated as an integral unit—is referred to as a *field*. A group or stack of related cards is termed a *deck*.

Two situations are common in laying out an IBM card: (*a*) the existence of more columns than necessary for recording the facts and data, and (*b*) an insufficient number of columns to accommodate the information in standard form. The first situation often calls for exercising restraint. So as to make full use of the card columns there may be a tendency to provide for information in excess of that which is sufficient, necessary, or desirable. Such action is to be avoided, because the consequences can be cumbersome and an unnecessary drag on the processing operations of the entire system. Instead provision should be made only for those facts and data that are germane to the program and can be justified in terms of their intrinsic worth.

If there are not enough columns to record the information in conventional terms on one card the solution may lie in any number of possibilities, such as the following:

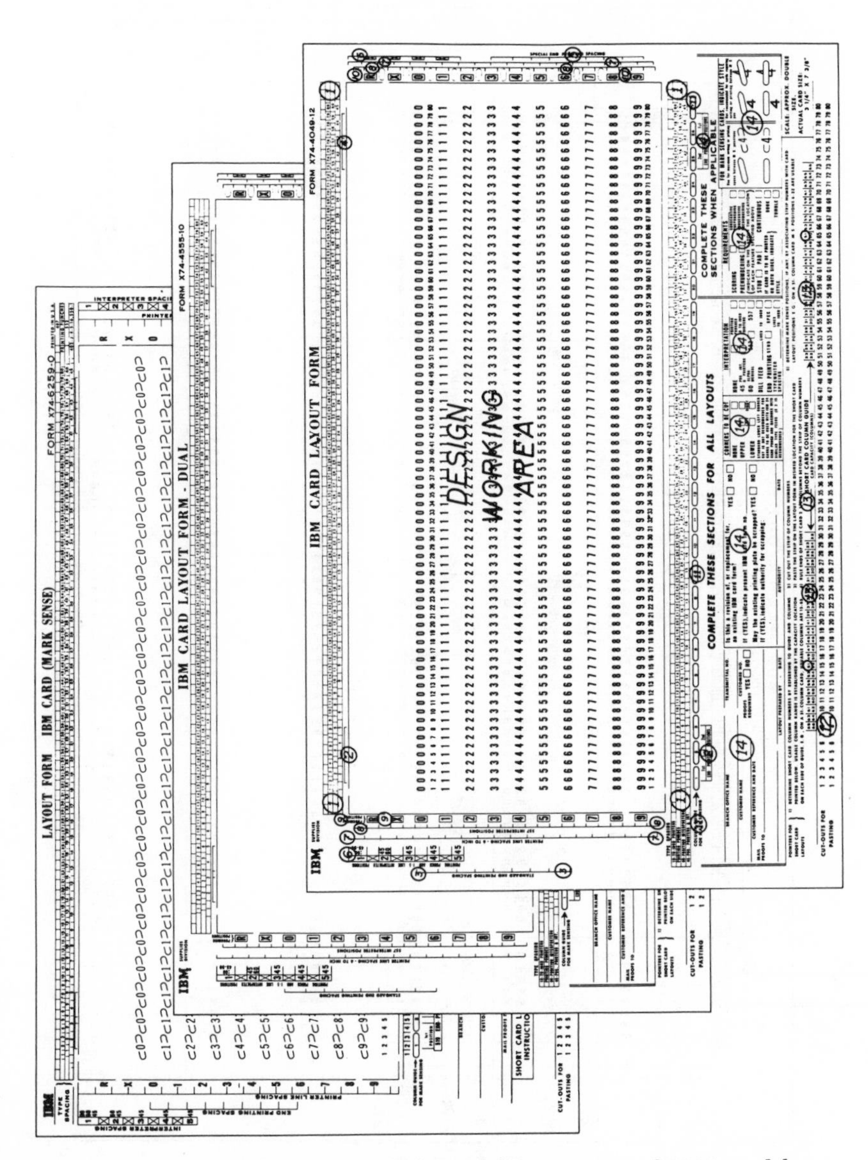

Figure 16-5. Forms for preparing three different types of IBM card layouts.

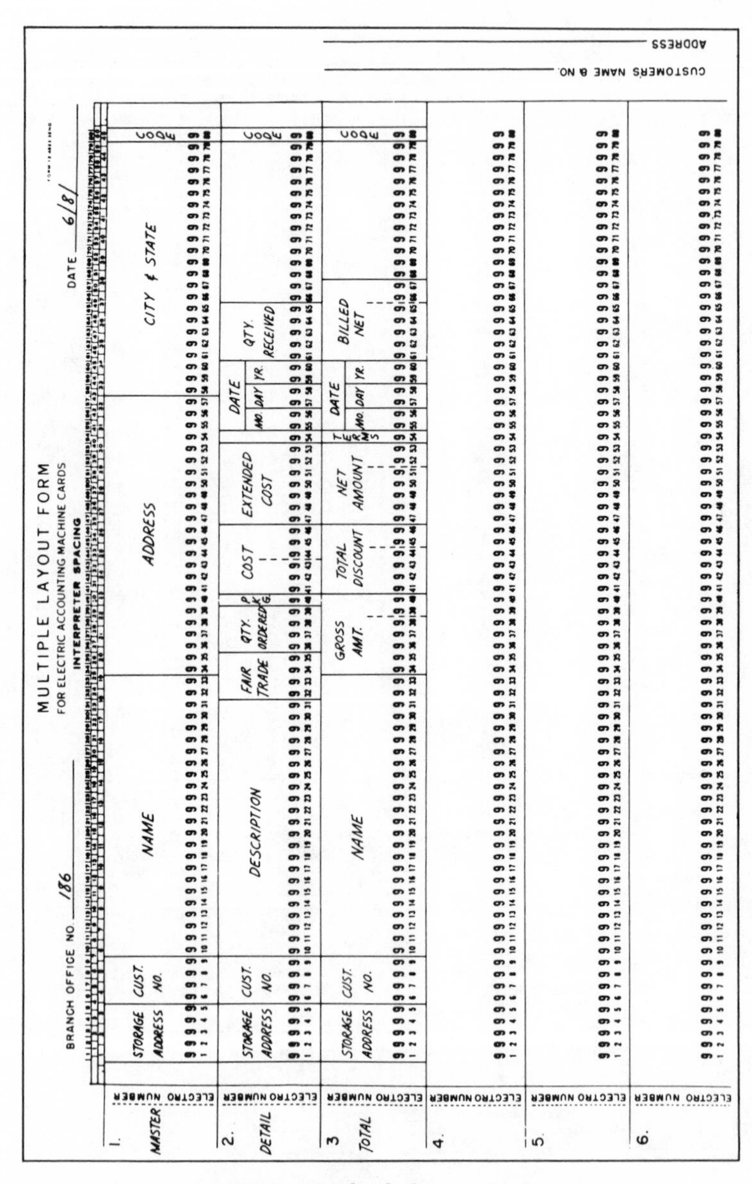

MULTIPLE LAYOUT FORM
FOR ELECTRIC ACCOUNTING MACHINE CARDS

BRANCH OFFICE NO. 186 INTERPRETER SPACING DATE 6/8/

1. MASTER: STORAGE ADDRESS | CUST. NO. | NAME | ADDRESS | CITY & STATE | CODE

2. DETAIL: STORAGE ADDRESS | CUST. NO. | DESCRIPTION | FAIR TRADE | QTY. ORDERED | PKG. | COST | EXTENDED COST | DATE (MO. DAY YR.) | QTY. RECEIVED | CODE

3. TOTAL: STORAGE ADDRESS | CUST. NO. | NAME | GROSS AMT. | TOTAL DISCOUNT | NET AMOUNT | TERMS | DATE (MO. DAY YR.) | BILLED NET | CODE

4.

5.

6.

ELECTRO NUMBER

CUSTOMER'S NAME & NO.

ADDRESS

Figure 16-6. Multiple layout form.

1. Reduce the number of characters in a field to the least possible number that is consistent with requirements.

2. Eliminate data that are redundant or in excess of requirements—for example, the recording of both an order number and an invoice number when one will serve the purpose.

3. Utilize the same field instead of two for recording alternative information; for example, by punching an X, or digit punch, in a selected card column meaning Dr. or Cr., it is possible to record either an invoice or credit transaction in a single field on an accounts-receivable card.

4. Use a single card column for recording a one-character code. Examples follow:

Example 1

Code	*Salesman*	*Code*	*Salesman*
1	Felix Ackerman	C	Leo J. Leary
2	Willard Atwood	D	Alexander P. Marshall
3	Harold J. Bond	E	Charles McCann
4	Robert P. Callahan	F	Franklin Paulson
5	Stephen Davies	G	Walter O. Rawlings
6	Anthony Dillon	H	Leland Robertson
7	Edgar Durfee	I	Leon Schultze
8	Ralph R. Galloway	J	Edmond O. Vaugn
9	Theodore Grace	K	Elwood Vickers
A	Philip M. La Point	L	Lee Wilson
B	Joseph G. Lanza		

Example 2

Quarterly Date Code

		Period							
	Quarter	1967	1968	1969	1970	1971	1972	1973	1974
January, February, March	1	A	E	I	M	Q	U	Y	3
April, May, June	2	B	F	J	N	R	V	Z	4
July, August, September	3	C	G	K	O	S	W	1	5
October, November, December	4	D	H	L	P	T	X	2	6

In addition to the above, the following more elaborate methods of coding may be used:

1. Block coding, in which groups, or blocks, of numbers are used to represent classifications. Examples are given below.

Example 1

Code	
1	Vitamin capsules— 30 to a bottle
2	Vitamin capsules— 50 to a bottle

3	Vitamin capsules—100 to a bottle
4	Vitamin capsules—225 to a bottle
5	Vitamin capsules—500 to a bottle
6	Seltzer tablets—10 to a jar
7	Seltzer tablets—15 to a jar
8	Seltzer tablets—25 to a jar

Example 2

Code	*Population*
11	1,000,000 or more
12	750,000 to 999,999
13	500,000 to 749,999
14	400,000 to 499,999

2. Group classification coding, in which major and minor classifications are represented by the succeeding digits of a code. The coding structure is based on digital position assignment, with the digits for the major class appearing on the left and the breakdown of that class on the right.

Example 1—Coding Structure

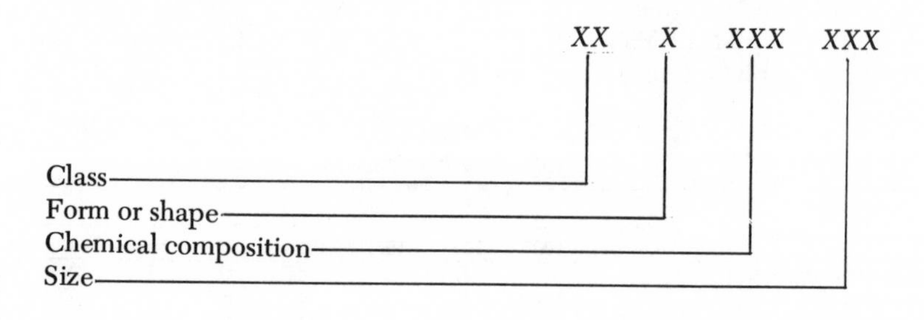

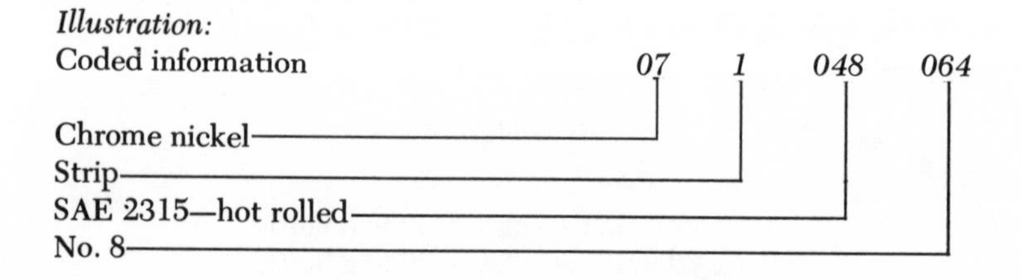

Example 2—Coding Structure for Cost Control

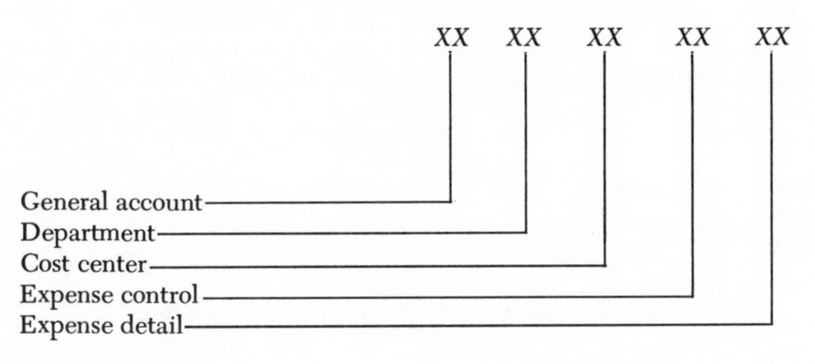

Illustration:

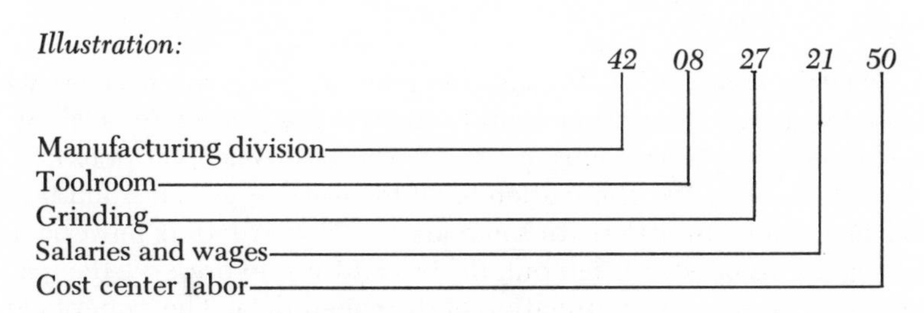

3. Significant digit coding, used mainly where it is advantageous to have built-in meaning that is readily identifiable. It is used chiefly for inventory-control systems. The coding structure may be based on size, weight, dimensions, etc.

Example 1

301.00.00	Electron tubes (size)
301.00.01	T–1
301.00.02	T–2
301.00.03	T–3
301.00.04	T–4

Example 2

18025	25 lb. sack (weight)
18050	50 lb.
18150	150 lb.

Example 3

240606	6 oz. 6 Pack, beverage (quantity and number)
241206	12 oz. 6 Pack
240612	6 oz. 12 Pack
241212	12 oz. 12 Pack

4. Final digit coding. This term is applied to a coding structure where the ending or final numbers designate certain additional information about the coded item. It is used only as an integral part of a more elaborate coding scheme and is not a code in itself.

Example 1

XXXX1 Piece Part
XXXX2 Subassembly
XXXX3 Assembly

Example 2—Coding for an Either-Or Situation

XXXX0 No–go gage
XXXX1 Go gage

5. Consonant coding. Consonant codes are great space savers where much alphabetic description would otherwise be necessary, such as nomenclature for material items or names of places. Wherever practical, the code is devised by the elimination of all the usual vowels beyond the first letter of surname or article. In some instances other letters must be omitted. Regardless of what is left out, the basic idea is to preserve the reading proficiency or visual interpretation of the coded data. The example given below is a detail part code for a five-column field.

Example—Detail Part Code

Noun	*Code*	*Noun*	*Code*
A frame	AFRAM	baffle	BAFLE
absorber	ABSBR	barometer	BRMTR
amplifier	AMLFR	bellcrank	BLCNK
arm	ARM	bushing	BSHNG
attenuator	ATNTR		
axle	AXLE		

Within the framework of the above schemes many different coding methods are devisable. The nature of the system and the characteristics of the data-processing equipment to be employed should be considered before selecting the type of code.

Where practicable, having all the information requirements on one card affords certain advantages; it alleviates sorting operations and simplifies other ADP tasks. Still, there are numerous times when it is necessary to use more than one card, and then thought must be given as to how the data are to be separated or assigned to each card. Techniques for accomplishing such divisions include the following:

1. Placing constant or stable information in one card and changing or variable information in a second card (see Figure 16-7).
2. Designing two cards for those instances where the information is derived from two separate source documents.
3. In cases where the system permits it, designing a card for each section of the source document (see Figure 16-8).

For the preparation of a report the master card and the trailer card can be machine-read so that the information on both cards is printed adjacent. A common arrangement is to have the data on the trailer card printed directly underneath the data provided by the master card. To accomplish this there must be a way of drawing the two cards together. Control numbers, account numbers, and the like serve this purpose. The idea is illustrated in Figure 16-9.

Depending on the purpose of the report and the characteristics of the machine that is to do the printing, the format of a report may be divided into a number of sections, and specified information may be printed in specified sections of the form; an example is given below.

SOLD TO JOHN WHITE DATE AUG 15
290 MAIN STREET
ANYTOWN

SHIP TO WHITE'S WAREHOUSE
40 BIRCH STREET
SAMETOWN

ITEM NO. 1
ITEM NO. 2
ITEM NO. 3

Types of Cards

IBM cards may be classified according to the manner in which they are prepared.

Transcript cards are for recording information derived from other documents. They may be plain, like the card shown in Figure 16-1, or they may be especially designed and printed, like those shown in Figure 16-8.

Dual cards are intended to function both as source documents and processing media. With this type, the cards themselves are subsequently keypunched directly from the information contained on them.

In designing a dual card care must be taken to see that the layout of the form permits continuous and unencumbered reading during the actual

ACCOUNT NO.	NAME	ADDRESS	CITY AND STATE
0 0 0 0 0 0 0 0	0 0 0 0 0 0 0 0 0 0 0 0 0 0 0 0 0 0 0 0 0 0 0 0	0 0 0 0 0 0 0 0 0 0 0 0 0 0 0 0 0 0 0 0 0 0 0 0	0 0 0 0 0 0 0 0 0 0 0 0 0 0 0 0 0 0 0 0 0 0 0 0
1 2 3 4 5 6 7 8	9 10 11 12 13 14 15 16 17 18 19 20 21 22 23 24 25 26 27 28 29 30 31 32	33 34 35 36 37 38 39 40 41 42 43 44 45 46 47 48 49 50 51 52 53 54 55 56	57 58 59 60 61 62 63 64 65 66 67 68 69 70 71 72 73 74 75 76 77 78 79 80
1 1 1 1 1 1 1 1	1 1 1 1 1 1 1 1 1 1 1 1 1 1 1 1 1 1 1 1 1 1 1 1	1 1 1 1 1 1 1 1 1 1 1 1 1 1 1 1 1 1 1 1 1 1 1 1	1 1 1 1 1 1 1 1 1 1 1 1 1 1 1 1 1 1 1 1 1 1 1 1
2 2 2 2 2 2 2 2	2 2 2 2 2 2 2 2 2 2 2 2 2 2 2 2 2 2 2 2 2 2 2 2	2 2 2 2 2 2 2 2 2 2 2 2 2 2 2 2 2 2 2 2 2 2 2 2	2 2 2 2 2 2 2 2 2 2 2 2 2 2 2 2 2 2 2 2 2 2 2 2
3 3 3 3 3 3 3 3	3 3 3 3 3 3 3 3 3 3 3 3 3 3 3 3 3 3 3 3 3 3 3 3	3 3 3 3 3 3 3 3 3 3 3 3 3 3 3 3 3 3 3 3 3 3 3 3	3 3 3 3 3 3 3 3 3 3 3 3 3 3 3 3 3 3 3 3 3 3 3 3
4 4 4 4 4 4 4 4	4 4 4 4 4 4 4 4 4 4 4 4 4 4 4 4 4 4 4 4 4 4 4 4	4 4 4 4 4 4 4 4 4 4 4 4 4 4 4 4 4 4 4 4 4 4 4 4	4 4 4 4 4 4 4 4 4 4 4 4 4 4 4 4 4 4 4 4 4 4 4 4
5 5 5 5 5 5 5 5	5 5 5 5 5 5 5 5 5 5 5 5 5 5 5 5 5 5 5 5 5 5 5 5	5 5 5 5 5 5 5 5 5 5 5 5 5 5 5 5 5 5 5 5 5 5 5 5	5 5 5 5 5 5 5 5 5 5 5 5 5 5 5 5 5 5 5 5 5 5 5 5
6 6 6 6 6 6 6 6	6 6 6 6 6 6 6 6 6 6 6 6 6 6 6 6 6 6 6 6 6 6 6 6	6 6 6 6 6 6 6 6 6 6 6 6 6 6 6 6 6 6 6 6 6 6 6 6	6 6 6 6 6 6 6 6 6 6 6 6 6 6 6 6 6 6 6 6 6 6 6 6
7 7 7 7 7 7 7 7	7 7 7 7 7 7 7 7 7 7 7 7 7 7 7 7 7 7 7 7 7 7 7 7	7 7 7 7 7 7 7 7 7 7 7 7 7 7 7 7 7 7 7 7 7 7 7 7	7 7 7 7 7 7 7 7 7 7 7 7 7 7 7 7 7 7 7 7 7 7 7 7
8 8 8 8 8 8 8 8	8 8 8 8 8 8 8 8 8 8 8 8 8 8 8 8 8 8 8 8 8 8 8 8	8 8 8 8 8 8 8 8 8 8 8 8 8 8 8 8 8 8 8 8 8 8 8 8	8 8 8 8 8 8 8 8 8 8 8 8 8 8 8 8 8 8 8 8 8 8 8 8
9 9 9 9 9 9 9 9	9 9 9 9 9 9 9 9 9 9 9 9 9 9 9 9 9 9 9 9 9 9 9 9	9 9 9 9 9 9 9 9 9 9 9 9 9 9 9 9 9 9 9 9 9 9 9 9	9 9 9 9 9 9 9 9 9 9 9 9 9 9 9 9 9 9 9 9 9 9 9 9
1 2 3 4 5 6 7 8	9 10 11 12 13 14 15 16 17 18 19 20 21 22 23 24 25 26 27 28 29 30 31 32	33 34 35 36 37 38 39 40 41 42 43 44 45 46 47 48 49 50 51 52 53 54 55 56	57 58 59 60 61 62 63 64 65 66 67 68 69 70 71 72 73 74 75 76 77 78 79 80

Figure 16-7 *a*. Master card.

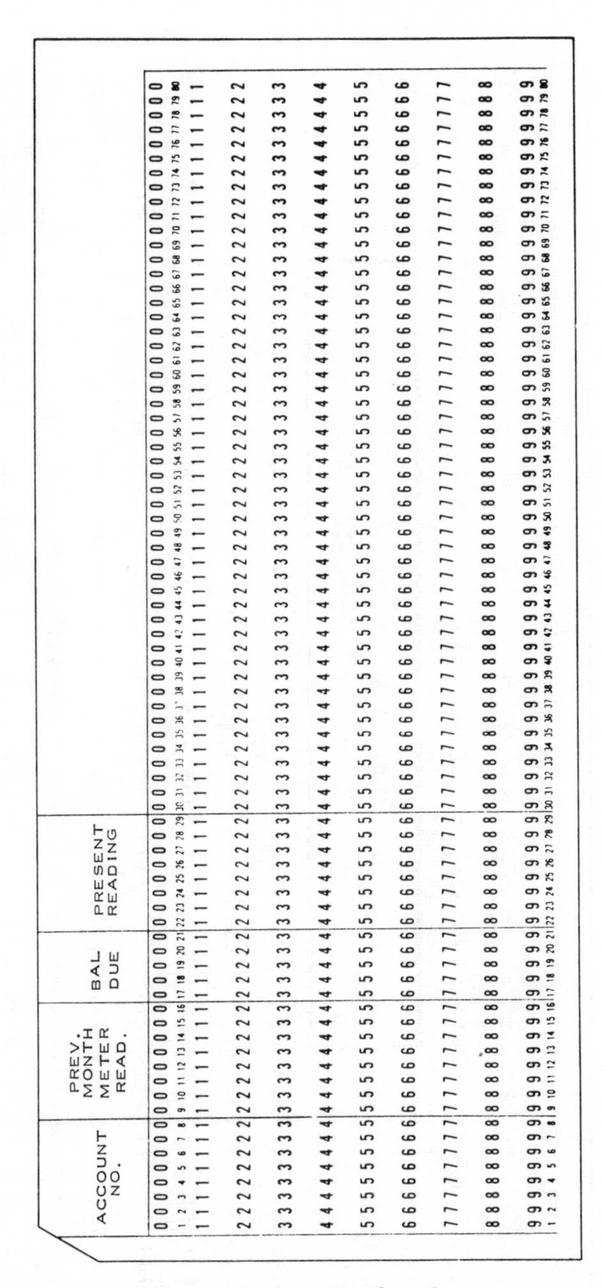

Figure 16-7 *b*. Detail card.

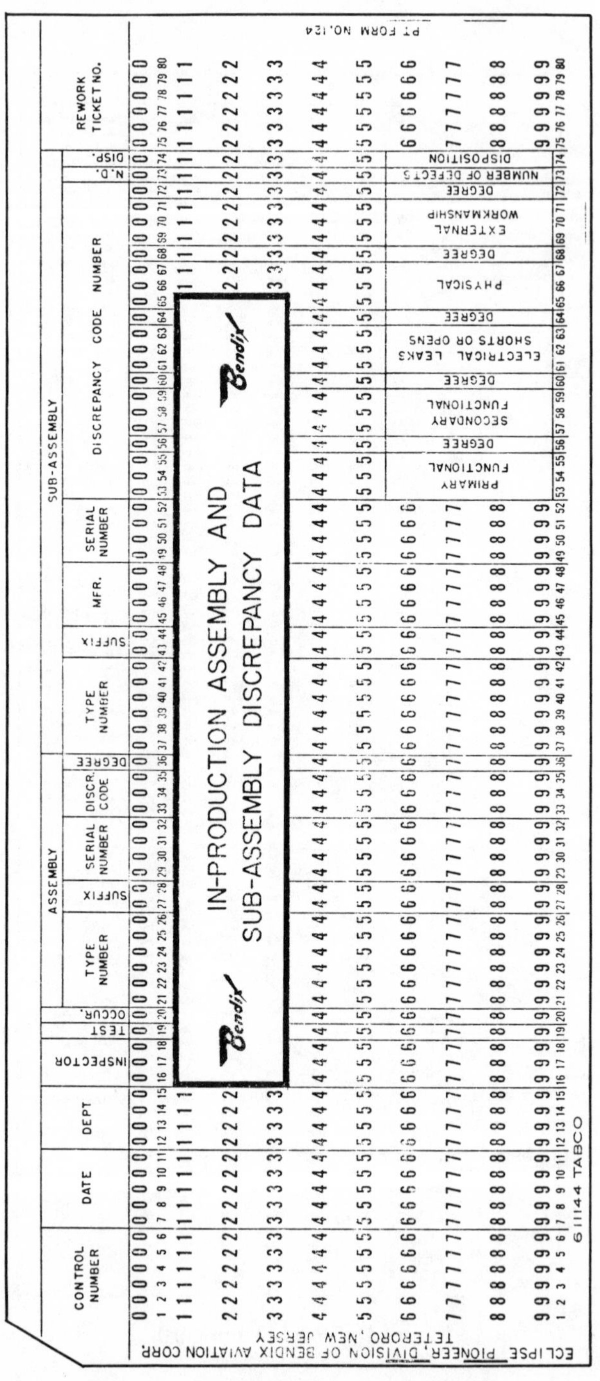

Figure 16-8 *a*. A master card for recording sub-assembly discrepancy data. (Information to be punched into this card is derived from a source document which is divided into two parts. The upper portion is for recording sub-assembly discrepancy data and the lower portion for recording piece-part failure information.)

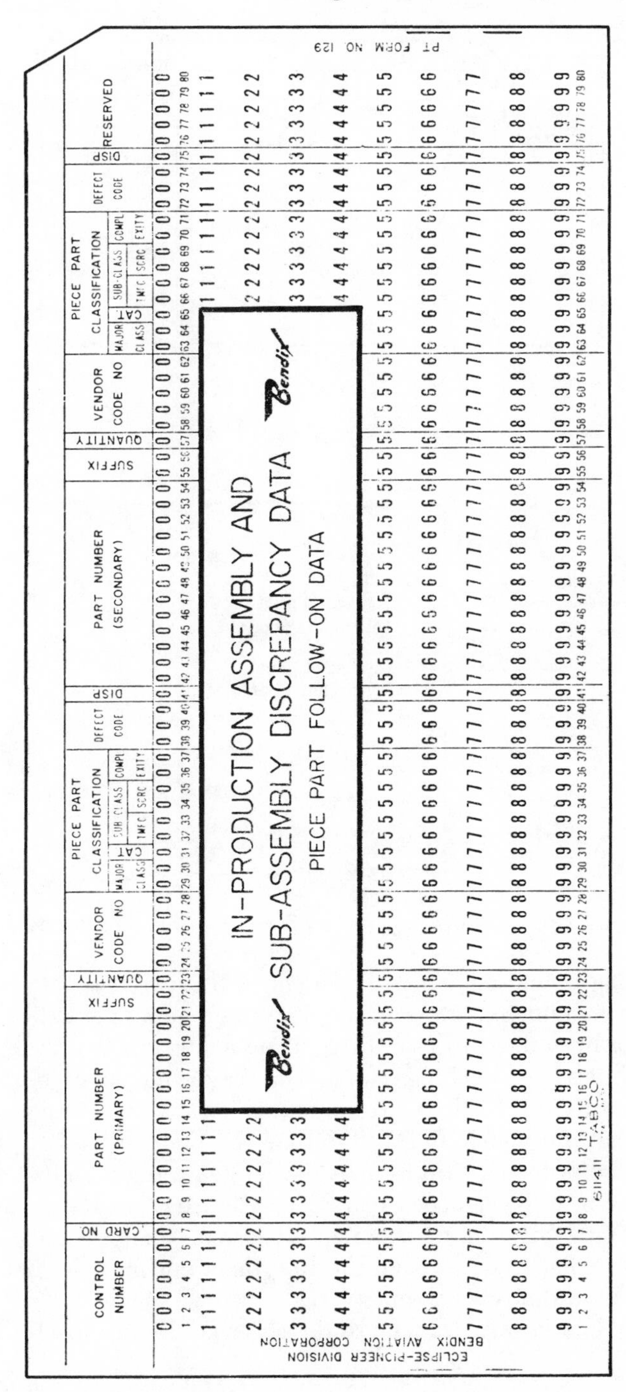

Figure 16-8 *b*. Piece-part trailer card. (For use with Figure 16-8 *a*. Note: Each master card may be followed by any number of trailer cards, depending on the number of piece-part failures.)

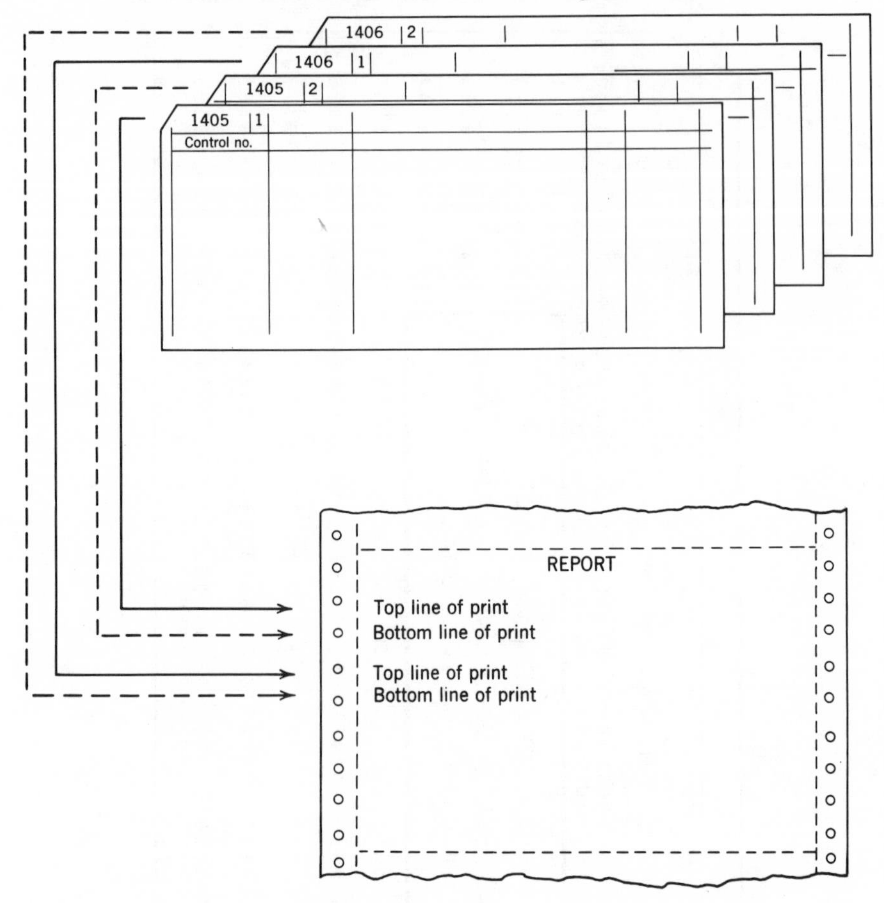

Figure 16-9. Preparing a report from master and trailer cards.

keypunching operation. There is danger of handwritten data being hidden under the punching-station bar when it is read for punching.

Just prior to the card's reaching the punching position the entire card is visible, but when punching is under way a maximum of nine columns are hidden (see Figure 16-10). Correspondingly, there are always at least 71 columns that are visible during the time the punching takes place. The important point is that card formats should be arranged so that each item of information to be punched can be seen by the operator at the proper moment. With this in mind experienced systems analysts make it a practice of preparing trial cards and testing them out on the keypunch machine before settling on a layout.

A dual tabulating card with stub is shown in Figure 16-11.

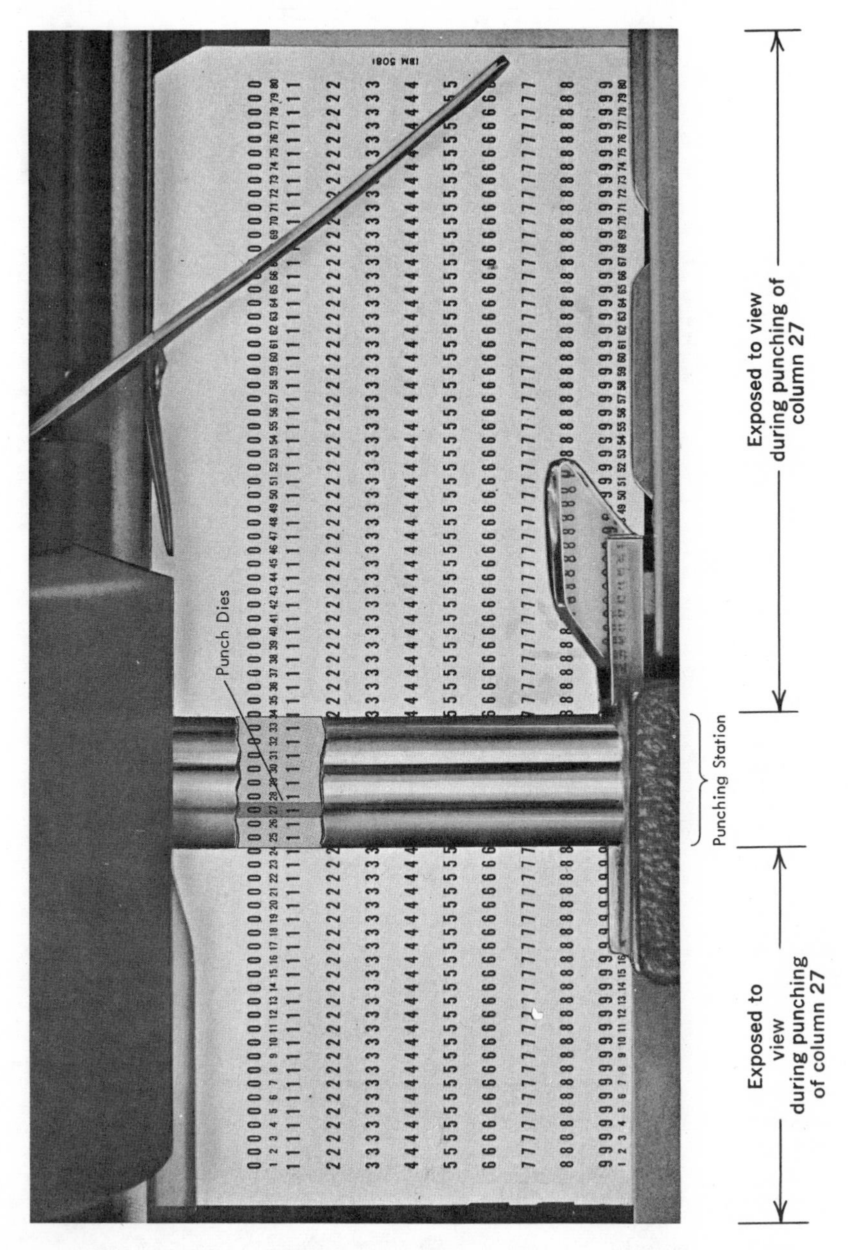

Figure 16-10. Tab card during keypunching operation. Note the columns that are hidden under the punching-station bar.

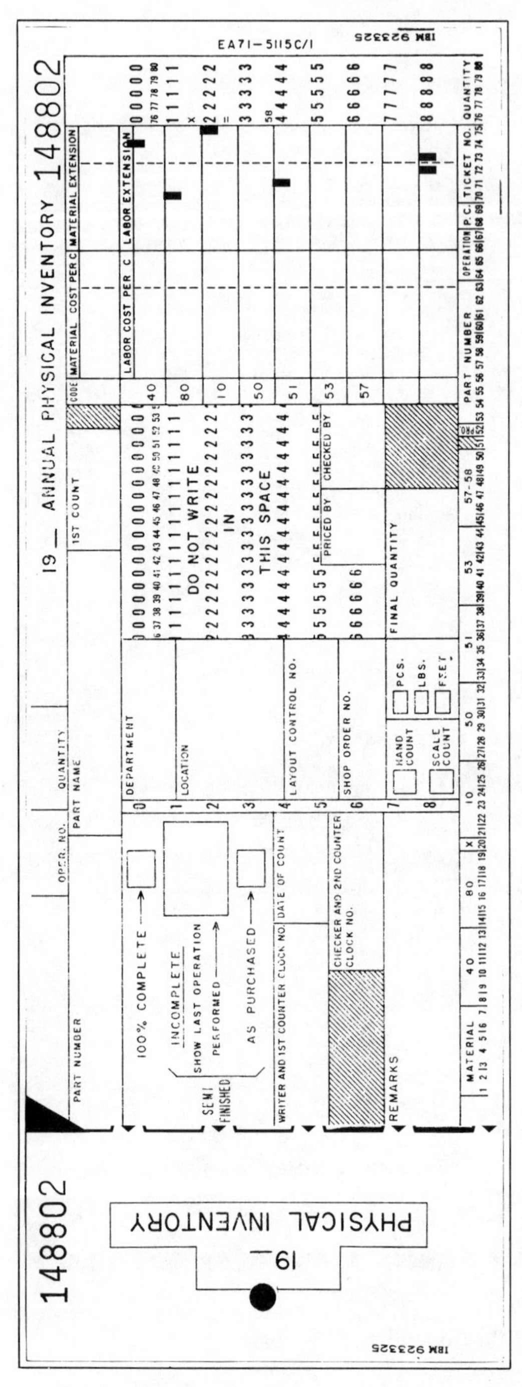

Figure 16-11. Dual tab card with stub.

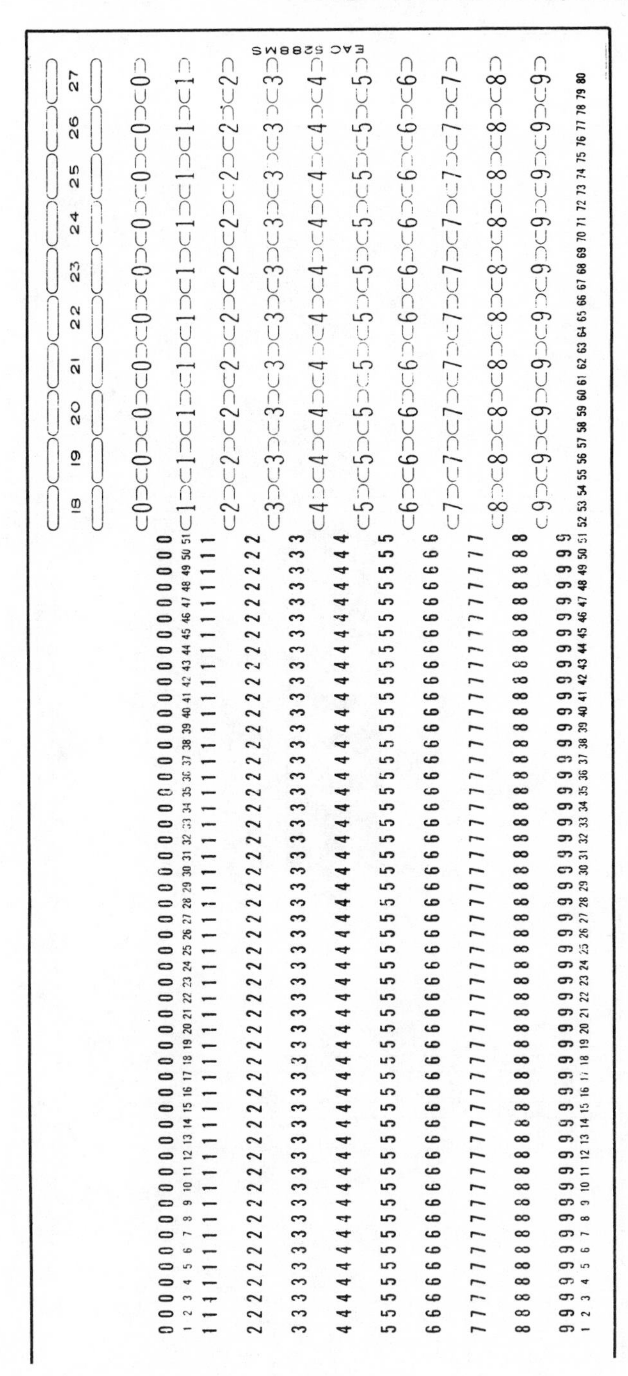

Figure 16-12. Mark sensing card.

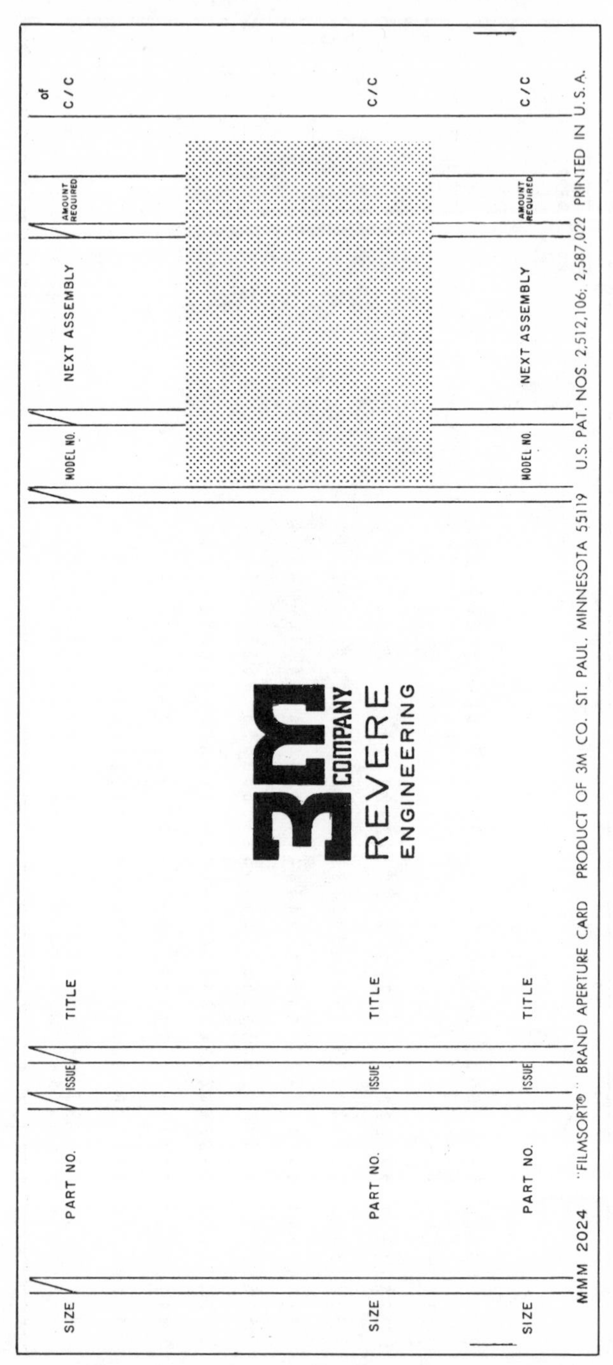

Figure 16-13. "Filmsort" aperture card. Use of such cards provides the advantages of both microfilm filing and punched-card processing. Technical drawings, specification sheets, and other documents may be retained in this manner.

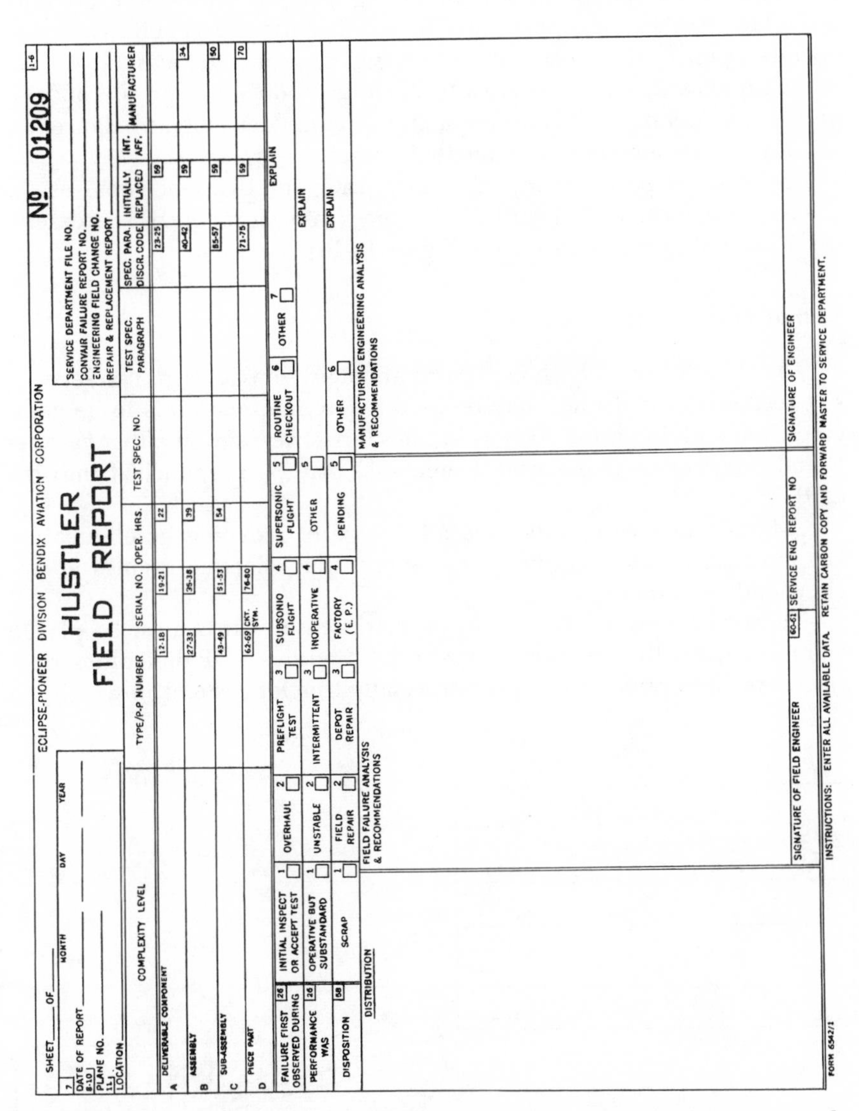

SHEET ___ OF ___

7 DATE OF REPORT 8-10 ___ MONTH ___ DAY ___ YEAR

PLANE NO. 11 ___

LOCATION ___

ECLIPSE-PIONEER DIVISION BENDIX AVIATION CORPORATION

HUSTLER

FIELD REPORT

SERVICE DEPARTMENT FILE NO. ___

CONVAIR FAILURE REPORT NO. ___

ENGINEERING FIELD CHANGE NO. ___

REPAIR & REPLACEMENT REPORT ___

№ 01209 1-6

COMPLEXITY LEVEL	TYPE/P-P NUMBER	SERIAL NO.	OPER. HRS.	TEST SPEC. NO.	TEST SPEC. PARAGRAPH	SPEC. PARA. DISCR. CODE	INITIALLY REPLACED	INT. APF.	MANUFACTURER
A DELIVERABLE COMPONENT	12-18	19-21	22			23-25	59		
B ASSEMBLY	27-33	35-38	39			40-42	59		34
C SUB-ASSEMBLY	43-49	51-53	54			55-57	59		50
D PIECE PART	62-69 CKT. SYM.	76-80				71-75	59		70

FAILURE FIRST OBSERVED DURING 26: INITIAL INSPECT OR ACCEPT TEST 1 ☐ OVERHAUL 2 ☐ PREFLIGHT TEST 3 ☐ SUBSONIC FLIGHT 4 ☐ SUPERSONIC FLIGHT 5 ☐ ROUTINE CHECKOUT 6 ☐ OTHER 7 ☐ EXPLAIN

PERFORMANCE WAS 26: OPERATIVE BUT SUBSTANDARD 1 ☐ UNSTABLE 2 ☐ INTERMITTENT 3 ☐ INOPERATIVE 4 ☐ OTHER 5 ☐ EXPLAIN

DISPOSITION 58: SCRAP 1 ☐ FIELD REPAIR 2 ☐ DEPOT REPAIR 3 ☐ FACTORY (E. P.) 4 ☐ PENDING 5 ☐ OTHER 6 ☐ EXPLAIN

DISTRIBUTION | FIELD FAILURE ANALYSIS & RECOMMENDATIONS | MANUFACTURING ENGINEERING ANALYSIS & RECOMMENDATIONS

SIGNATURE OF FIELD ENGINEER | 60-61 SERVICE ENG REPORT NO | SIGNATURE OF ENGINEER

INSTRUCTIONS: ENTER ALL AVAILABLE DATA. RETAIN CARBON COPY AND FORWARD MASTER TO SERVICE DEPARTMENT.

FORM 6542/1

Figure 16-14. Conventional source document from which information is derived for the keypunching operation. The small numbers relate to the card-punching locations and are meant to assist the keypunch operator.

Mark-sensed cards are automatically punched from electrographic pencil marks that are recorded in especially designated places on the face of the card (see Figure 16-12). To be properly recorded these marks must be made with a special pencil containing a highly conductive graphite lead. The brushes on a punch machine will sense the pencil marks and operate a punch-die mechanism, converting the marks to holes.

Output cards are automatically machine produced as a result of data-processing operations. These frequently contain data such as balance forward, summary amounts, and updated figures.

Special purpose cards make up a category that encompasses (*a*) checks in the form of IBM cards, (*b*) invoices in the form of IBM cards, and (*c*) "Filmsort" aperture cards (see Figure 16-13).

Source Documents

As already noted, both dual cards and mark-sensed cards can serve as source documents. Either may be used where information is to be gathered from outlying areas such as in sales territories, and in either case selected information (e.g., control numbers) may be prepunched into the card.

Mark-sensed cards have the advantage of reducing or eliminating keypunching and key-verifying operations, thus effecting savings in machine and operator expenses.

However, because the marks on mark-sensed cards must be placed and recorded with exceptional care to avoid serious ADP difficulties, dual cards are often preferred. This is especially true for systems in which the

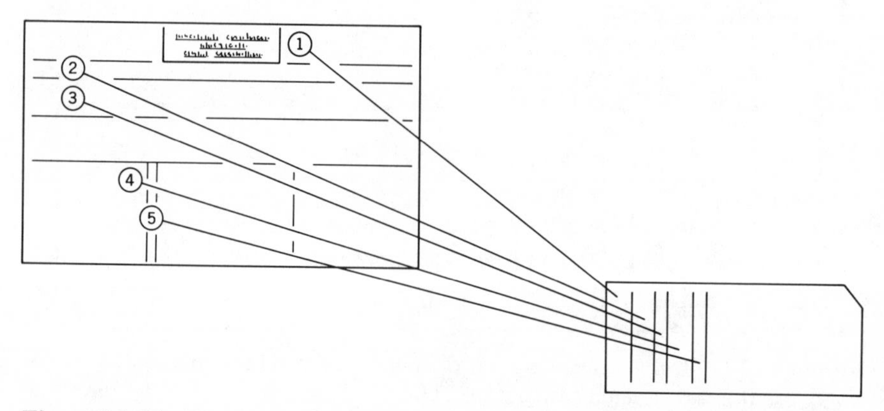

Figure 16-15. Forms involving transcriptions should be keyed to each other.

card undergoes considerable handling and is in danger of having the graphite markings smudged.

Dual cards and mark-sensed cards are in themselves machine processable. However, depending on the system, another type of form may be preferred for capturing data at its source; for example, an 8½- by 11-inch multipart form can be designed so that the information contained on one copy is easily read for keypunching into a transcript card. Figure 16-14 shows such a form. When this method is used both forms, the regular form and the card form, should be keyed to each other with respect to the location of information (Figure 16-15). For efficient keypunching the sequence of items and their relative arrangement on the forms must be orderly and harmonious.

Observe, in Figure 16-14, the small numbers in the various boxes on the "Hustler Field Report." These figures relate to the columns on the card into which the recorded information will be punched, and are for the purpose of assisting the keypunch operator.

17

Basic Machines and Their Functions

Each unit of EAM equipment is designed to perform a specialized function or several closely related functions, so that it takes a battery of interrelated machines to process the cards. The machines that are normally part of EAM installations are discussed below.

PUNCHING—KEYPUNCH MACHINES

Card punching, as implied by the term, is the act of perforating cards with the small holes that represent the information desired. The two most widely used machines for this purpose are the IBM 024 and the 026, which are the same in general appearance, features and function, except that the 026 can print characters above the columns punched. As an option either type of machine may be equipped with a numerical keyboard or a combination numerical and alphabetical keyboard, both kinds of keyboards being movable to facilitate manipulation and provide for the comfort of the operator.

Figure 17-1 shows a typical printing card punch. This particular machine possesses a combination keyboard, which is used to register alphabetical, numerical, and special-character information on tab cards. Basically, here is what occurs: the operator visually reads a character on the source document, touches the appropriate key, and thereby activates the punching mechanism, which cuts one, two, or three holes in the one column, depending on the key depressed. If the print switch—the third functional control switch on the upper part of the keyboard—is in the "on" position, the character will also be printed above the column punched. It

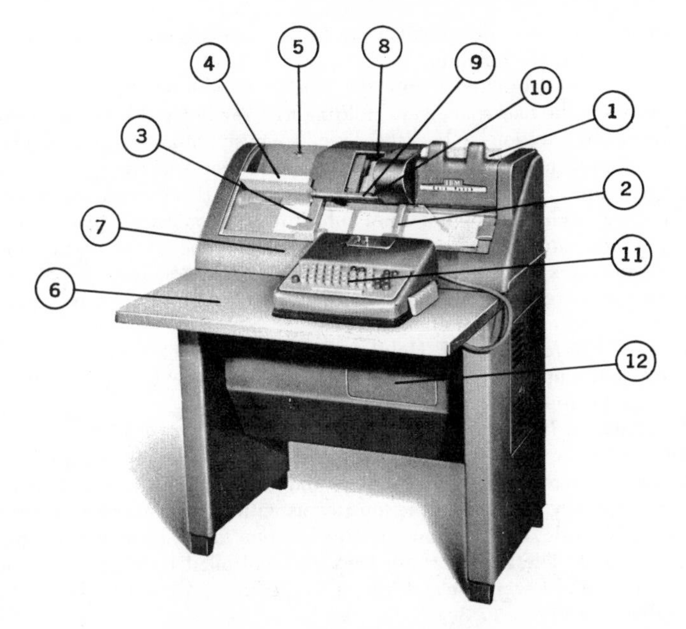

Figure 17-1. Printing card punch.

1. *Card Hopper.* Blank cards are fed from this position. The hopper holds about 500 cards and dispenses them, one at a time, either automatically or as a result of depressing a feed key.

2. *Punching Station.* This is the first of two stations in the card bed through which the cards move during processing. While the cards traverse from right to left, the card columns are punched in the opposite direction: from left to right. Ordinarily, an operation is begun by feeding two cards into the card bed at the right of the punching station. Synchronous action occurs: the feeding in of the second card automatically positions the first card at the punching station. During the punching of the first card, the second one awaits its turn on the right side of the card bed. After column 80 of the first card passes the punching station, moving on to the next station, the second card is brought up to the punching station, and another card is fed from the hopper to fill the space made vacant by the previous movement.

3. *Reading station.* Located about one-card length to the left of the punching station is the reading station. Normally, one card is passing through the reading station while the succeeding one is being punched. Movement of the two cards is synchronous, column by column, to permit information to be transcribed from the first card to the second. Transfer of information from one card to another is controllable so that only the desired information is duplicated.

4. *Card stacker.* From the reading station each card is fed into the stacker on the upper left side of the machine. The cards are accumulated at an angle, row 12 facing downward, and fall into their original sequence. A pressure plate holds the cards in position.

5. *Main-line switch.* Located above the stacker is the main-line switch. Before the machine is put into operation a period of about one minute is needed to warm up the electronic tubes. Filling the stacker results in throwing the switch off automatically.

6. *Reading platform.* The movable keyboard rests on the reading platform. Both

the keyboard and the source documents from which the cards are punched can be positioned to accommodate the operator.

7. *Backspace key.* Situated below the card bed between the reading and the punching stations is the backspace key. Holding this key down causes the cards at the punching and reading stations to backspace, continuously. Moreover, the program card that controls skipping and duplication is simultaneously backspaced.

8. *Program unit.* This is where the program card is mounted on the program drum. A clamping mechanism on the drum holds the program card in place.

9. *Column indicator.* This shows the operator the number of the next column to be punched. The numbers are listed around the base of the program drum holder, which rotates as the cards are being punched. Reference to the indicator makes it easier to space or backspace to a particular column.

10. *Pressure-roll release lever.* Holding down on this lever permits the removal of a card caught at the punching or reading station. Such cards can usually be set free without damage if care is exercised.

11. *Combination keyboard.* The keyboard, shown in Figure 17-2, is in many respects similar to those found on standard typewriters. The arrangement of the letter keys is identical to permit operation by the standard typewriter touch method. However, the numerical keys are not on the top row as with a typewriter keyboard, but are placed on the right side of the keyboard. This location facilitates numerical punching with the right hand. The dual-purpose keys are controlled by shift keys, similar in operation to upper and lower case shifting on a typewriter. A chart of the combination keyboard is shown in Figure 17-3.

12. *Chip box and fuses.* The receptacle for the punched-out chips is located under the reading platform. Removal of this box affords access to the fuses of the machine.

is possible, through program-card coding, to suppress printing for one or more columns of the card, even though the print switch is on. After the information for a column has been punched, the card is automatically brought to the next position.

The components of the printing card punch may be briefly described as follows:

Cards are said to be *interpreted* when the punched holes are correspondingly recorded in print.

Both the 024 and the 026 machines are equipped with a control unit that controls automatic skipping, automatic duplicating, and shifting from one mode to the other (from the numerical mode to the alphabetical, or vice versa). The unit consists mainly of a program card mounted on a program drum and inserted in the machine. Punched holes in this card, the card itself being a regular IBM card, control the automatic operations for the corresponding columns of the cards being punched. A different program card is prepared for each different punching application, and it may be used over and over.

The speed at which data can be initially keypunched into cards is subject to many variables, such as the number of columns to be punched, the experience of the operator, and the nature of the information. As a guide-

Figure 17-2. Alphabetic and numeric keyboard.

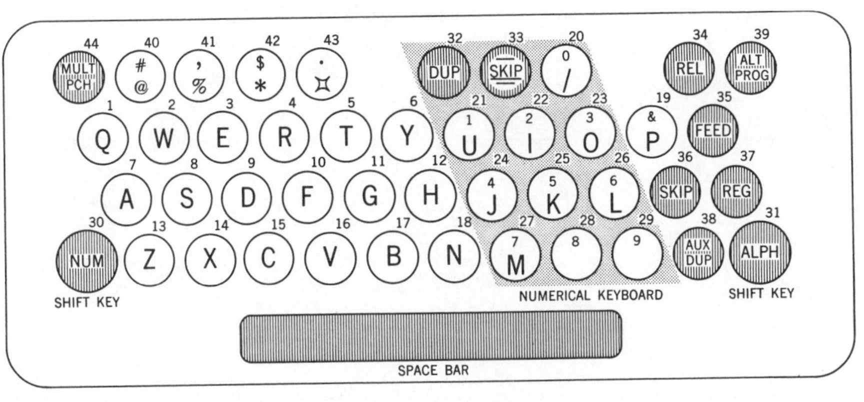

Figure 17-3. Combination-keyboard chart.

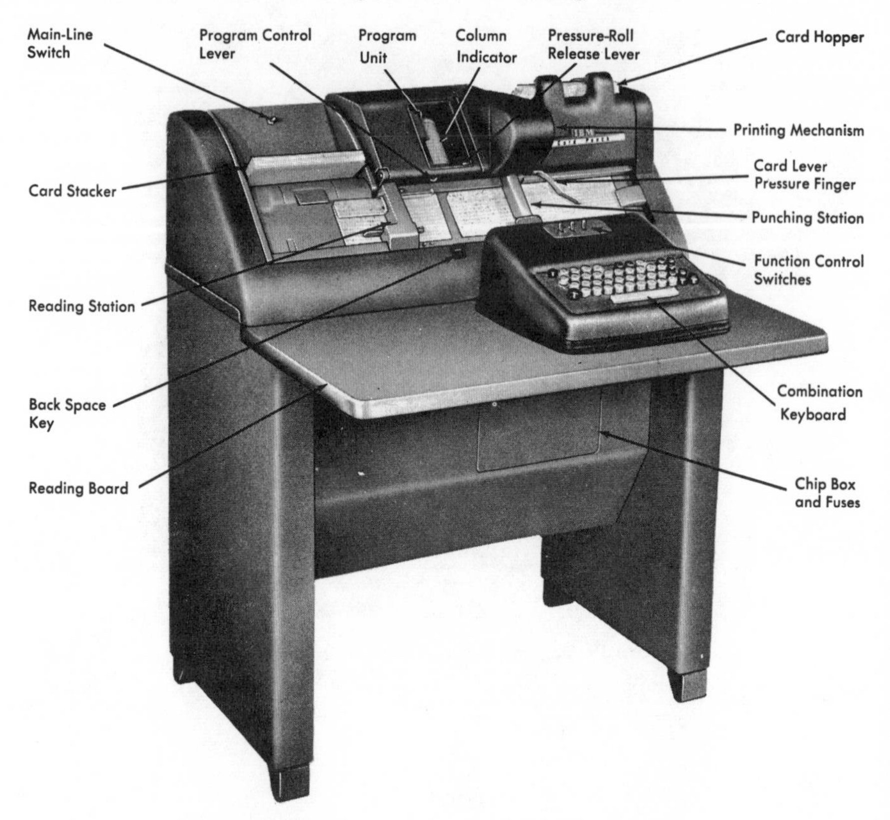

Figure 17-4. IBM 026 Printing card punch.

post, however, an experienced operator can punch cards in all columns at the rate of approximately 100 cards per hour. An IBM 026 printing card punch is shown in Figure 17-4.

A different type of card punch is the portable numerical unit shown in Figure 17-5. It is lightweight, movable, and manually operated. For these reasons it has a variety of applications at places away from the central tabulating facility, such as in field sales offices and selected locations throughout the plant. The IBM Corporation markets a machine similar to this one, called a Porta-Punch.

CARD VERIFIER

The introduction of errors during card-punching operations is always a possibility, especially when large quantities are being punched on either

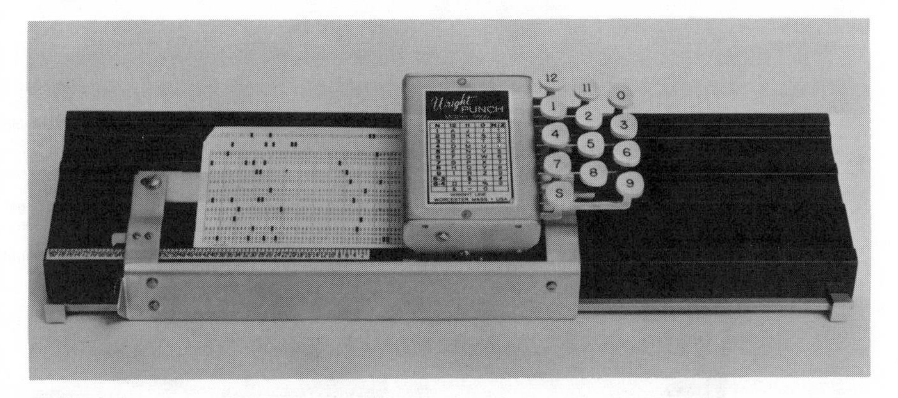

Figure 17-5. Portable card punch.

the 024 or 026 card-punch machine. Therefore some way is needed to determine if the punched information is in accordance with the source document. A machine known as a verifier performs this function (see Figure 17-6). In appearance and in general operation (except for the error-detection routine) it resembles the 024 and 026 card-punch machines.

Verification is accomplished by processing the cards over again as if they were being keypunched for the first time. Instead of punching holes, however, the verifier senses them.

The operator of the verifier places the punched cards into the card hopper, reads from the same source document from which the material was originally punched, and proceeds as in the manner of punching new cards. Should the verifier sense all of the punched columns and find no discrepancies, the verified card will receive an "OK" notch (Figure 17-7) in the right end of the card and then be ejected to the stacker.

On the other hand, should there be an inconsistency between what is punched in the column of a card and the key depressed by the verifier operator, the machine will stop and an error light will go on. When this occurs the operator is given the opportunity to find out whether the error is in the verification or in the original punching. If the card was properly punched originally and the verifier operator depressed the wrong key, the light will go off when another attempt at verification is made correctly. Actually, the verifier operator has three opportunities to check the accuracy of a punched column. If, after the third attempt, there is still an indicated discrepancy, the verifier will cut a notch directly over the affected column (Figure 17-8) and permit the operator to go on with the verification. Such cards are not cut with a final OK notch.

In a deck of verified cards it is easy to sight check and remove all

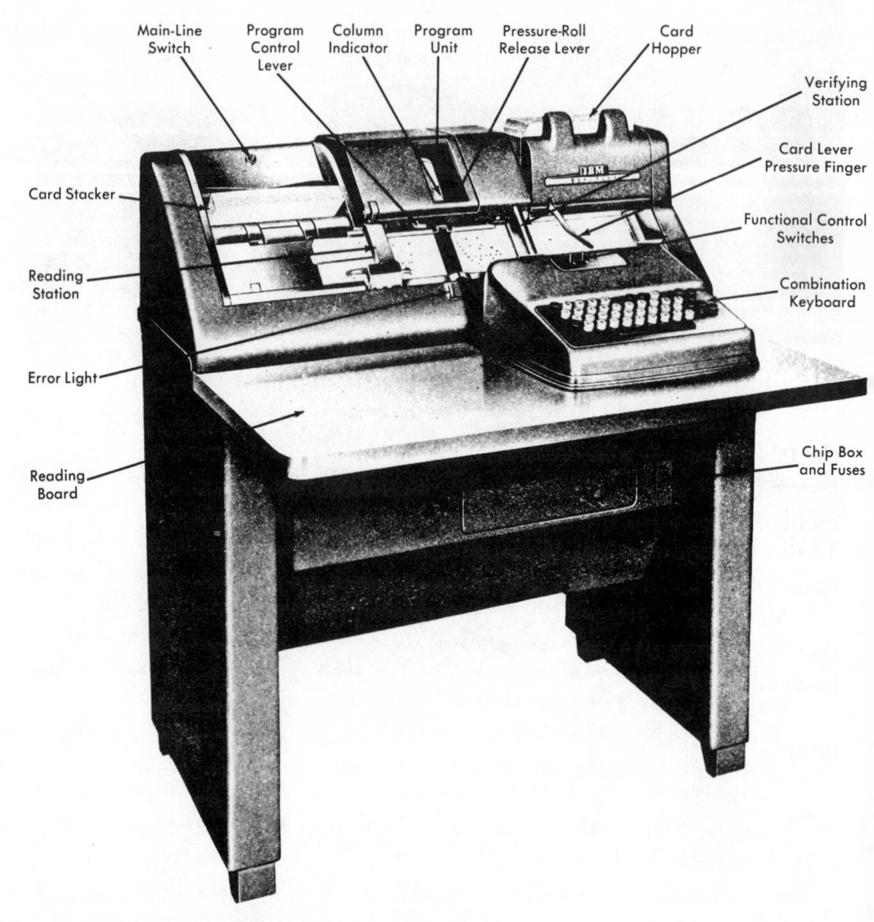

Figure 17-6. Card verifier.

cards that do not contain the right-hand notch. These cards should be repunched and then reverified. In the process of doing the repunching all columns except the ones containing an error notch will be duplicated on a new card, and only those columns that were incorrect are repunched. Thus time is saved by eliminating the need for repunching all columns. This also eliminates the chance of error through incorrectly repunching a column that was initially correct.

By intent, verification may be either partial or complete. In most cases numerical data should be completely verified. In certain cases the verification of alphabetic information can be omitted—a calculated risk being preferred to taking the time and effort necessary to assure near perfection.

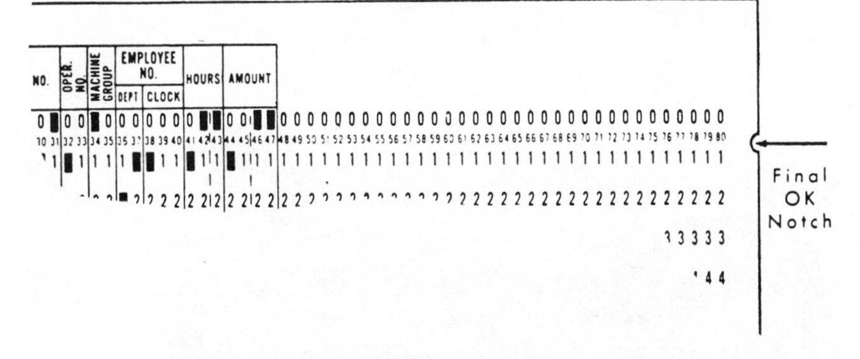

Figure 17-7. Punched card with "OK" notch.

Wherever feasible a different operator should perform the verifying operation. This makes it less likely that the same error will be repeated.

CARD SORTER

To prepare various management reports it is necessary that the data on the same cards be presented in many different ways. Consider, for example, sales reports. If distributions of sales by individual territories, products, and salesmen are wanted, then it is necessary to arrange the cards for the particular sequence or grouping of the information desired on the report. A sorter such as the model shown in Figure 17-9 performs this function. Punched cards placed in this machine, or a similar one, can be arranged in numerical or alphabetical sequence according to any classification punched in them.

It will be observed from the illustration that the machine is designed with 13 compartments, or pockets. From left to right, facing the machine,

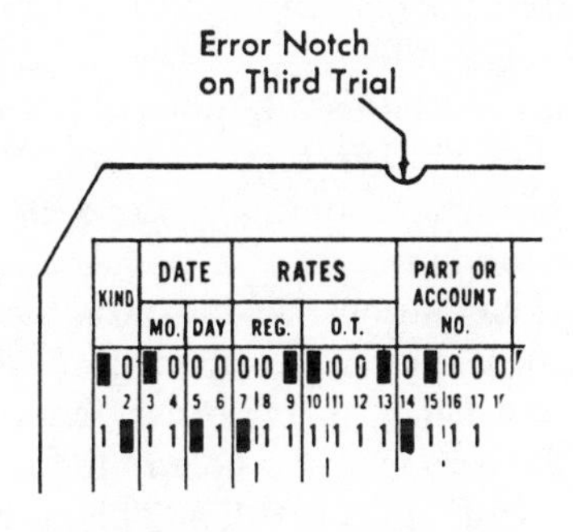

Figure 17-8. Error notch.

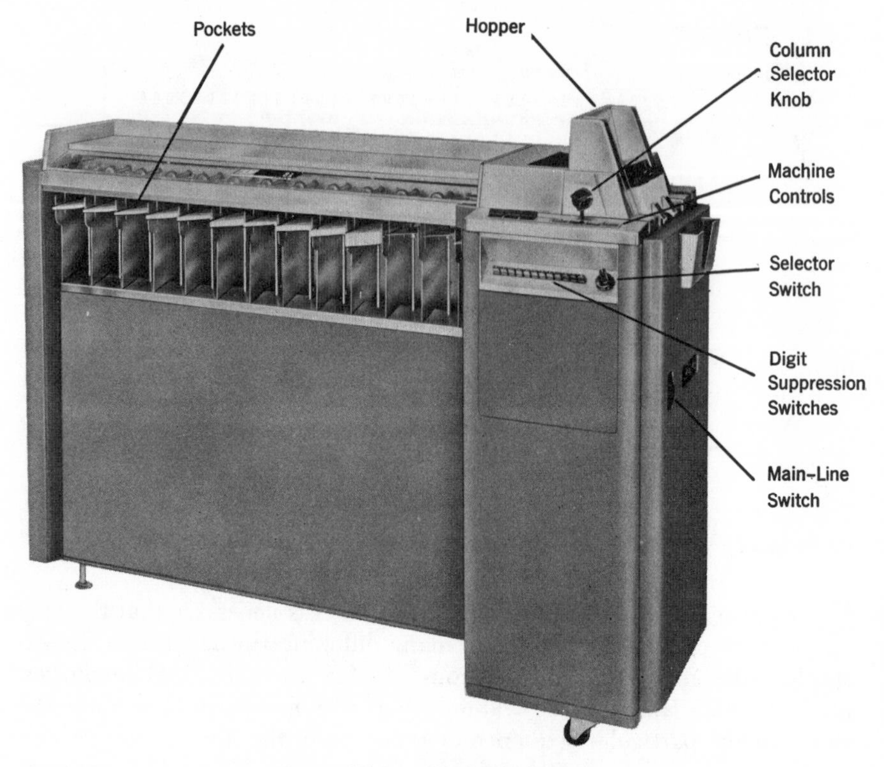

Figure 17-9. IBM 083 Sorter.

these pockets are designated: 9, 8, 7, 6, 5, 4, 3, 2, 1, 0, 11, 12, and R. Each of the first 12 pockets corresponds to one of the 12 punching positions in a card column. The R, or reject, pocket receives cards that are unpunched in the column being sorted. By means of the selection switch it is possible to separate from a file all cards containing any specified punch, without disturbing the remainder of the file.

Cards are placed into the feed hopper of the machine face down, 9-edge first. They are then read in sequence from the bottom of the deck. The sorting at any one time is normally done according to one vertical column; that is, for a unit number, letter, or special character.

Pressing the start key automatically starts the cards feeding in from the bottom of the stack and advances them on feed rolls to a reading station. Here they pass between a sorting brush, made of fine steel wires, and a metal roller. With the presence of a hole in the column of a card contact is made between the brush and the roller. This results in the closing of an electrical circuit, which causes the opening of a chute blade that corresponds to the punched hole, and the card is directed by the feed rollers and the chute into the appropriate receiving pocket. Accordingly, for the column being sorted, every card with a *1 position* punched will drop into the *1 pocket,* every card with the *2 position* punched will drop into the *2 pocket,* and so on. In the absence of a punched hole, the card will drop into the R pocket.

Sorting is normally carried out from right to left within the card columns of a field; for example, the number 269453, in one field that terminates in card column 20, would be sorted on columns 20, 19, 18, 17, 16, and 15—in that order.

Whenever the volume of cards is so great as to be unwieldy, a technique known as *block sorting* may be used. Under this plan the cards are separated into groups so that each group can be handled independently of the others; for example, it may be desirable to divide a file into 10 blocks by sorting according to the left-hand column of a five-digit number. Then, each block can be handled separately and sorted normally, from right to left, through the unsorted columns of the field.

When operations involve more than one field, blocking can be accomplished by first sorting the major field. Each major division can then be treated independently.

Block sorting permits the simultaneous use of more than one sorter, not only for establishing the blocks but for sorting each block as well. It also makes it possible to process completed blocks on other machines while the remaining blocks are still in the process of being sorted. Block sorting thus affords the advantages of saving time and speeding up the preparation of reports.

In numerical sorting it is sufficient to make one sort on each pertinent column, because a number is distinguished by a lone hole in the one column. To state it differently, the cards must be passed through the machine once for each column of the field being sorted. Alphabetic sorting, however, requires two complete passes, because a letter is recorded by two holes in a single column. Inasmuch as the cards are read starting with the ninth row, it follows that the first sort is a digit sort, and the second sort a zone sort.

Sorters that are available operate at speeds ranging from 250 to 1000 cards per minute. These use the most common reading mechanism—the brush. Newer and faster machines, however (some sorting up to 2000 cards per minute) utilize a photoelectric cell for reading the cards. Even though the reading method is different, the basic technique of sorting is the same.

Card-counting units may be used on sorters to ascertain the total number of cards processed and to count the number of cards going into each pocket. These devices provide fast and dependable totals, and are especially useful for obtaining statistical compilations.

To demonstrate how a sorter operates let us look at Figure 17-10, which shows the sorting operation in diagrammatic form.

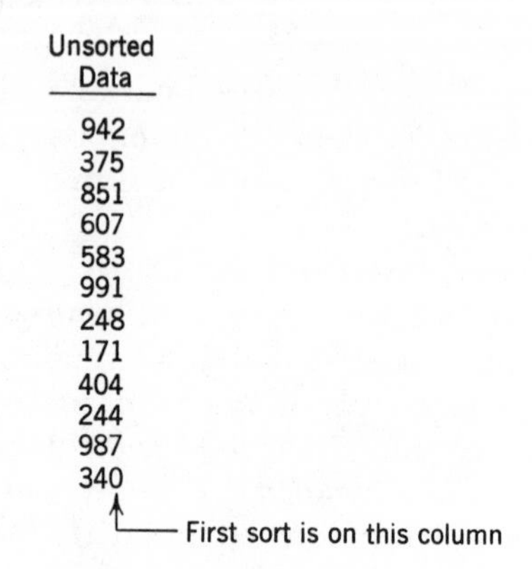

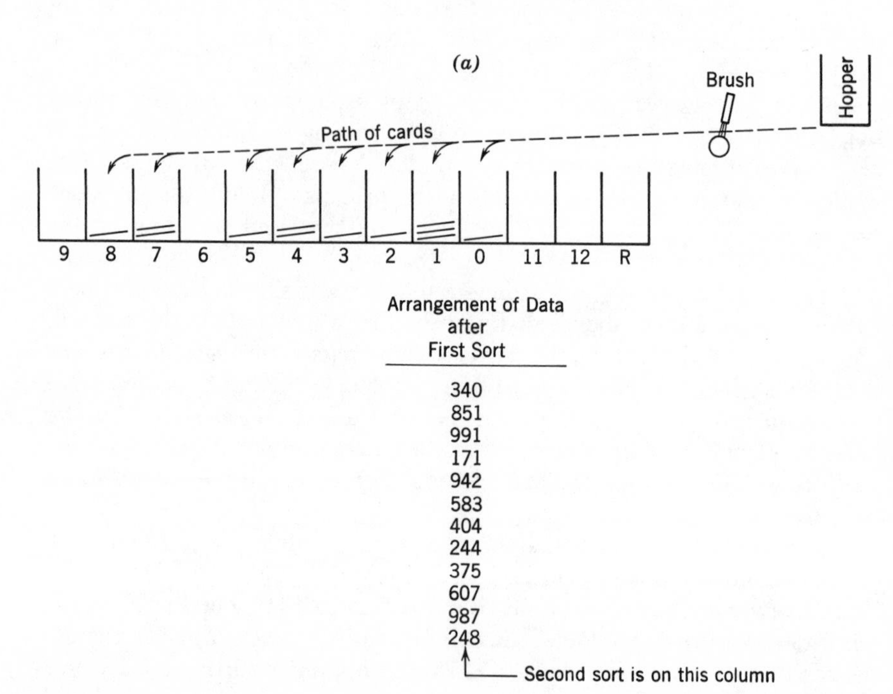

Figure 17-10. Schematic representation of the sorting operation: (*a*) first sort; (*b*) second sort; (*c*) third sort.

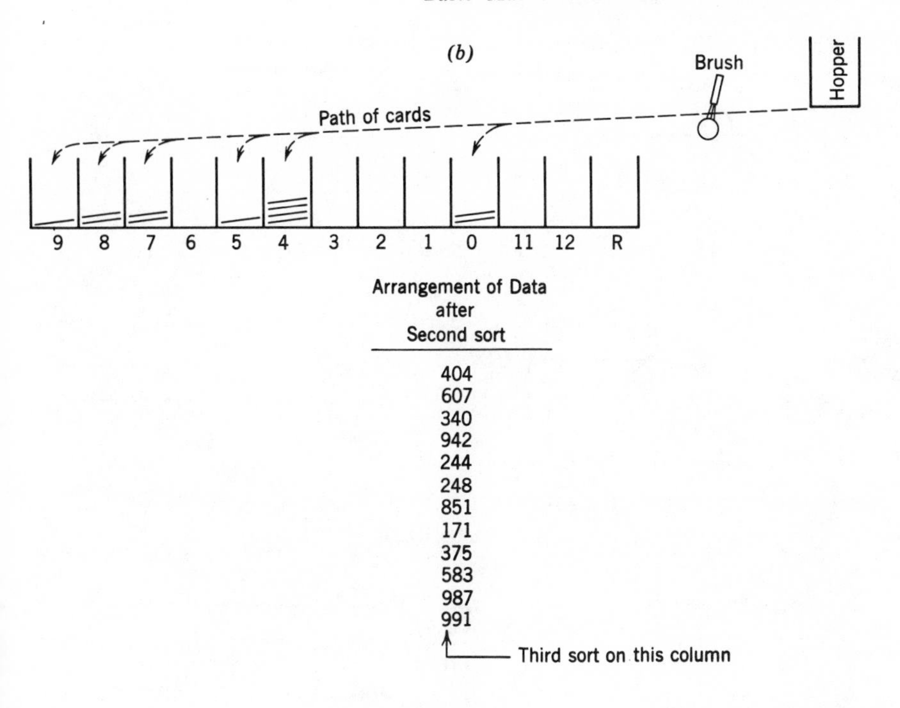

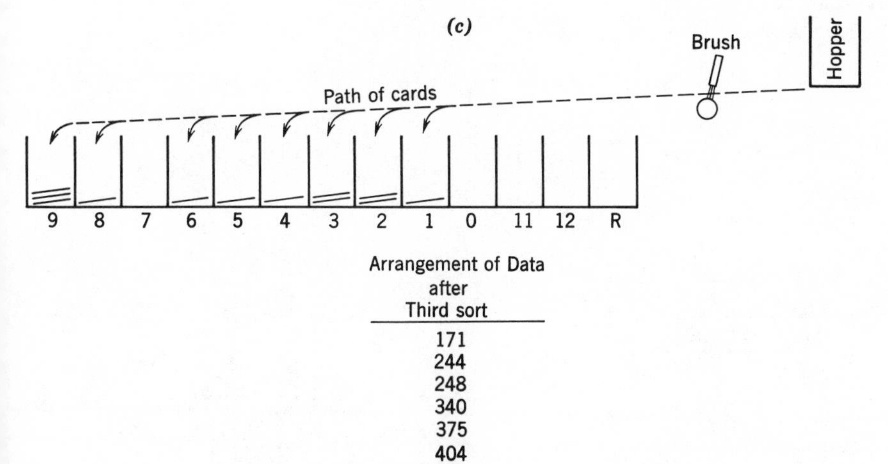

Figure 17-10. *Continued*

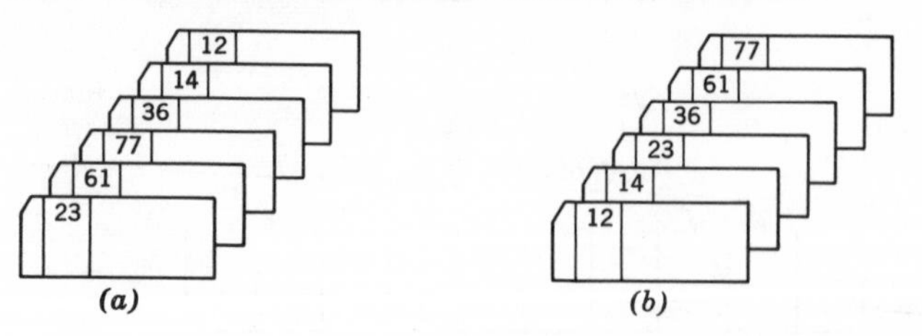

Figure 17-11. Sequencing: arrangement of data (*a*) before; (*b*) after.

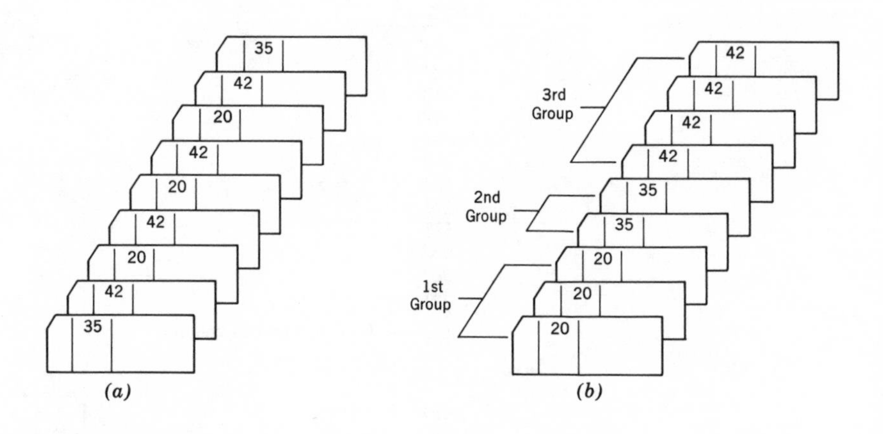

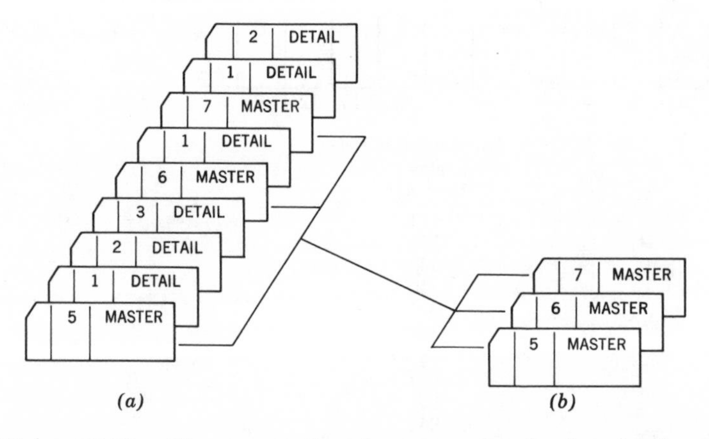

Figure 17-12. Two types of sorting—grouping (top) and selecting (bottom). In both cases the arrangement of data before sorting is shown in (*a*); and after, in (*b*).

The type of sorting shown in Figure 17-10 is called *sequencing*. Two other types of sorting, with names that are equally appropriate, are known as *grouping* and *selecting*. Examples of all three are shown in Figures 17-11 and 17-12.

COLLATOR

The basic operations of the collator are selecting specific cards from a file, checking the card sequence in a file, combining two decks of cards into one (with or without the selection of certain cards), and matching two sets of cards while selecting any unmatched cards from each set.

There are a number of different makes and models of collators. The

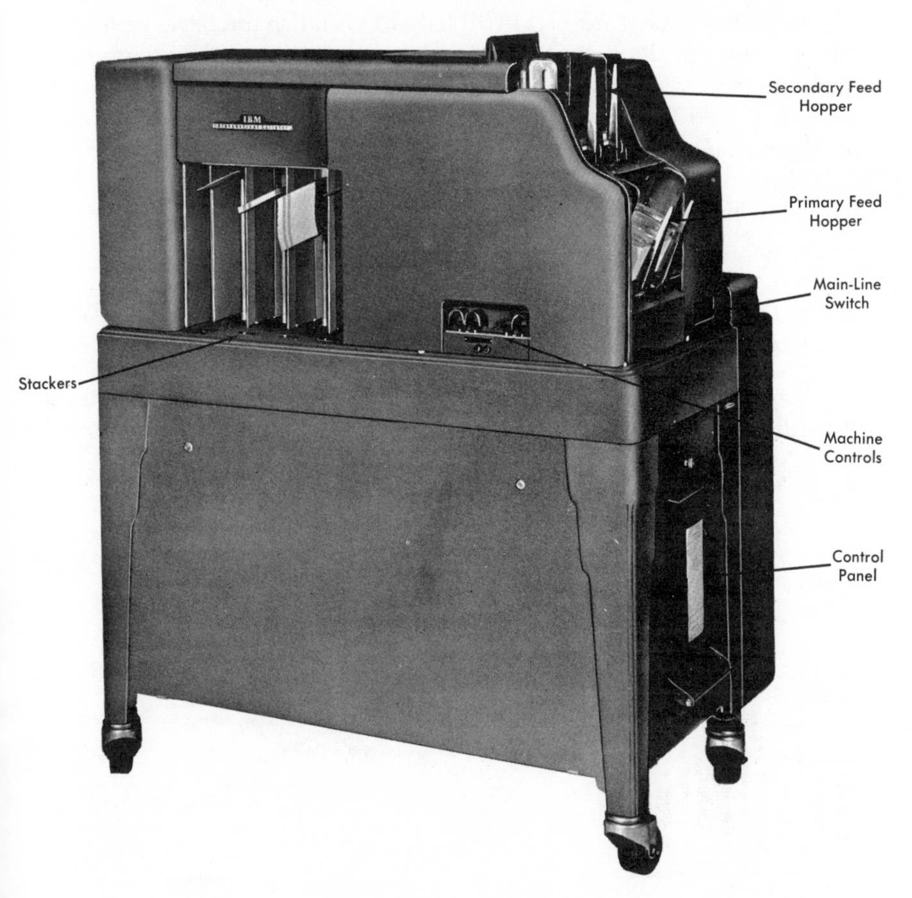

Figure 17-13. Collator.

IBM 77, 85, and 88 are numerical, while the 87 and 89 are both alphabetical and numerical. Figure 17-13 shows the IBM 89.

In construction the collator has two feeding stations, called a *primary feed hopper* and a *secondary feed hopper*, and four stackers for receiving processed cards. The reason for the dual feeding stations is to permit two sets of cards to be fed simultaneously into the machine for the purpose of matching or merging them.

The stackers, or pockets, are situated on the left side of the machine and are numbered from right to left, 1 through 4. Pocket 1 is for selected primary cards, 2 is for merged cards, and 3 and 4 are for selected secondary cards. Primary cards can be directed to either pockets 1 or 2. Secondary cards can be directed to pockets 2, 3, or 4. In merging operations the two sets of cards are brought together in pocket 2.

Reference to Figure 17-14 will help to visualize the basic principle of the operation of the collator. It may be observed from this diagram that there are three read stations, two on the primary feed track and one on the secondary feed track. In order for all 80 columns to be read there are 80 brushes at each read station.

Machine functions that the collator is capable of performing fall into the following four classifications:

1. Sequence checking
2. Selecting
3. Merging
4. Matching

Sequence Checking. As the name infers, sequence checking is the act of verifying the filing order of a group of cards. This is normally done by

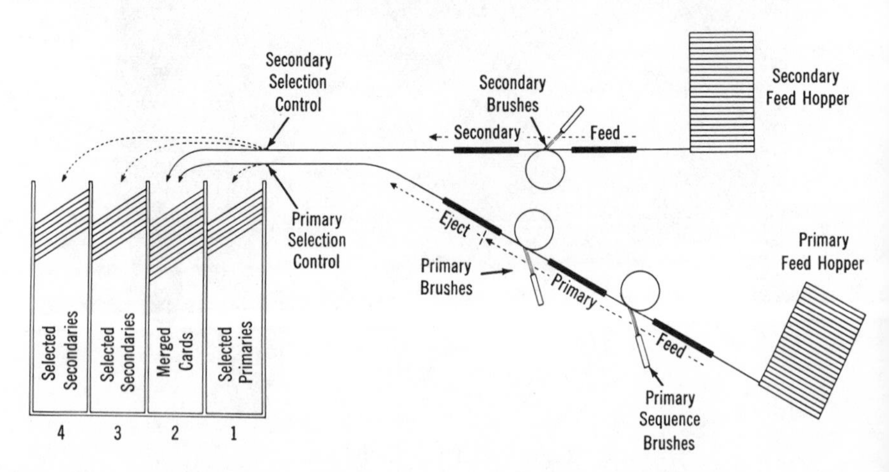

Figure 17-14. Schematic of card-processing channels.

placing the file to be checked in the primary feed hopper, from which they are fed through the machine and deposited in the second pocket. Along the way each card is compared with the one ahead to determine whether the card is higher, lower, or equal to the card ahead of it. When an error in sequence is detected, card feeding automatically stops and the error light goes on. Cards from both the hopper and the stacker are removed. After the error-reset key has been pressed the cards within the machine are passed through by pressing the run-out key. The second card that is run out is called the step-down card; it may or may not be the card out of sequence. A visual check must be made to locate the precise card or cards that are out of order so that they can be properly filed by hand.

Selecting. The process of extracting one or more cards from the file placed in either feed hopper is called selecting. The nature of the selection is governed by the instructions wired into the control panel. Kinds of possible selections include (*a*) all cards with a particular number; (*b*) single cards that are not part of any group; (*c*) the first or last of a group of cards punched with the same control number; or (*d*) all cards punched with numbers between an upper and a lower limit—for example, 1530 and 4685.

In any of these selection operations selection usually involves the primary feed; and selected cards are stacked in the first pocket, whereas the cards that are to remain in the file are stacked in the second pocket. However, since the secondary feed permits the use of three stackers, it is sometimes used for the purpose of making a double selection. When this is the case selected cards can be directed to pockets 3 or 4, and the cards that are to remain in the file are stacked in pocket 2.

Merging. This is the process of interfiling two sets of cards, which are in similar order, into one set in that same order. Usually the cards to be merged are in ascending order, and therefore the resultant merged file will be in ascending order.

In merging a card from one file is compared to a card in the other file to determine which of the two should be moved first to the combined file. The comparison indicates one of three possibilities: the card in the first file is low, the card in the second file is low, or the cards in both files are equal. Low cards are merged in front, so the file is arranged in ascending order; when the cards are equal the card in the first file is merged in front of the card in the second file.

To accomplish the merging operation one file of cards is placed into the primary feed hopper and the other into the secondary feed hopper. Primary cards are merged ahead of secondary cards whenever the cards from both hoppers are equal. Accordingly the cards that should appear first in the merged file should be placed in the primary feed hopper. The stacking of the combined file takes place in the second pocket of the machine.

Merging is an operation that can be conducted in conjunction with both sequence checking and selection.

Matching. The term "matching" is used to describe the checking or comparison of two similar sets of cards to find out if each card or group of cards in one file has a counterpart in the other file. Unlike in the merging operation, the two sets of cards remain apart throughout the processing and end up being stacked in separate sets in the same order in which they were initially fed into the machine. The primary cards are collected in the second pocket, whereas the secondary cards are collected in the third pocket. Unmatched cards from either or both sets can be selected or withdrawn from the files. Thus all four machine pockets may be utilized—two for the separate sets that match, and the others for two groups of selected cards.

ACCOUNTING MACHINE

Known also as tabulators, machines of this classification automatically read the information contained in punched cards and prepare reports that are both readable and meaningful to management. Information punched in cards can be read, and the machine will add, subtract, and print out any required combination of totals. The printed information is usually recorded on a continuous form while it is automatically fed through the machine carriage. Figure 17-15 shows the IBM 407 accounting machine, one of the most widely used of its type.

Accounting machines can be regulated to print on one or several lines from each card or on one line from a number of cards. They can print by "detail" or by "group." Detail printing is the printing of selected information from every card as it is run through the machine. The technique is used to prepare comprehensive reports wherein the enumeration of particulars is desired. Group printing, on the other hand, involved summarizing the information contained on groups of cards and printing the totals on a report. The totals may involve additions, subtractions, or crossfooting. To accomplish this the accounting machine reads the information from the punched cards and enters it into automatic counters. If programmed for totals at the end of each group of cards, the machine will read out the totals and list them in report form. Descriptive information may be provided for all totals.

In terms of the arrangement of information the layout of the tab card does not need to coincide with that on the printed form. They should be

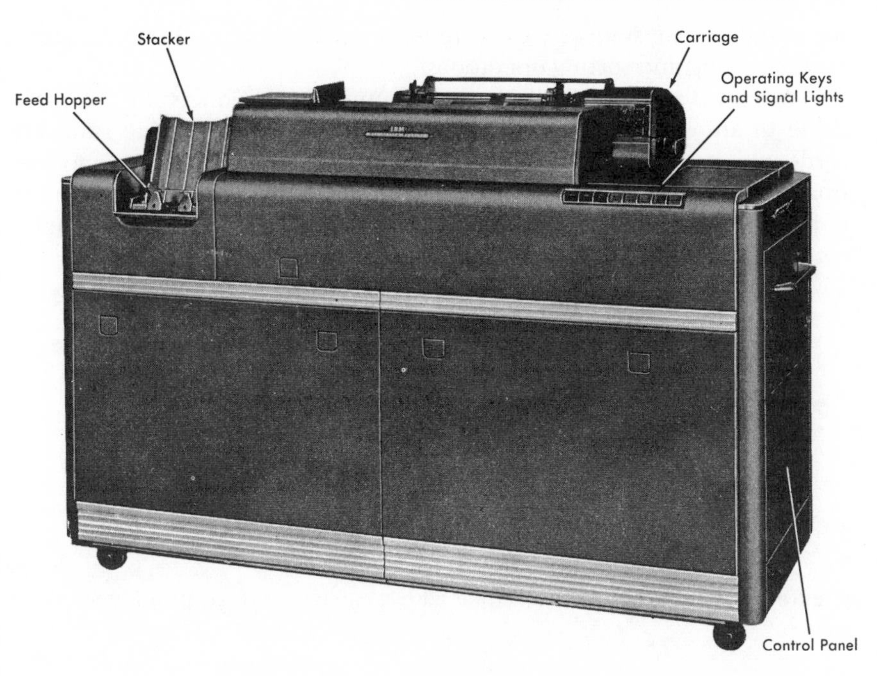

Figure 17-15. Accounting machine.

adapted to each other, of course, but they do not have to correspond exactly. Like most EAM equipment the accounting machine has a control panel, or "brain" as it is sometimes called, and this panel instructs the machine where to print the information it reads. Accordingly information sensed in any given field on the card can be transferred to selected columns on the printed form. The panel also functions to control other operations as the cards are fed through the machine.

The control-panel feature provides flexibility. Since the panel is readily removable, wiring changes can be made with ease; also, one panel may be conveniently replaced with another that is set up for accomplishing an operation of a different kind. In this way operational programs are preserved for future use.

SUMMARY PUNCH

This is a punch-card machine that may be connected to another machine to produce punched cards bearing summarized calculations or totals from the other machine. This information may be used for subsequent operations; for example, to carry totals for producing a report the following month. Using summary cards reduces card volume and thereby facilitates handling and sorting operations.

A typical application of a summary punch is to connect it by way of a cable to an accounting machine for the purpose of obtaining summary cards with punched totals (see Figure 17-16). The cable carries impulses from the various accounting-machine circuits to the punching mechanism of the summary punch. Control over these impulses is exercised through a control panel located in the base of the punch.

Figure 17-16. Accounting machine with a cable-connected printing summary punch.

Illustrative of a project for an accounting machine and summary-punch hookup is the preparation of inventory reports. The accounting machine can be set up to produce a report showing for each item the opening balance on the first of the month, the receipts and issues during the month, and the balance on hand at the end of the month. While the report is being prepared the accounting machine will also automatically transmit the information on totals for each class of items to the summary punch. Cards are thus made available for use in the preparation of the next report.

Except for the summary-punching feature summary punches perform the same functions and operate in the same manner as standard card punches, such as the printing card punch shown in Figure 17-1. When the summary punch is not being used for summary punching it may therefore be used as a keypunch for regular card punching.

REPRODUCER

In manual systems it is often necessary to transfer information, in whole or in part, from one record to another. In some instances information generated at one point may be needed at another, requiring duplication of the same basic data. At times, too, totals must be accumulated and transcribed from certain documents to others.

When punched-card systems are used these repetitious tasks can be performed automatically by machines known as reproducers.

Machines of this classification are basically punching units designed to perform some or all of the following functions:

1. Card reproduction
2. Gang punching
3. Summary punching
4. Mark-sensed punching
5. End printing

Not every reproducer can perform all of these functions, but most models can perform the first three.

Card Reproduction. Reproducing is a duplication process wherein information on one set of cards is automatically punched into another set of cards. The punched cards are placed in the read-feed hopper of the machine; those that are set to be punched are placed in the nearby punch-feed hopper. Both are fed through the machine synchronously so that the previously punched cards can be sensed in the *reading channel,* while blank cards are correspondingly perforated in the *punching channel,* to become copies of the originals. Throughout the processing the two stacks of cards remain apart, and each set is accumulated in a separate stacker. A typical reproducing punch is shown in Figure 17-17, and a schematic of card-processing channels is given in Figure 17-18.

The layout of both card forms need not be identical. Information contained in the original set of cards can be punched into the same or entirely different columns of blank cards, as suits the requirements of the system. Positioning of the information is accomplished by means of control-panel wiring.

Built into the machine is a comparing unit for verifying the punched information as it is reproduced.

Gang Punching. This is the automatic duplication of punched information from a master card into any number of individual or detail cards that follow it. Cards are fed from the punch-feed hopper; as they pass through the punch unit the master, or lead, card is read, and the desired information is transferred to the next card. Feeding of the cards is continuous, and the punching process itself is repetitive. When more than one detail card follows a master card the machine reads the first detail card, which has just been punched, and transfers the information to the next detail card. This action is repeated until every detail card that follows a master card is punched with the identical information. The card in front is

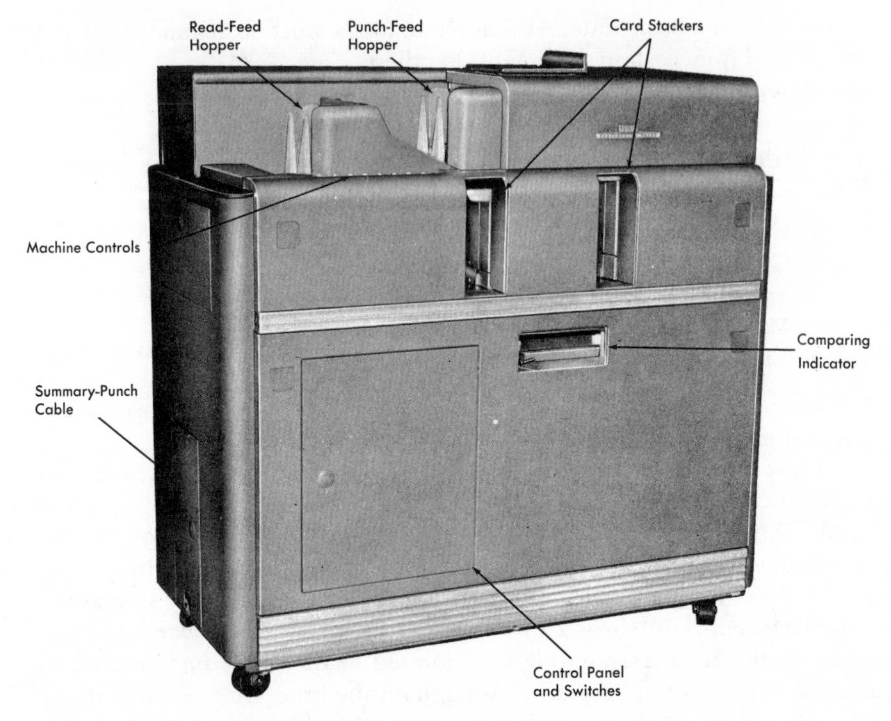

Figure 17-17. Reproducing punch.

always responsible for the information punched into the card immediately following.

Two methods of gang punching are common—single master-card gang punching and interspersed master-card gang punching. In the former method the same information is punched into every card that is processed through the machine. The master card is placed in front of the cards to be gang punched, and the entire set of cards is fed into the punch-feed hopper. As a result all of the cards end up having identical information punched into them.

The interspersed master-card method uses multiple master cards. Information in the master card is punched into all cards following it until the next master card is reached. With a change in master cards there is then a change in the information that is subsequently punched. This method of gang punching is used where the common information varies by groups within the file.

Summary Punching. The automatic punching of totals that have been accumulated or summarized by machine is known as summary punching. For this purpose the reproducer may be connected by cable to an accounting machine and receive data from specified locations of the accounting-machine circuits, as in the case of the summary punch previously described (see Figure 17-16). Another application, which is possible with the IBM 528 accumulating reproducer, is to use it independently to prepare summary cards at high speed from detail cards.

Mark-sensing is an important operational feature of some reproducers. By sensing the electrically conductive marks on a card these machines automatically convert the markings into regular holes in selected columns on the card. Once cards have been translated into punched-hole form, they can then be read by other machines in the usual manner.

End Printing. This process converts punched information into bold printing across the end of the card. In the IBM system this operation is a function of the 519 document-originating machine (Figure 17-19), a major type of reproducer. Cards can be printed with as many as eight digits of information on one line in a single pass through the punch unit. An additional line may be printed by repositioning the printing unit and passing the cards through the machine a second time. The information that is recorded on each card may be read from the card itself, a process called interpreting; from a card in the read unit at the comparing brushes, a process called transcribing; or from an emitter.

End printing is useful where visual referencing is important—such as in the case of employee time cards, which are stored in racks for ready reference and selection, and in inventory-control systems where cards are stored on end in a tub file.

As an operation end printing can be done concurrently with reproducing, gang punching, summary punching, and mark-sensed punching.

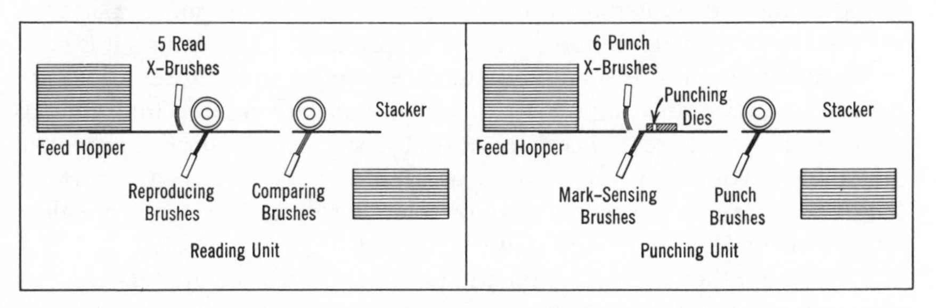

Figure 17-18. Schematic of card-processing channels.

CALCULATING PUNCH

The accounting machine will add or subtract, but a class of machines known as calculating punches will also multiply and divide. In the devel-

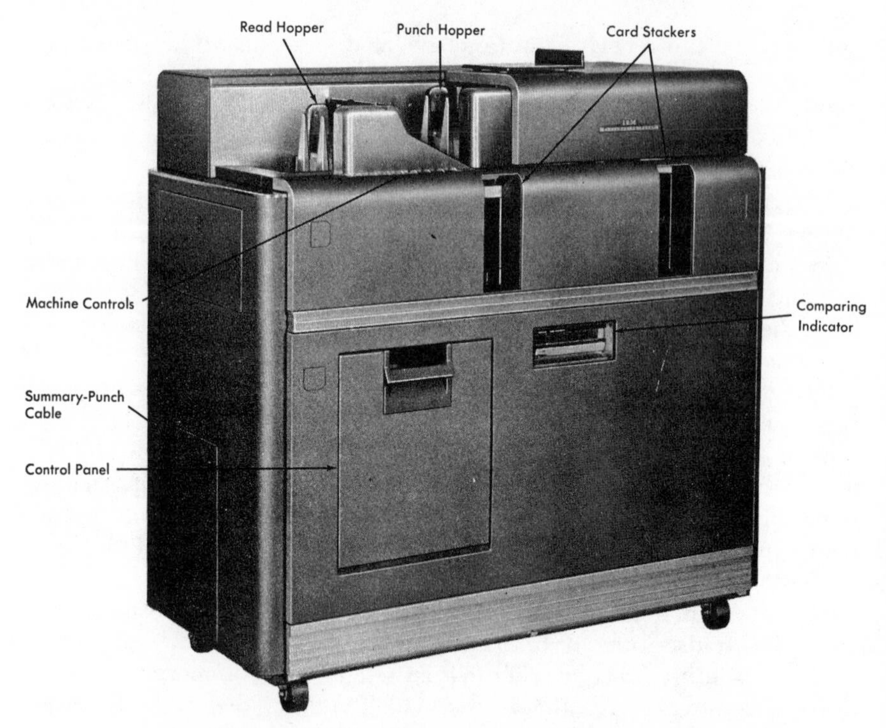

Figure 17-19. Document-originating machine.

opment of automatic-data-processing equipment they are considered to be successors of key-driven rotary-dial calculators and forerunners of electronic computers. One model, the electronic calculating punch, performs its operations with electronic devices and circuitry. Even though it is electronic and computes, it is not an automatic computer, because it does not have stored programming where all instructions are entered into the machine in the proper sequence to perform the steps necessary to complete a given application or problem from data entered. Neither can it perform the operations of sorting, merging, or selecting data. Two types of calculating punches are shown in Figure 17-20.

Although they are not automatic in the sense that computers are, calculating punches can perform some highly sophisticated functions. In addition to the operations of addition, subtraction, multiplication, and division, calculators can be programmed to *group calculate* or *summary punch,* either as separate operations or in combination. Group calculation is a method that allows a common factor to be stored in the machine so that it may be applied against all cards in a group. This factor is usually read from a master card in front of a group of detail cards. An example of

Figure 17-20. Calculating punches. The three units on the left comprise the 607 electronic calculating punch (604 is similar). Pictured on the right is the 602A calculating punch. With this model the calculating and punching operations are performed by the same unit.

this would be to read into the machine a unit cost from a master card and then to read the factors in each detail card and multiply them by that cost. The result is punched into each detail card. Summary punching on the calculator is accomplished by accumulating figures from each card in a group and punching the results in a summary card at the end.

Like most IBM machines, the calculator utilizes a control panel to provide instructions on how to perform various programs.

Calculating punches are used to compute payrolls, agents' discounts and commissions, interest charges, invoice amounts, insurance premiums, etc.

CARD INTERPRETER

Information may be printed onto cards in one of three ways: by the accounting machine or tabulator, by the printing card punch, and by the card interpreter. Of these three the last is the only one that provides for sensing holes already punched in a card and printing the equivalent in readable form on the face of the same card. Interpreters can also take information from master cards, hold it in the print unit, and repeat it into as many successive detail cards as is desired.

Figure 17-21 shows the IBM 557 interpreter. This machine reads information punched into a card and prints what it reads at the rate of 100 cards per minute. It is possible to print as many as 60 characters on one of the 25 lines on each card. These lines are identified in Figure 17-22. Observe that they fall between the rows on the card and that the data being interpreted are not directly above the columns punched for the data. The location of the printing is controlled by a control panel in the base of the machine. A single pass of the cards through the machine is required for each line to be printed.

Selective stacking is a feature of some interpreters. Such units come equipped with as many as four stackers, the purpose of which is to permit a limited amount of sorting along with interpreting.

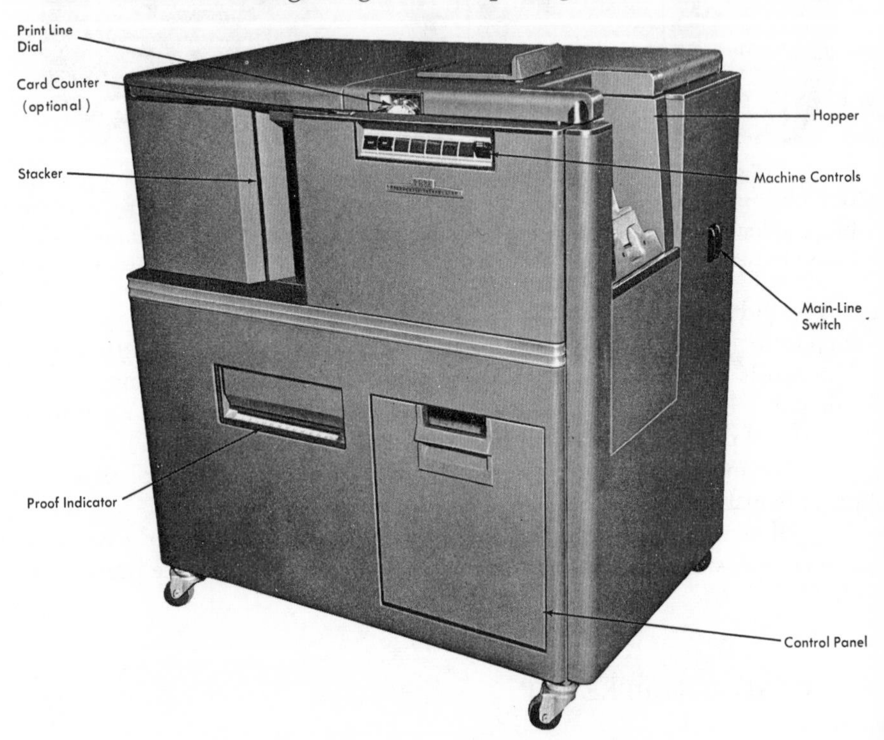

Figure 17-21. Card interpreter.

Being able to read the tab card as one would read a typewritten record enhances its use as a data-processing medium. It can be used as a document as well as for printing reports by machine. Applications of this kind include uses of original source documents such as address cards, payroll cards, tab-card checks, invoice cards, time tickets, and automobile licenses.

ELECTRONIC STATISTICAL MACHINE

In earlier times the use of statistical methods in business and industry was very limited. Now an immense amount of statistical work is carried

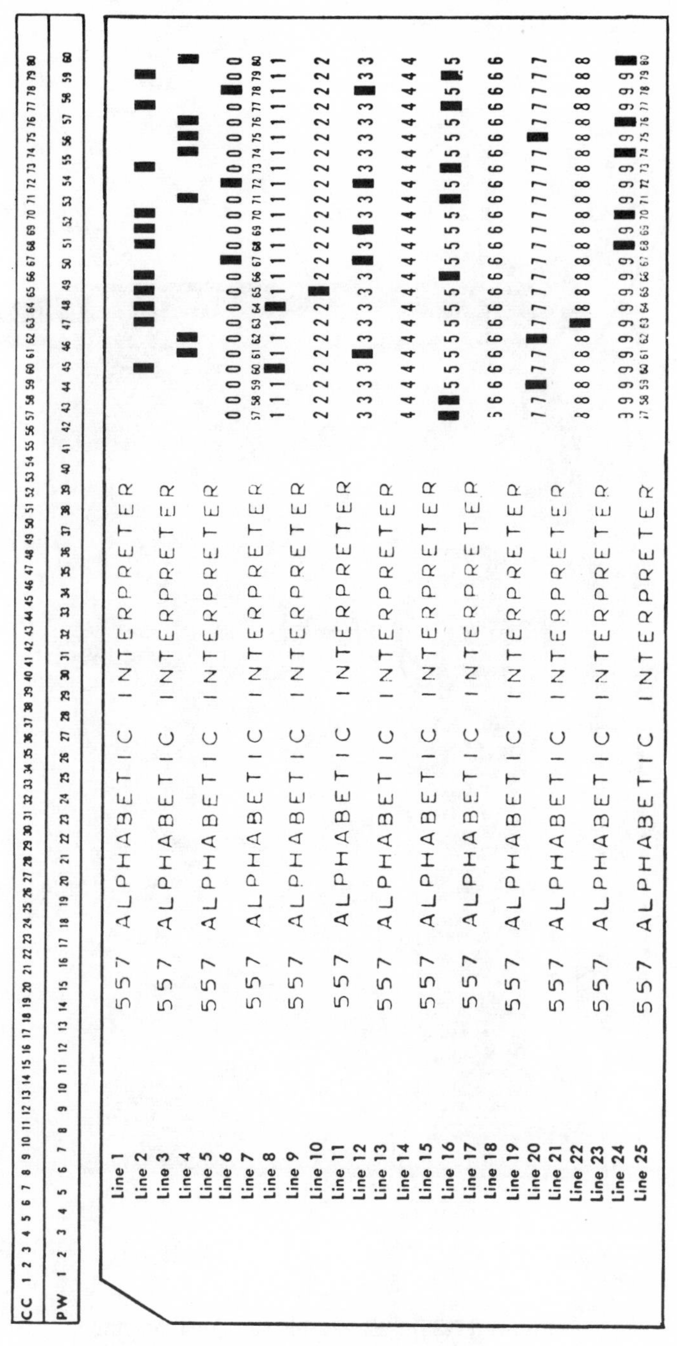

Figure 17-22. Interpreted card.

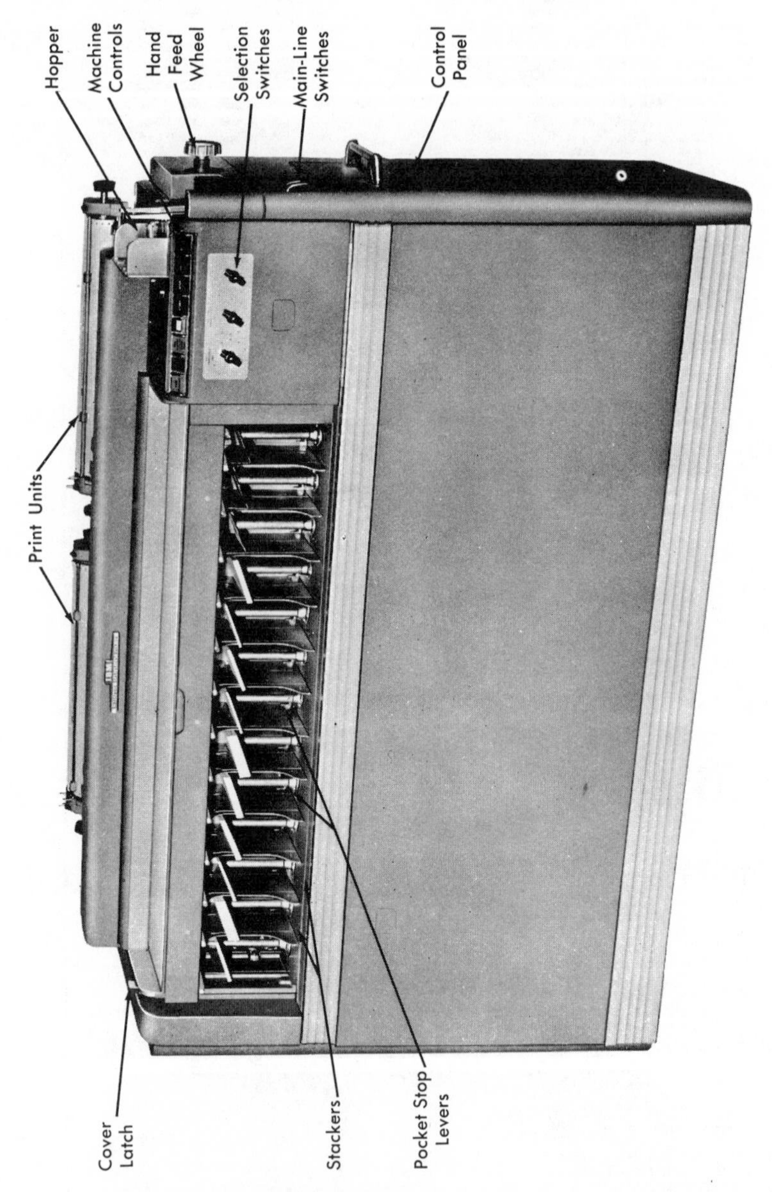

Figure 17-23. Electronic statistical machine.

out by all but the smallest enterprises. Typical areas of statistical application include production, financial planning, quality control, reliability engineering, and cost analyses.

Statistical work generally requires a great deal of sorting, counting, accumulating, balancing, and editing. The electronic statistical machine, shown in Figure 17-23, can perform all of these functions and, in addition, produce printed reports.

APPENDIX

APPLICATION OF PUNCHED CARD FUNCTIONS TO UTILITY BILLING OPERATIONS

Procedures Related to Billing

This illustration on the use of punched cards is in utility record keeping and billing for a municipal water company. The tabulating card illustrated in Figure 17-24 is adapted to a number of purposes. It is (*a*) a post card prepared completely on ADP equipment, (*b*) a medium for billing customers, (*c*) a delinquency notice when all three parts are mailed, and (*d*) a means for providing both the customer and the company with a record of the transaction.

Figure 17-25 illustrates another tabulating card that is a basic part of the system. This is the service billing card; among other data, it shows the route and folio numbers.

Since billing procedures are cyclical operations, it is necessary for reasons of explanation to break in on the cycle at some point—in this case, after the preparation of the service billing cards. The cards at this stage have already been mark-sensed by a meter reader that uses a special type of graphite pencil; they have been punched for the pencil-recorded data by a reproducer (which translated the marks into holes); they contain the previous reading, which was punched into the card on the previous cycle date, and the result of subtracting that figure from the present reading; they also contain the product of having multiplied a prepunched rate by the amount of water used.

Figure 17-26 illustrates a third type of tabulating card that is necessary for the billing operation. This is a master card containing the name, street address, city and state, and account number. Name-and-address cards are merged with their service billing cards when needed by using the number that appears in both as common information (account number, including folio designation).

The tabulating card shown in Figure 17-25 is termed "current" be-

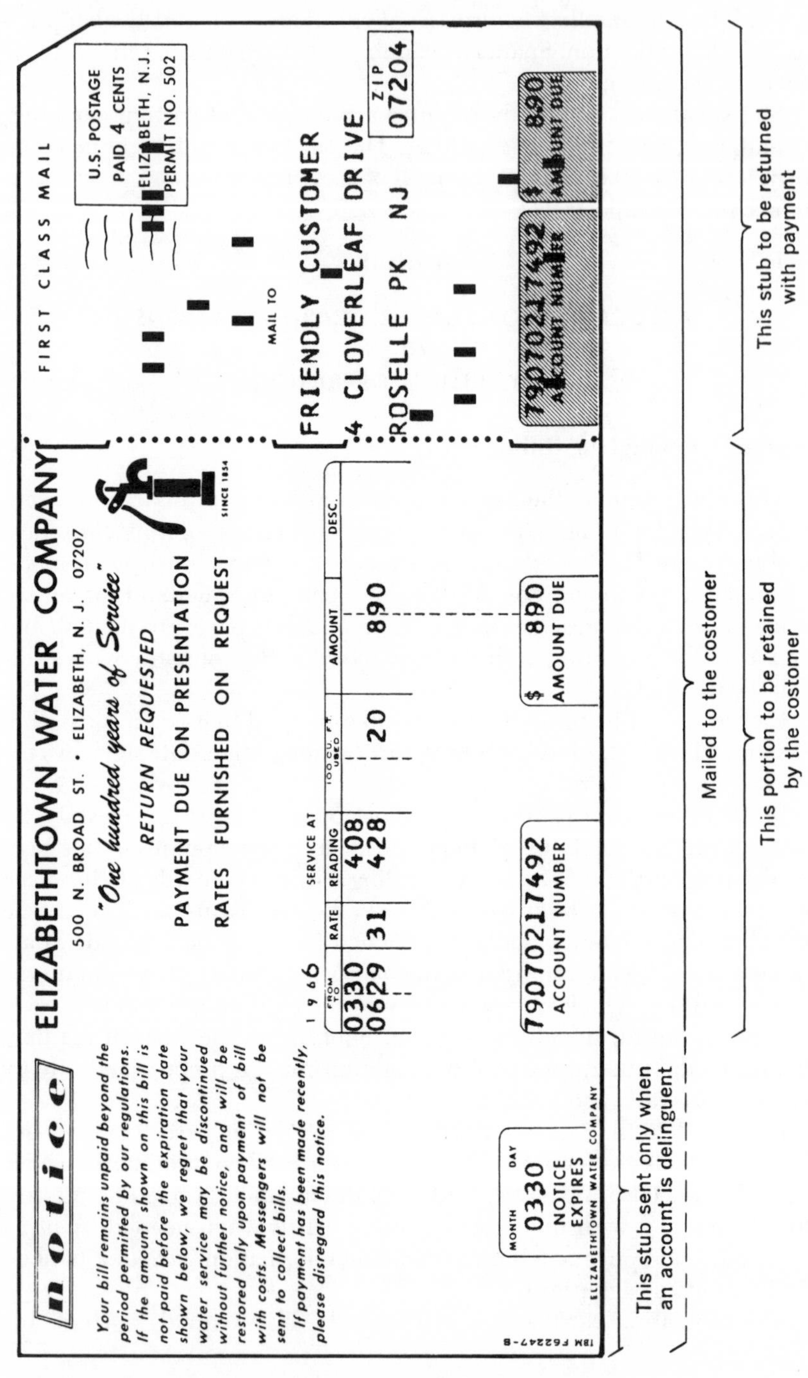

Figure 17-24 *a*. Billing card (front).

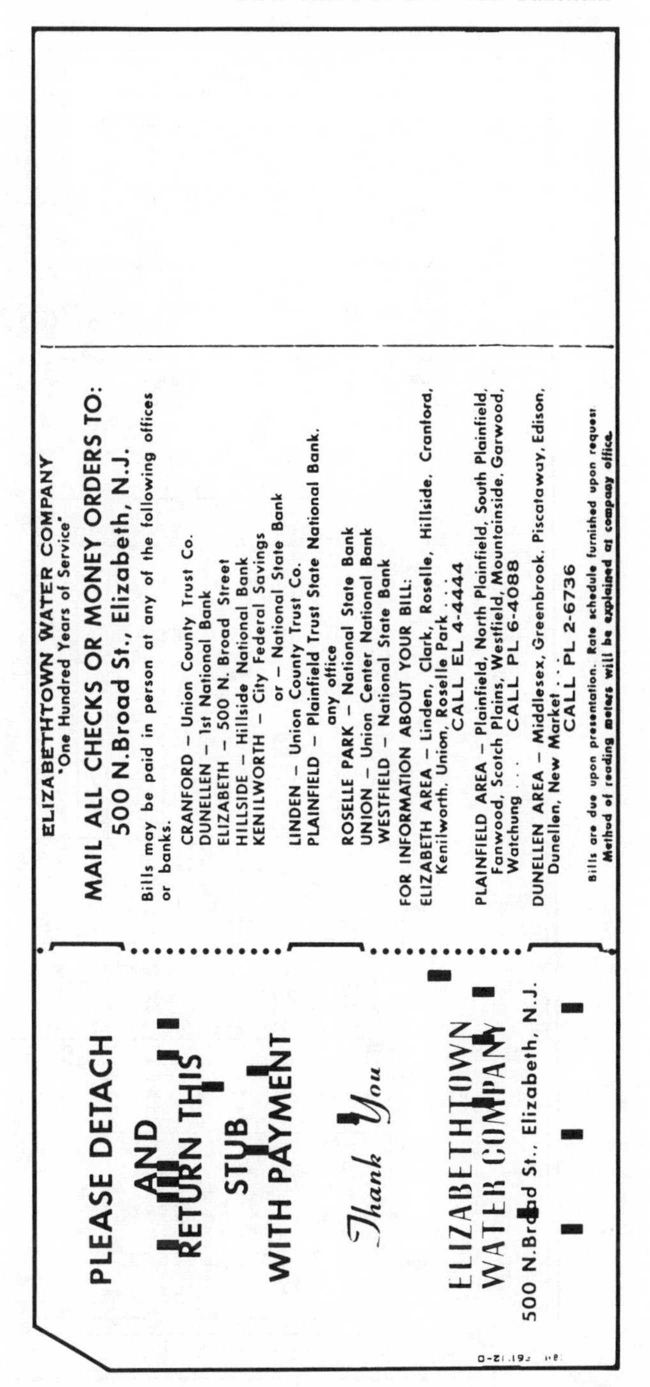

PLEASE DETACH
AND
RETURN THIS
STUB
WITH PAYMENT

Thank You

ELIZABETHTOWN
WATER COMPANY
500 N.Broad St., Elizabeth, N.J.

ELIZABETHTOWN WATER COMPANY
"One Hundred Years of Service"

MAIL ALL CHECKS OR MONEY ORDERS TO:
500 N.Broad St., Elizabeth, N.J.

Bills may be paid in person at any of the following offices or banks.

CRANFORD – Union County Trust Co.
DUNELLEN – 1st National Bank
ELIZABETH – 500 N. Broad Street
HILLSIDE – Hillside National Bank
KENILWORTH – City Federal Savings
or – National State Bank
LINDEN – Union County Trust Co.
PLAINFIELD – Plainfield Trust State National Bank.
any office
ROSELLE PARK – National State Bank
UNION – Union Center National Bank
WESTFIELD – National State Bank

FOR INFORMATION ABOUT YOUR BILL:
ELIZABETH AREA – Linden, Clark, Roselle, Hillside, Cranford, Kenilworth, Union, Roselle Park . . .
CALL EL 4-4444
PLAINFIELD AREA – Plainfield, North Plainfield, South Plainfield, Fanwood, Scotch Plains, Westfield, Mountainside, Garwood, Watchung . . . CALL PL 6-4088
DUNELLEN AREA – Middlesex, Greenbrook, Piscataway, Edison, Dunellen, New Market . . .
CALL PL 2-6736

Bills are due upon presentation. Rate schedule furnished upon request.
Method of reading meters will be explained at company office.

Figure 17-24 *b*. Reverse side of the punched card shown in Figure 17-24 *a*.

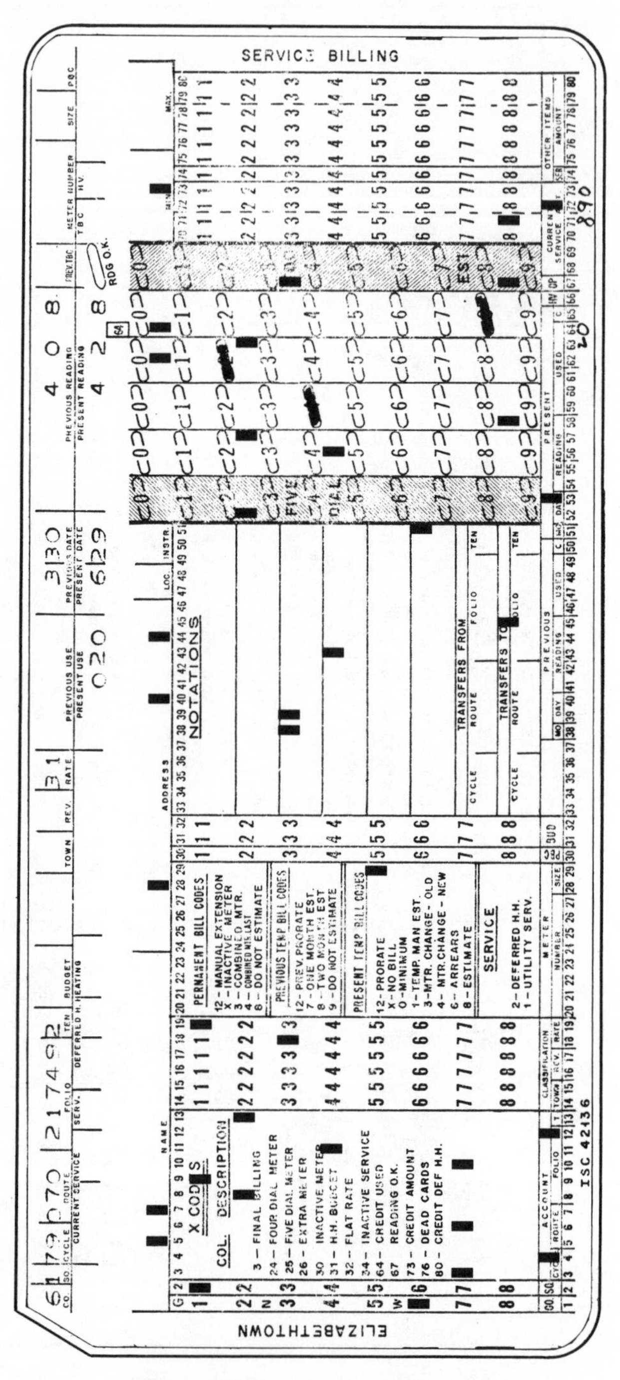

Figure 17-25. Service billing card.

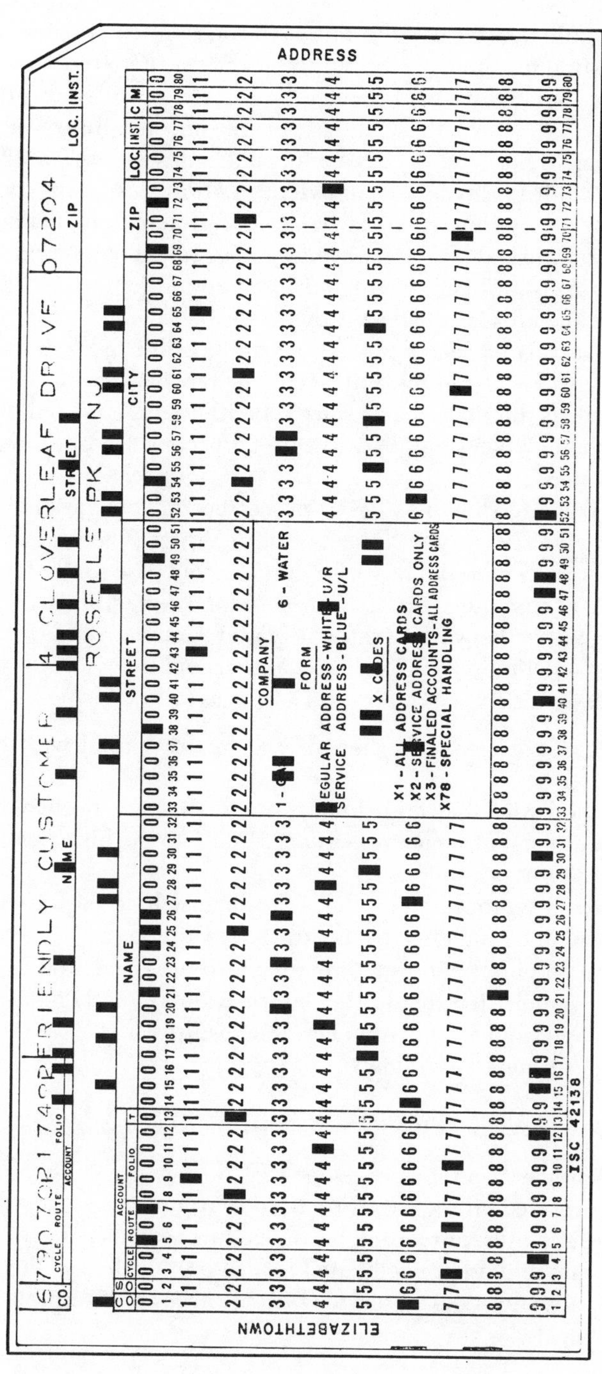

Figure 17-26. Name-and-address card used in billing.

cause it is punched with up-to-date information. As already noted, the present reading for the current month has been compared with the previous reading and the units of service have been calculated. For the next cycle this month's present reading will become the future month's previous reading. Thus it is necessary to punch this month's reading into a similar card, but in the field designated for "previous." This reproduction of data from the existing set of cards into a different field of a new deck of cards is accomplished by the reproducer. Placement of the punches is controlled by the control panel, which is inserted in the front of the machine. The new set of cards is called an "advanced" deck.

Two interrelated sets of procedures are possible by means of the cards thus far mentioned—billing and accounts receivable. These operations are depicted in the flow charts shown in Figures 17-27 and 17-28. They have been somewhat simplified in order to focus attention on the main procedural steps.

The cards are shown as being processed on conventional tabulating equipment, since that is the subject of this chapter. These same three card types can, however, be the bases for carrying out essentially the same procedures on a computer installation instead of electric accounting machines. Actually, the Elizabethtown Water Company does, in practice, make use of a computer for its billing and accounts-receivable operations.

The sequence shown in Figure 17-27 is as follows:

1. Meters are read and the cards are mark-sensed with the present reading.

2. In the event that a customer's meter has not been read, perhaps because he was not at home, reference is made to the historical record file, and the reading is estimated.

3. Cards are processed on the reproducer where they are mark-sense punched, and gang punched to record the reading date.

4. After having been put through the reproducer, the cards go to a calculating punch that reads the information from them, performs the arithmetic operations, and punches the results on the same cards. The charges are computed as follows:

$$\text{present reading} - \text{previous reading} = \text{water used} \times \text{rate} = \text{charge}.$$

5. Cards are then interpreted.

6. The service billing cards are used to reproduce an advance deck of cards where the present reading becomes the previous reading on the new cards. The rate and other fixed data are also prepunched at this stage.

7. The advance deck of service billing cards are interpreted, and held for the next cycle.

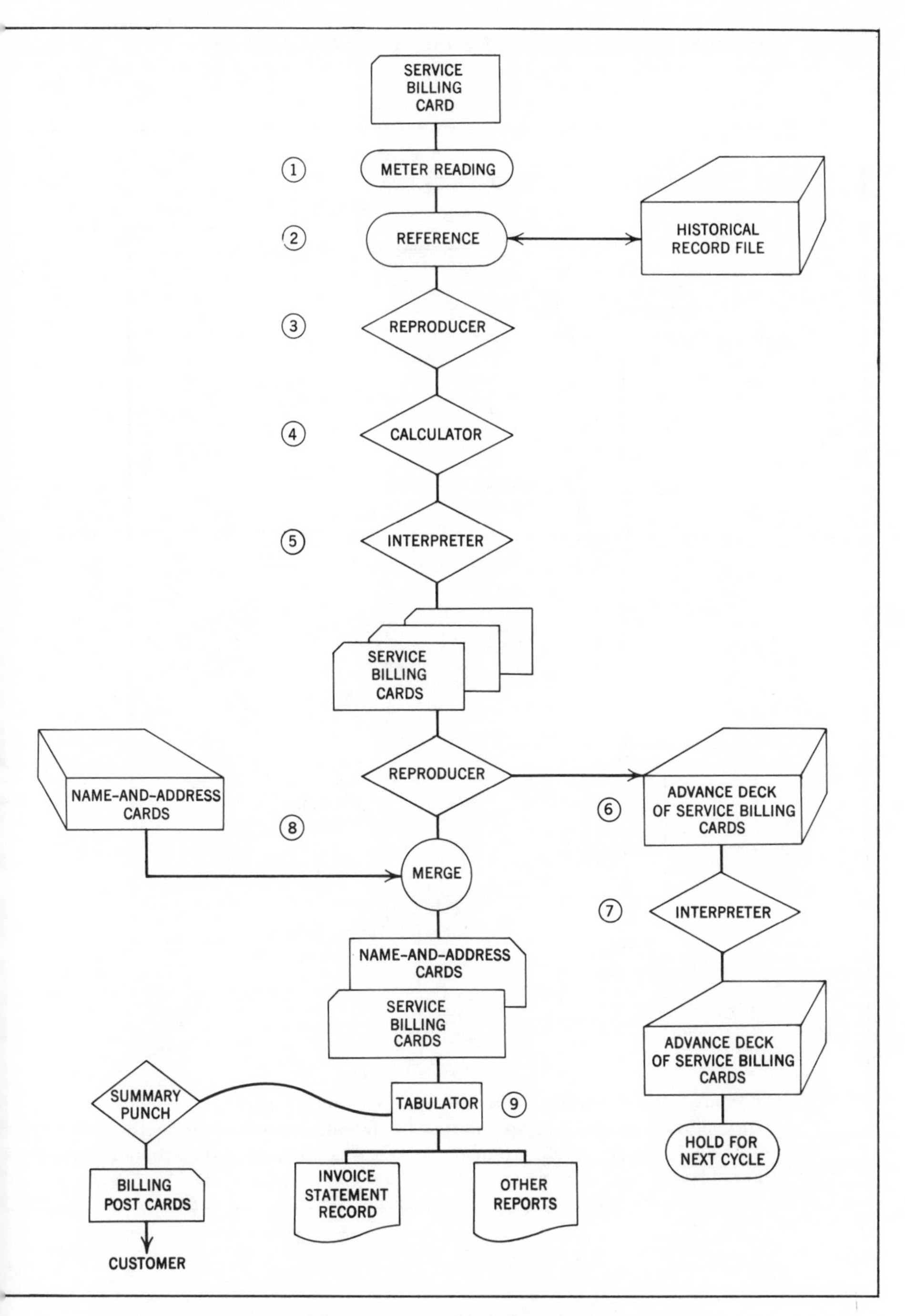

Figure 17-27. Steps in the billing cycle.

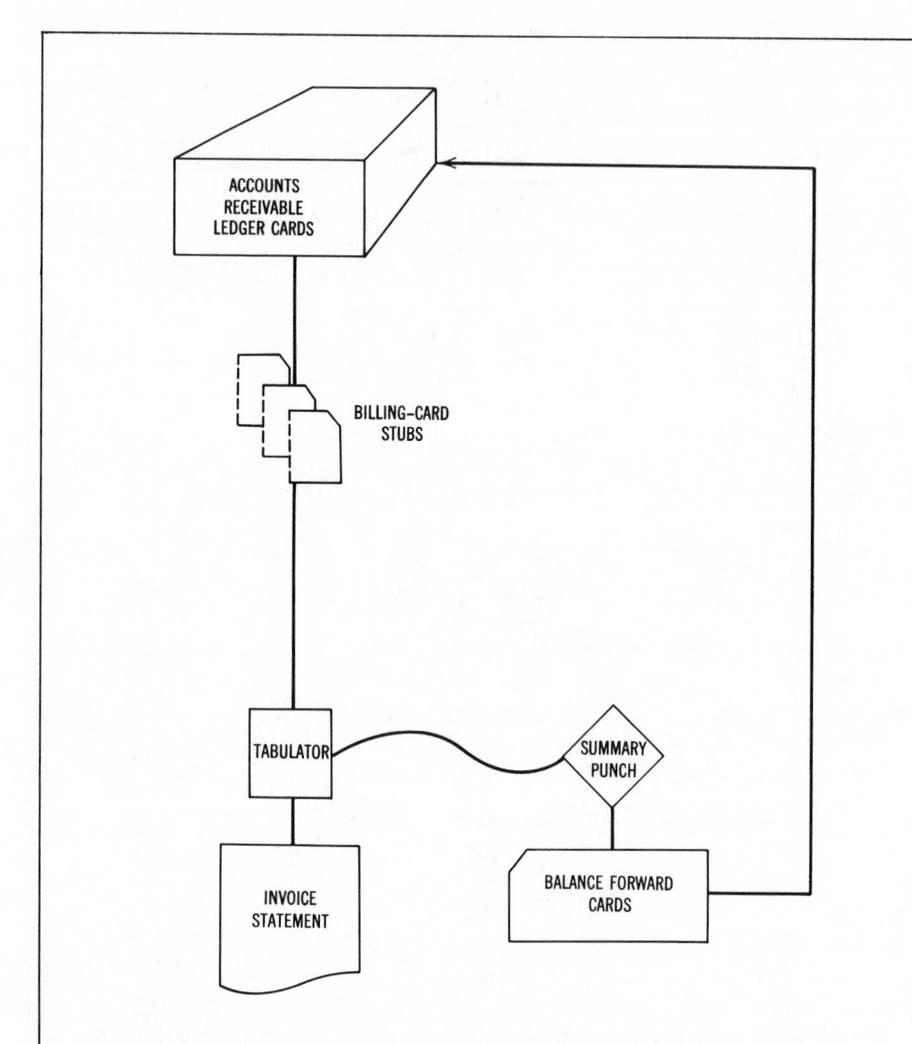

1. Prepare invoice statements on tabulator (accounting machine).
2. In conjunction with the preparation of statements, balance-forward cards are prepared on an auxiliary summary punch. These cards are filed in the accounts receivable ledger file. Old cards are separately filed.

Figure 17-28. Machine processing steps for accounts receivable.

8. By means of the collator the current service billing cards are merged with the name-and-address cards in preparation for the actual billing operation.

9. The service billing cards and the name-and-address cards are run through the tabulator to produce reports. Billing cards are simultaneously generated on a cable-connected summary punch.

Procedures Related to Accounts Receivable

Cards generated for use in a billing operation may be used for other operations such as accounts receivable, inventory control, and sales analysis. Accounts receivable here provides a good example not only of the use of machines but also of the utilization of cards for more than one purpose.

In accounts-receivable operations a debit card may be punched and filed for each sale. Among other data this shows the account number, date, invoice number, and amount due. (The billing service card may serve as the debit card or selected information may be punched from it into a new card to serve as the debit card.) Similarly, a credit card may be punched and filed for each cash receipt on account. These cards show date of payment received, amount of payment, account number, name of customer, and gross credit. During the invoice-preparation process a new balance-forward card may be produced as a by-product operation; for example, if the billing is done on a tabulator, a balance-forward card for each statement may be produced on a summary punch connected with the tabulator. The procedures in the monthly processing of accounts receivable, which incorporate the use of a tabulator and summary punch, are portrayed in Figure 17-28. The procedure shown is as follows:

1. Prepare invoice statements on tabulator (accounting machine).

2. In conjunction with the preparation of statements, balance-forward cards are prepared on an auxiliary summary punch. These cards are filed in the accounts-receivable ledger file. Old cards are filed separately.

part 6

Automatic Data Processing—Integrated

18

Integrated Data Processing

On occasion an expression emerges in our language that captures the imagination and fancy of a large number of people in the business community. Such has been the case with the phrase "integrated data processing." It admittedly has a pleasant ring to it—a ring that seems to promise much. Integrated data processing does, indeed, have behind it a substantial record of achievement.

Initially the concept of integrated data processing, or IDP, implied the utilization of certain types of equipment. However, inasmuch as this concept was implied and not expressed in the phraseology itself, the term was borrowed by many and liberally applied to all kinds of procedural operations, some of which are intrinsically manual.

In this chapter IDP will be discussed as a part of automatic data processing. Within this framework IDP may be defined as the treatment of information on a network basis—from point of origin through successive stages—so that it is recorded at its source in mechanical form for subsequent machine applications. Fundamentally, as we shall see later on in this chapter, much IDP equipment is an adaptation of office machines such as typewriters, adding machines, calculators, cash registers, and bookkeeping machines. The important difference is that these machines can produce and read common-language media, and special cards and/or tape. Furthermore, since they can be used in sets, each succeeding step in information processing is mechanized, and therefore accomplished quickly and accurately with a minimum of human intervention. This provides a direct and efficient means for communicating with the more complex automatic-data-processing equipment—tabulating machines and electronic digital computers.

CHARACTERISTIC FEATURES OF IDP

The following three features provide a means and a basis for identifying IDP installations:

1. Information can be recorded at the point of origin on automatic equipment that produces an original document and simultaneously produces a by-product tape or card containing all or part of the main recording;
2. Once it is in machine language form, the information thus recorded is used for subsequent data-processing operations on the same machine or on common-language machines; and
3. The different kinds and models of equipment involved possess a basic and systematic compatibility.

The basic concept of IDP is the automatic perpetuation of repetitive data. This self-perpetuating principle is based on the ability of IDP equipment to be activated by the same common language contained on punched paper tape, edge-punched cards, or tab cards.

Typically, at the point of origin data are stored in common-language form as a by-product of preparing the source document. This stored information, which contains all or selected portions of the writing, can be used without recopying for subsequent data-processing operations. Often, too, even the original preparation of records is made partially automatic through the prerecording of repetitive information. This is considered a feedback feature of the equipment. As an illustration, in the preparation of sales orders constant or nonvariable information regarding repeat customers (such as names and addresses, manner of shipment, and particular instructions) can be prerecorded on tapes or cards—either as an initial by-product operation or solely by preparing the tapes or cards beforehand—for the semiautomatic preparation of future sales orders.

ADVANTAGES AND BENEFITS OF IDP

The evaluation of benefits accruing from the use of IDP is a problem. This complexity arises from the fact that there are both direct and indirect benefits, and that the indirect benefits are, in many cases, more important than the direct ones. Both direct and indirect benefits may be further classified as to whether they are tangible or intangible. The money saved may be calculated, held in the hand, so to speak, and spent. The value of improved customer service on the other hand is intangible.

In general some of the advantages of IDP are the following:

1. Speed. Savings in time can assume exceptional importance in a variety of business situations, and the fast preparation of source documents is a step in that direction. To cite an obvious situation: Quick service is frequently an important factor in the order-billing-shipping cycles of a business.

2. Increased accuracy. Once information is recorded on a tape or card and verified to be correct, it can be used over and over again with virtual assurance of its accuracy. Of course, even basic information changes from time to time (for example, the addresses of customers), and provision must be made to keep such information current.

3. Improved management control through the preparation of records and reports that are current and therefore more meaningful.

4. Additional information. There is the advantage of obtaining information that previously was too costly to acquire or too time consuming to accumulate. A fine example of this is provided by the case of an IDP procurement system generating by-product tab cards that are used to produce an up-to-date record of purchases—a record that was previously not obtained. Here is how one such system operates: An automatic writing machine, Flexowriter, is electrically connected to a card punch. As the purchase order is being prepared on the Flexowriter the equipment is regulated so that selected information about the purchase is simultaneously transferred over the cable and punched into a card. At designated periods the accumulated cards are sent to the central data-processing installation and used to produce valuable reports on purchasing commitments.

5. Direct savings through the elimination of retranscriptions of data.

THE COMMON-LANGUAGE MEDIA OF IDP

Integrated data processing is most efficient when there is a high degree of direct compatibility between the machine units comprising a system and the related information-carrying media. As inferred, compatibility in this respect is a relative thing. In some cases both the physical units and the code media vary, and conversion operations are necessary. Machines are available that will convert divergent information from one form to another; for example, there are tape-to-card punch machines that will transform the coded information on a tape into that of a card. More will be said about equipment of this sort later on in the chapter.

As mentioned earlier the common-language media of IDP equipment are of three major types: tab cards, punched paper tape, and edge-punched cards. These may be further classified as follows:

- I. Tab cards
- II. Channel code instruments
  - A. Punched paper tape
    1. Five-channel
    2. Six-channel
    3. Seven-channel
    4. Eight-channel
  - B. Edge-punched cards
    1. Five-channel
    2. Six-channel
    3. Seven-channel
    4. Eight-channel

Tab Cards

In the previous chapter the layout and function of the tab card was discussed in connection with the use of conventional tabulating equipment. They are also widely used in both IDP and EDP installations as a major means for storing and reproducing information in common-language form.

On the early equipment of Hollerith cards were read by the mercury electrical contact method—as previously explained. Now, on modern automatic-data-processing equipment, holes may be read not only by electromechanical means, such as feeler pins and electric brushes, but also by photoelectric sensing. The method used depends on the equipment.

More will be said about the use of tab cards as an IDP medium later on in this chapter, in connection with IDP equipment.

Punched Paper Tape

A widely used alternative method for storing and processing information is paper tape. Its use is not new. In the field of communications punched paper tape has long been a dependable means for the automatic communication of typewritten messages over long distances. From this background a natural outcome was (*a*) the application of the same idea to the automatic transmission of information between machines in the same general location, and (*b*) the extensive utilization of the feedback principle wherein punched tapes are used as master tapes to automatically record repetitive information on the same machine.

As the value of tape-operated equipment became more and more apparent, engineers and designers kept making improvements and adaptations. What commenced as a rather specialized application of basic principles for the transmission of messages by wire, ultimately furnished the groundwork for a large array of machines for the automatic processing of information in connection with the storage and transmission of facts and data on this equipment.

The use of punched tape has other benefits. Among these are the following: Paper tape is narrow and thin and thus stores information more compactly than do punched cards; it is of light weight and therefore lends itself to inexpensive mailings; it is easy to file and can be made readily available for repeated use; and it records and processes information in a continuous manner.

Persons who are unfamiliar with its use may consider paper tape to be quite fragile and unsuitable for applications requiring considerable handling. Although it is not as strong a medium as the tab card, it is strong enough for applications of nearly every kind. A way to alleviate this handicap is to use tape-card carriers of the type discussed in Chapter 12 (see, in particular, Figure 12-28) or as an alternative, form-tape envelopes (see Figure 18-5). Where a more rugged medium than tape is needed, the edge punched card may be used, because it will accommodate the same channel codes.

A more important limitation of paper tape lies in its inability to function as a source document. Unlike the tab card, it does not have the space or type of surface for recording information in longhand. Here, again, this handicap can be surmounted by the use of tape-card carriers or form-tape envelopes.

Punched paper tape can be machine read in one of two ways: electromechanically or photoelectrically. In the former method a typical tape unit moves tape over a metal plate and onto the reading heads. Wherever there is a hole in the tape a wire brush makes contact with the plate and thereby produces a pulse that in turn activates the equipment. In the latter method an electric eye is employed. As previously mentioned in connection with tab cards, coded information in the form of holes can be read by photoelectric sensing of the holes. Here is how the technique functions: At the reading station light is concentrated on the moving tape, and where it is perforated, the light shines through and strikes a series of photoelectric cells on the other side whose electrical properties produce the desired action.

Places where holes may or may not appear on a tape in accordance with a coding scheme are known as positions. These positions are said to contain bits or no-bits of information. The word "bit" is an extraction from

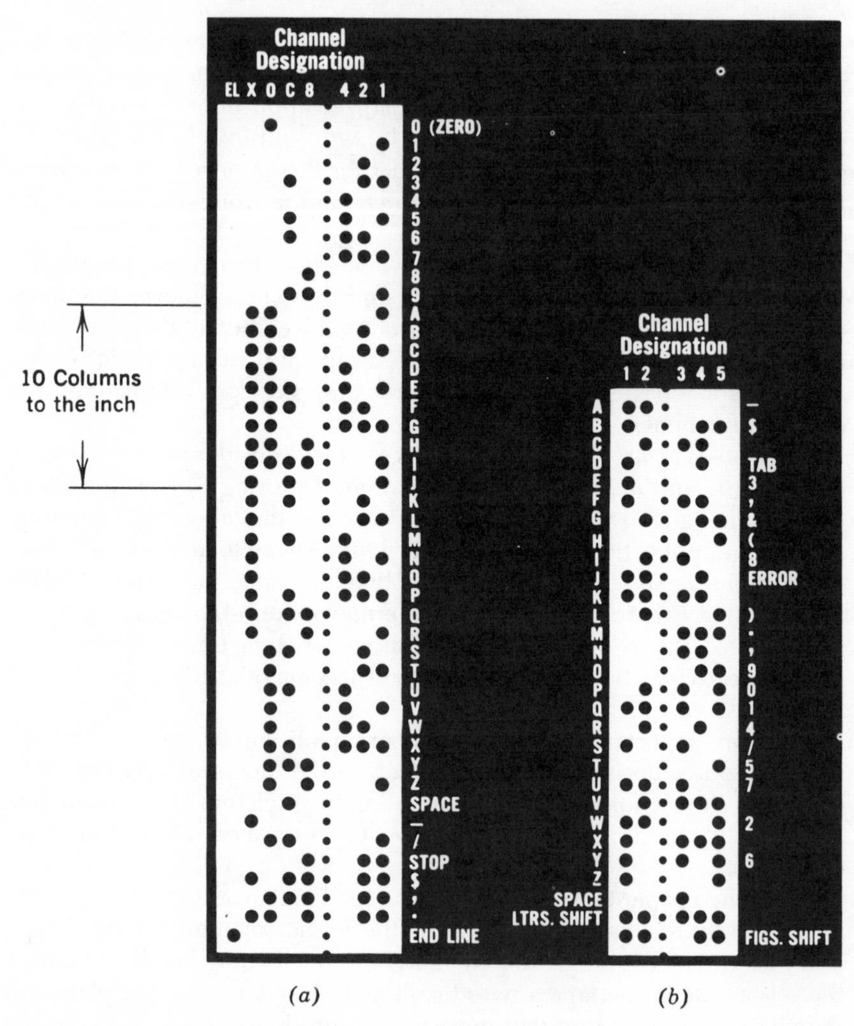

Figure 18-1. Paper-tape codes: (*a*) eight-channel type; (*b*) five-channel type.

*b*inary dig*it;* it may be defined as a unit of information represented by a figure 0 (nothing) or a 1; a hole or no hole, as in a tape or card; or a spot or no spot, as on magnetic tape. Bits are established singly and in combination to represent coded characters. Along the length of a regular tape there are normally 10 columns of positions to the inch, each of which might contain one or more bits of information.

One widely used coding method for paper tape employs no more than five bits to designate all individual letters, numbers, and other characters. How is this made possible with so few bits? Algebraically, we know that the number of ways of arranging two situations (a hole or no hole), five in a series, is 2^5 or 32. Thus it is that 32 combinations of characters are possible for each column. But this is not sufficient to accommodate the 26 letters of the alphabet, plus 10 digits, and the several special characters that are needed. Greater capacity is obtained by employing two modes, as in the shifting arrangement of a typewriter. One mode is primarily for the letters; and the other, primarily for numerals and special characters. By this technique an additional 32 coding places become available. However, not all of them may be used for letter, numerals, and special characters. Two places are required for the letters shift and the figures shift for each mode (see Figure 18-1*b*); thus $64-4=60$. These codes cause the machine (e.g., Teletypewriter or Flexowriter) to shift from one mode to the other so that either letters or numerals and special characters are printed until the alternative code is read and restores the machine to the previous mode of operation. Out of the remaining 60 coding places only 52 are commonly used for characters because several coding places are required to control machine operations—spacing, carriage return, and so on.

Turning again to Figure 18-1*b*, it will be observed that each letter is represented by one hole or a combination of holes. The letter A is produced by holes 1 and 2; the letter B is produced by holes 1, 4, and 5; the letter C is produced by holes 2, 3, and 4; and so on. In the other mode numbers and special characters are similarly represented. The alphanumeric code thus established is used nationally and internationally. Since this code is directly compatible with wire communications equipment, it entails no conversions or modifications when it comes to the transmittal of information produced by equipment generating five-channel tape.[1]

The format of eight-channel tape is, understandably, different—simpler, as a matter of fact. Looking at Figure 18-1*a*, we find that the channel designations, reading from left to right, are decipherable in the following terms:

EL	End of line (carriage return)
X	For alphabetic characters
0	For alphabetic characters
C	For check (or parity) bit
8	Numerical channel
o	Feed holes

[1] The five-channel code is sometimes referred to as the Baudot code, after the name of the man who introduced it, Jean Maurice Baudot, about 1871. The code gained international acceptance and has been in use continually since that time. Baudot's having been French probably explains the basic coding format, because the frequency of occurrence of a letter in the French language appears to be the reason for the number of coded holes used to designate letters.

4	Numerical channel
2	Numerical channel
1	Numerical channel

The EL channel provides for the carriage return on the equipment. Bits in the X and 0 channels are combined with bits in the numerical channels so as to produce the letters of the alphabet. Combinations of positions in these channels are also used to form special characters and to convey certain signals. The next channel, called the parity-bit channel, serves a technical function; it does not, as we shall see, form a part of the basic coding scheme. Parity checking is the automatic sensing by the equipment of each column for a given number of bits of information. With eight-channel tape an odd number of bits is used to represent each character. The odd number is assigned in the check column to those characters that would otherwise have an even number of bits. With this code system, when the equipment detects an even number of bits while automatically making a parity check, it stops until the error is corrected. Some equipment, using a different code format, makes use of an even-parity method of checking. Under either plan the checking operation is done automatically within the machine as the information is being processed.

The reader may wonder why five of the channels are identified by the digits 0, 8, 4, 2, and 1. The reason ties in with the basic coding scheme. Here, as with the coding format for five-channel tape, the characters are formed by the placement of the bits—but in a radically different manner. Consider the numerical code first, because it is predicated on the values assigned to specified locations. Notice from the illustration, Figure 18-1*a*, that the figure 0 is designated by a single bit in the 0 channel, the 8 by a single bit in the 8 channel, the 4 by a single bit in the 4 channel, and likewise for 2 and 1. The remaining digits (namely 3, 5, 6, 7, and 9) are represented by bits that are the sum of their location values; for example, 3 is comprised of a bit in the 2 channel and a bit in the 1 channel (2+1=3). To take another example: 7 is comprised of a bit in the 4 channel, another bit in the 2 channel, and still another bit in the 1 channel (4+2+1=7). What we have been talking about is really a binary-coded decimal (BCD) format, which is the *language* of high-speed computers. Additional treatment is given the subject of binary codes in Chapter 20.

The binary-coded decimal format also provides for coding the alphabet and special characters. The X and 0 channels, similar to the X and 0 zones on IBM cards, are used in conjunction with the numerical channels to record these characters. This similarity is clearly shown in Figure 18-2. The logic of the arrangement is also apparent: A is comprised of bits in channels X, 0, and 1; B is comprised of bits in channels X, 0, and 2. Accordingly, for each of the first nine letters of the alphabet the characters are formed by bits in both the X and 0 channels, plus bits in the numerical channels. Except for the elimination of bits in the 0 channel, the next nine letters are formed in the same way. The last eight letters are made up of combinations of bits in the 0 channel and bits in the numerical channels.

Although eight-channel tape cannot be used for direct transmission and reception over standard communication channels, this presents no insurmountable problem. By employing appropriate equipment such as the Teledata tape transmitter-receiver (a product of Friden, Inc.) it is possible to transmit, receive, and check information encoded in more than five channels. Different models are available and correspond to the capacity of the tape to be transmitted: five-, six-, seven-, or eight-channel tape. Briefly, here is how the equipment functions: While the tape is being processed on the reader of the transmitter-receiver unit at one station, it is simultaneously reperforated and checked on the punch of a similar unit at some other location. Designed into the equipment is a code-parity method of checking the accuracy of the transmission.

In using tapes, regardless of the number of channels, provision must be made for their filing so that they can be conveniently withdrawn and easily refiled. Of the many filing techniques the most common are (*a*) coil the tapes and store them in specially designed trays with compartments for ready identification, (*b*) insert tapes in vertical tape racks with holders containing index labels for easy identification, and (*c*) file the tapes within labeled envelopes behind well-marked dividers.

Edge-Punched Cards

Holes may be punched along the edge of cards in accordance with the same format that is used for punching holes in tape. The edge-punched cards thus produced are of two types—unit record cards and fanfold continuous cards. (See Figures 18-3 through 18-6.)

Unit record cards are used when the required punched information can be contained on a single card. Two opposite edges of a card may be used (as in Figure 18-4) or just one edge.

Fanfold cards lend themselves to situations in which the punched information is normally of sufficient length to warrant the use of at least three cards but not long enough to require an extensive tape.

EQUIPMENT FOR INTEGRATED DATA PROCESSING

At this point let us consider the different types of IDP equipment that utilize one or more of the common-language media. For convenience we

STANDARD TAB CARD PUNCHING POSITIONS												CARD PUNCH	STANDARD TAPE CHANNEL NUMBERS									FLEXOWRITER 2201	FLEXOWRITER MODEL SPS/SPD
12	11	0	1	2	3	4	5	6	7	8	9		8	7	6	5	4	FEED	3	2	1		
		I										0			●			•				) — 0 (ZERO)	(ZERO) 0 —)
			I									1						•			●	! — 1	1 — !
				I								2						•		●		@ — 2	2 — @
					I							3				●		•		●	●	# — 3	3 — #
						I						4						•	●			$ — 4	4 — $
							I					5				●		•	●		●	= — 5	5 — =
								I				6				●		•	●	●		¢ — 6	6 — ¢
									I			7						•	●	●	●	? — 7	7 — ?
										I		8					●	•				* — 8	8 — *
											I	9				●	●	•			●	(— 9	9 — (
I			I									A		●	●			•			●	a — A	a — A
I				I								B		●	●			•		●		b — B	b — B
I					I							C		●	●	●		•		●	●	c — C	c — C
I						I						D		●	●			•	●			d — D	d — D
I							I					E		●	●	●		•	●		●	e — E	e — E
I								I				F		●	●	●		•	●	●		f — F	f — F
I									I			G		●	●			•	●	●	●	g — G	g — G
I										I		H		●	●		●	•				h — H	h — H
I											I	I		●	●	●	●	•			●	i — I	i — I
	I		I									J		●		●		•			●	j — J	j — J
	I			I								K		●		●		•		●		k — K	k — K
	I				I							L		●				•		●	●	l — L	l — L
	I					I						M		●		●		•	●			m — M	m — M
	I						I					N		●				•	●		●	n — N	n — N
	I							I				O		●				•	●	●		o — O	o — O
	I								I			P		●		●		•	●	●	●	p — P	p — P
	I									I		Q		●		●	●	•				q — Q	q — Q
	I										I	R		●			●	•			●	r — R	r — R
		I		I								S			●	●		•		●		s — S	s — S
		I			I							T			●			•		●	●	t — T	t — T
		I				I						U			●	●		•	●			u — U	u — U
		I					I					V			●			•	●		●	v — V	v — V
		I						I				W			●			•	●	●		w — W	w — W
		I							I			X			●	●		•	●	●	●	x — X	x — X
		I								I		Y			●	●	●	•				y — Y	y — Y
		I									I	Z			●		●	•			●	z — Z	z — Z
												SPACE				●		•				SPACE	SPACE
	I													●				•				" — -	- — "

12	11	0	1	2	3	4	5	6	7	8	9		EL	X	O	CH	8	FEED	4	2	1		
		▮	▮									/			●	●		·			●	: — /	/ — :
					▮					▮		#					●	·		●	●	F1 (MNP STOP)	STOP
	▮				▮					▮		$		●		●	●	·		●	●	% — —	% — —
		▮			▮					▮		,			●	●	●	·		●	●	, — ,	, — ,
▮					▮					▮		.		●	●		●	·		●	●	. — .	. — .
						▮				▮		("				●	●	·	●			F2	NONPRINT (AUX. SPACE)
		▮				▮				▮		%			●		●	·	●			F3	PRINT-RESTR. (AUX. ZERO)
	▮					▮				▮		'		●			●	·	●			F4	ON 1—ON 2 (U.C.)
▮						▮				▮		□		●	●	●	●	·	●			UPPER CASE	UPPER CASE
▮												&		●	●	●		·				; — &	& — ;
		▮						▮		▮		SKIP			●	●	●	·	●	●		TAB	TAB
								▮		▮		END CARD 1					●	·	●	●		F9	CONTROL (AUX. 2)
		▮							▮	▮		END CARD 2			●		●	·	●	●	●	F6	PUNCH OFF
									▮	▮		COR. TAB				●	●	·	●	●	●	F10	DATA SELECTOR (AUX. 3)
	▮								▮	▮		ERROR		●			●	·	●	●	●	F11	FORM FEED (AUX. L)
				▮						▮		PI 1				●	●	·		●		± — +	PI 1 (2, 8, SP)
	▮			▮						▮		PI 2		●			●	·		●		' — °	PI 2 (2, 8, -)
		▮		▮						▮		PI 3			●		●	·		●		BACKSPACE	PI 3 OR BS (2-8-0)
		▮					▮			▮		PI 4			●	●	●	·	●		●	F13	PI 4 (AUX. /)
▮							▮			▮		PI 5		●	●		●	·	●		●	¼ — ½	PI 5 (AUX. A)
	▮						▮			▮		PI 6		●		●	●	·	●		●	F12	ADDRESS IDEN. (AUX. J)
							▮			▮		PI 7					●	·	●		●	F8	SKIP RESTORE (AUX. 1)
▮				▮						▮		SP-1		●	●	●	●	·		●		LOWER CASE	LOWER CASE
▮								▮		▮		SP-2		●	●		●	·	●	●		F5	ON 2
	▮							▮		▮		CR		●		●	●	·	●	●		F7	FC ON
▮									▮	▮		TAPE FEED		●	●	●	●	·	●	●	●	TAPE FEED	TAPE FEED
									▮		▮	END LINE	●					·				CARRIAGE RETURN	CARRIAGE RETURN
X	0	10																					
ZONES													CHANNEL NUMBERS										

Figure 18-2. Comparison of codes. In the last two columns the codes for such functions as "UPPER CASE," "LOWER CASE," and "BACK SPACE" are so-called "open" or "blank" coding places, since their particular assignment and function varies with the equipment. The codes for the numbers, letters, and most special characters, however, are standard on all eight-channel tape.

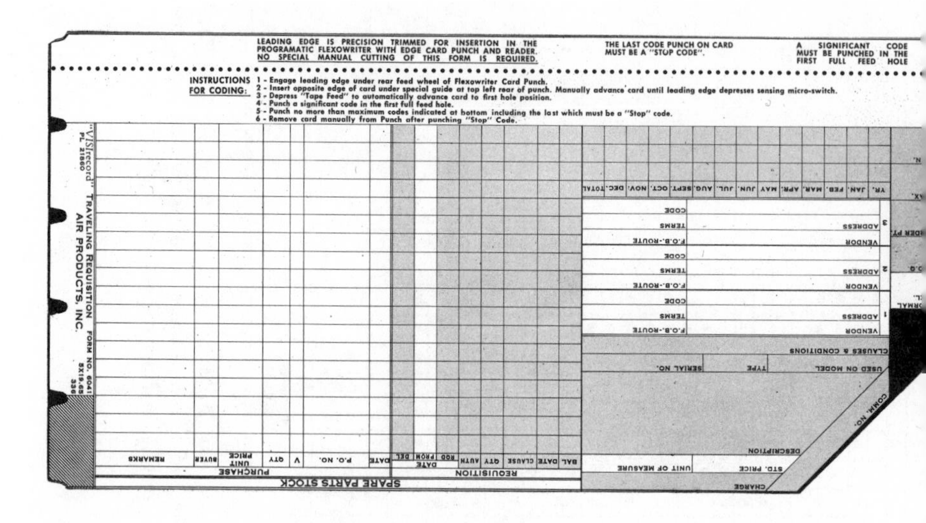

LEADING EDGE IS PRECISION TRIMMED FOR INSERTION IN THE PROGRAMATIC FLEXOWRITER WITH EDGE CARD PUNCH AND READER. NO SPECIAL MANUAL CUTTING OF THIS FORM IS REQUIRED.

THE LAST CODE PUNCH ON CARD MUST BE A "STOP CODE".

A SIGNIFICANT CODE MUST BE PUNCHED IN THE FIRST FULL FEED HOLE

INSTRUCTIONS FOR CODING:
1 - Engage leading edge under rear feed wheel of Flexowriter Card Punch.
2 - Insert opposite edge of card under special guide at top left rear of punch. Manually advance card until leading edge depresses sensing micro-switch.
3 - Depress "Tape Feed" to automatically advance card to first hole position.
4 - Punch a significant code in the first full feed hole.
5 - Punch no more than maximum codes indicated at bottom including the last which must be a "Stop" code.
6 - Remove card manually from Punch after punching "Stop" Code.

"V/S record" PL 21660 "TRAVELING REQUISITION" FORM NO. 6041 AIR PRODUCTS, INC.

YR. JAN. FEB. MAR. APR. MAY JUN. JUL. AUG. SEPT. OCT. NOV. DEC. TOTAL

VENDOR F.O.B.-ROUTE ADDRESS TERMS CODE

CLAUSES & CONDITIONS

USED ON MODEL TYPE SERIAL NO.

COMM. NO. DESCRIPTION

CHARGE STD. PRICE UNIT OF MEASURE

BAL DATE CLAUSE QTY AUTH ROD FROM DEL DATE

REQUISITION

DATE P.O. NO. V QTY UNIT PRICE BUYER REMARKS

PURCHASE

SPARE PARTS STOCK

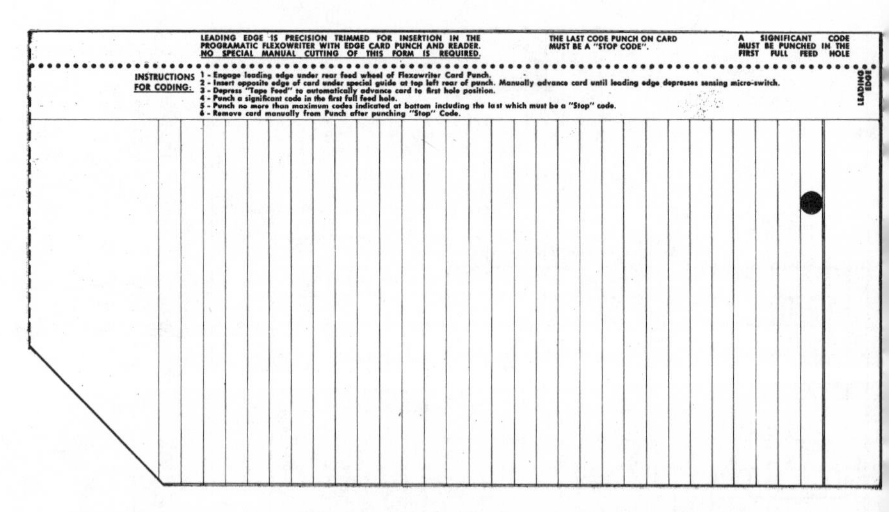

LEADING EDGE IS PRECISION TRIMMED FOR INSERTION IN THE PROGRAMATIC FLEXOWRITER WITH EDGE CARD PUNCH AND READER. NO SPECIAL MANUAL CUTTING OF THIS FORM IS REQUIRED.

THE LAST CODE PUNCH ON CARD MUST BE A "STOP CODE".

A SIGNIFICANT CODE MUST BE PUNCHED IN THE FIRST FULL FEED HOLE

INSTRUCTIONS FOR CODING:
1 - Engage leading edge under rear feed wheel of Flexowriter Card Punch.
2 - Insert opposite edge of card under special guide at top left rear of punch. Manually advance card until leading edge depresses sensing micro-switch.
3 - Depress "Tape Feed" to automatically advance card to first hole position.
4 - Punch a significant code in the first full feed hole.
5 - Punch no more than maximum codes indicated at bottom including the last which must be a "Stop" code.
6 - Remove card manually from Punch after punching "Stop" Code.

LEADING EDGE

Figure 18-3. Two-part record card.

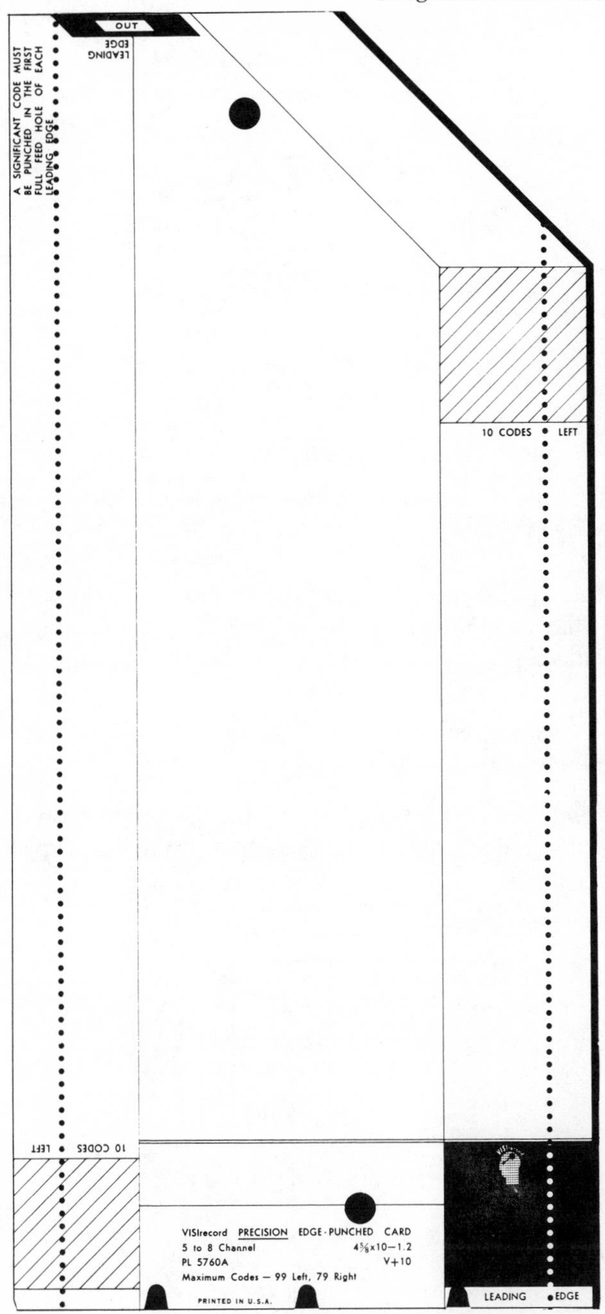

Figure 18-4. Card utilizing opposite edges for containing information.

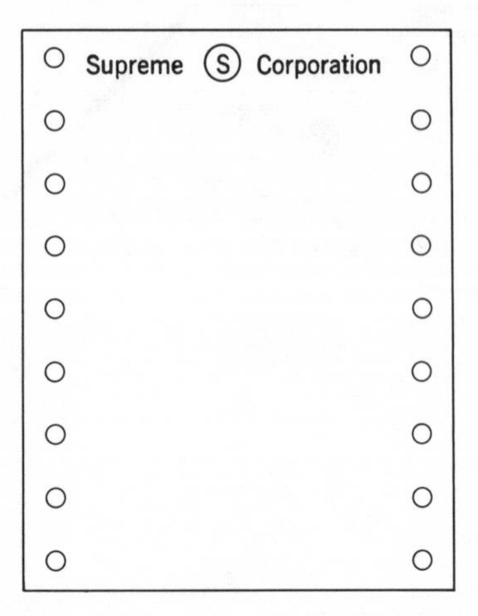

Figure 18-5. Form-tape envelope actual size in inches: 8½ × 11. Tape is stored within the envelope.

shall treat the equipment in terms of the various types falling within these categories; that is, modified electric typewriters, calculators, adding machines, bookkeeping machines, cash registers, and plate-making machines.

Modified Electric Typewriters

The Teletypewriter (marketed by the Teletype Corporation), the Flexowriter (manufactured by Friden, Inc.), and the Typewriter Tape

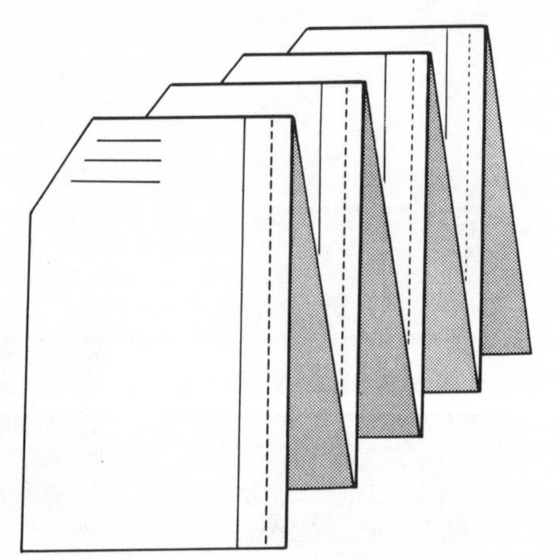

Figure 18-6. Fanfold continuous cards.

Punch (put out by IBM), are typical of equipment belonging to this classification. Each of these units is an automatic writing machine with basically the same keyboard as on a standard typewriter. As a document is being prepared any of the information being typed can be recorded on tape as a by-product of the typing operation. The completed tape may contain all of the typed information or only selections, depending on how the equipment is regulated. Afterwards the tape may be transported to other locations as the common-language link to subsequent operations. The locations may be near at hand or at some distance. Economy is a factor for consideration. If the locations are far apart and the need for the information is not truly urgent, the tape can be mailed, and its arrival at the destination point in a day or two will be in sufficient time for the purpose.

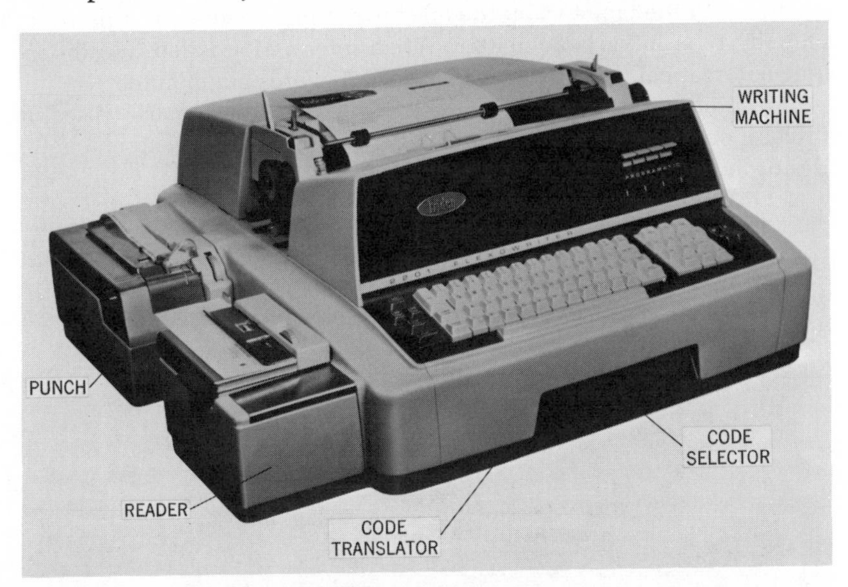

Figure 18-7. Flexowriter showing the reading of an edge-punched card and the simultaneous creation of a code-punched tape as a by-product operation.

Where wire transmittal of the information is desired it may be feasible to hold the tape for transmission during the evening when lower rates are in effect.

Figure 18-7 shows the Flexowriter equipment. Noteworthy of the Flexowriter is the number of built-in customized features afforded by the various models. Units are obtainable with one or more of the following: tape reader, edge-punched card reader, or tab-card reader; single- or double-case keyboard; a selection of type faces and substitute characters;

special carriages and platens; and an electric linefinder. A wide variety of associated equipment may also be added; for example, auxiliary tape or card punches.

One of the most interesting applications of the Flexowriter is when it is cable connected to a card punch, such as the IBM 026, via a small code-converting unit known as a TCPC (Tab-Card Punch Control). With this arrangement the Flexowriter punches data simultaneously into tape and tab card as a by-product of either manual or automatic typing. In this way the need for keypunching or tape-to-card conversion is eliminated.

Only new information—information being recorded for the first time—need be entered by the Flexowriter keyboard. Information of a repetitive nature can be automatically typed from tapes or cards. Stop codes are punched into the tape to stop the machine at places where it is desired to manually type in variable matter. When information is fed into the machine automatically it types at a speed of 100 words per minute.

Machines of the modified-electric type are customarily used for purchase-order writing, preparing receiving reports, inventory control, sales-order writing, and billing.

Figure 18-8. A common language machine of the calculator class: the Computyper.

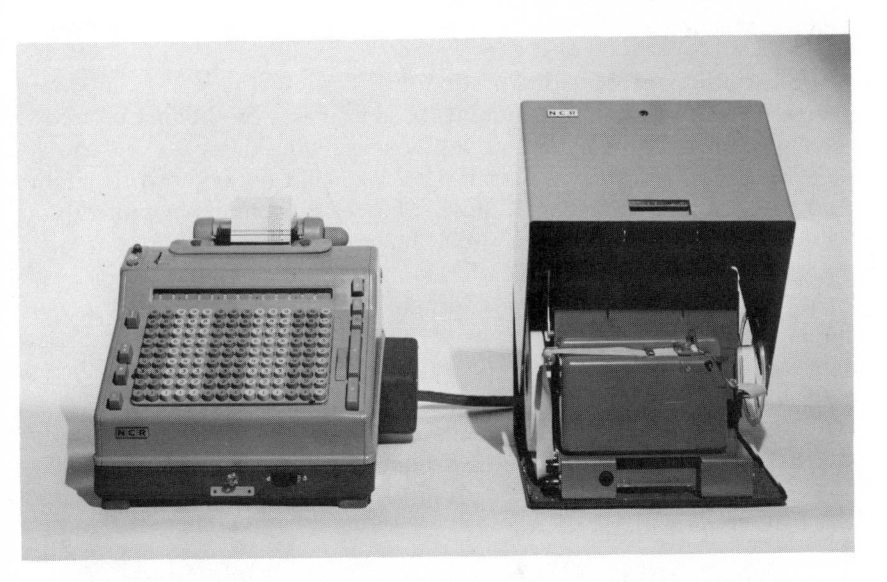

Figure 18-9. Adding machine with cable-connected tape recorder.

Calculators

The IBM Cardatype accounting machine and the Friden Computyper (shown in Figure 18-8) are representative of the equipment comprising this category. These units have much in common with equipment of the modified-electric-typewriter class: They can produce several copies of a document from a single typing; they can produce a by-product tape and/or set of punched tab cards; they can read from previously punched cards or tapes and print the information; and they can be the master of any number of slave units (such as additional tape punches or auxiliary typewriters) that are programmed to produce related documents.

What makes equipment of this type different from that of the modified-electric-typewriter class are its calculating capabilities.

By means of an automatic computing unit, equipment of the calculator category can perform the arithmetic operations of addition, subtraction, multiplication, and division. Instructions may be given the machine either through the input medium (tapes, cards) or at the command of program keys activated by the operator. Tabulations and decimal placements are made quickly and with maximum accuracy.

An illustration of the use of the Computyper, with auxiliary equipment, is given in the appendix to this chapter.

Adding Machines

The adding machine that produces a by-product tape is an important device for capturing information at its source as a by-product of a basic accounting or add-listing operation. Depending on the make and the design, the equipment may be a self-contained unit or, as shown in Figure 18-9, comprised of two coupled units. Models are obtainable with either a full-bank or 10-bank keyboard, and a choice of carriage widths.

Equipment of this classification function as regular adding machines for all kinds of work involving additions and subtractions—such as sales audits, production and inventory control, and the tabulation of statistical data. As the figures are printed on a roll or form selected data may be automatically punched for subsequent automatic processing.

Like the Flexowriter, certain common-language adding machines can be used in combination with a card-punch machine. All items, subtotals, and totals appearing on the adding-machine tape may be punched automatically into tabulating cards.

Bookkeeping Machines

These machines, known also as accounting machines, come in a variety of models to meet specific needs. Categorically they fall into two main classes; that is, numerical, or nondescriptive, machines and alphanumeric, or descriptive, machines. Numerical units have a single bank of digital and special character keys that produce figures, arithmetical signs, asterisks, and various other marks. Alphanumeric units, on the other hand, commonly possess two banks of keys—one above the other (see Figure 18-10). We use the word "commonly" because there are exceptions; that is, machines of the alphanumeric class having single banks containing alphabetical, numerical, and special keys. In two-bank machines the lower bank is primarily composed of letter keys and the uppermost bank is primarily composed of numerical keys. Both banks contain special character keys and operational activators.

Applications requiring distribution and totals accumulation are especially adaptable to the bookkeeping machine with a paper-tape punch. Uses of the machine typically include tasks related to accounts receivable, stock records, payroll accounting, and general ledger record keeping. Other applications include expense distributions, sales distributions, cost distributions, and budgetary accounting.

The machine shown in Figure 18-10 has a mechanical tape punch that is an integral part of the machine. Some manufacturers design their equipment so that the punch unit is separate and is connected to the bookkeeping machine by means of a cable.

Figure 18-10. Alphanumeric accounting machine with tape punch as an integral part of the unit.

Bookkeeping machines of the common-language type may also be linked directly to a card punch so that information can be simultaneously captured on tabulating cards.

Cash Registers

The common-language cash register is an excellent instrument for obtaining merchandising and point-of-sale information in punched-tape form, with little or no effort on the part of the salesperson. By capturing information in this manner and using it as the input for further processing it is possible to produce timely and meaningful merchandise and financial reports. A cash register designed to produce a by-product tape is shown in Figure 18-11.

Plate-Making Machines

Recording information on metal or plastic documents (e.g., charge plates) is accomplished by a process called embossing. Plate-making ma-

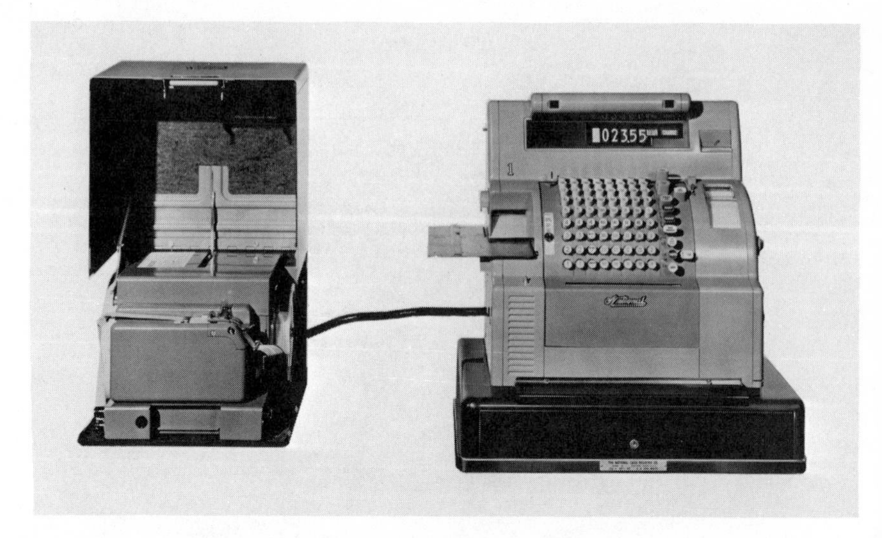

Figure 18-11. Cash register designed to produce a by-product tape.

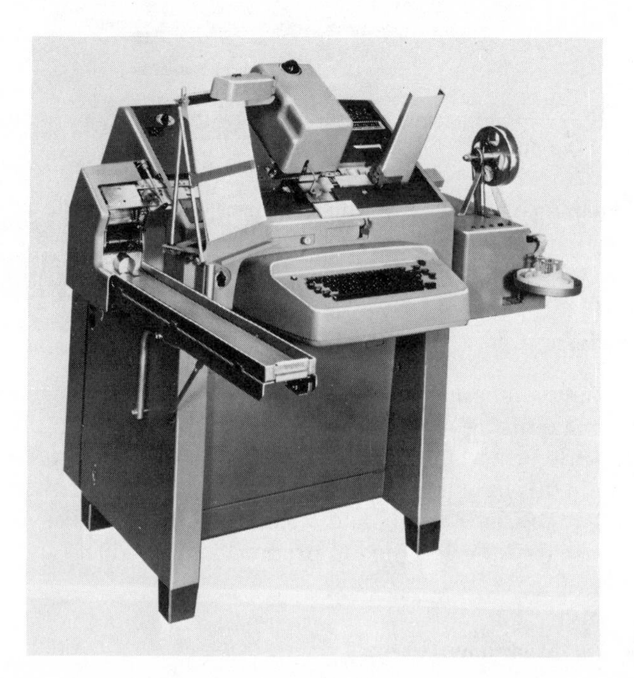

Figure 18-12. Plate-making machine. Known as the Graphotype, this particular machine is the product of the Addressograph Multigraph Corp.

chines perform this function. The particular unit shown in Figure 18-12 is a high-speed automatic machine that can accept input from fully electric keyboards, paper-tape readers, or punched card readers.

ASSOCIATED EQUIPMENT

Plant Data Collection

The Friden Corporation's Collectadata data-collection system, the IBM 357 data-collection system, and the factory data collector manufac-

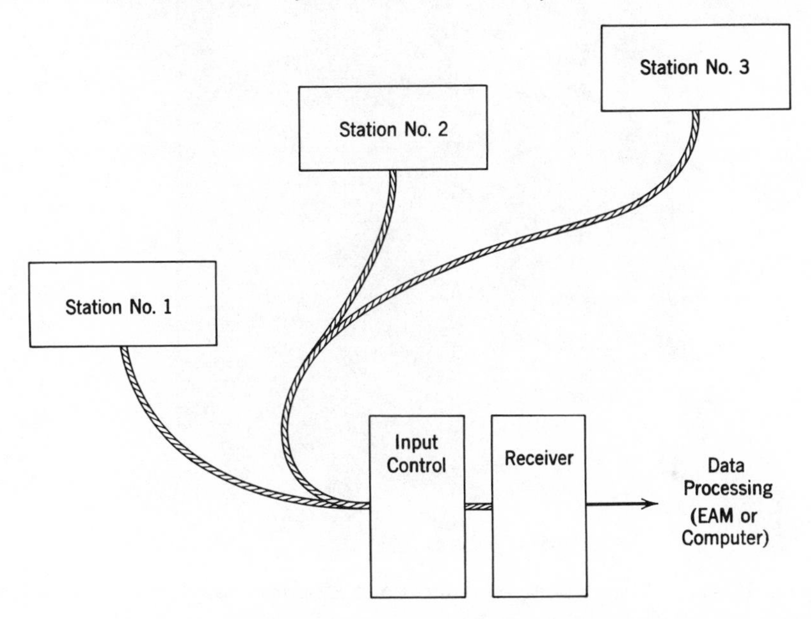

Figure 18-13. Data collection system.

tured by the Radio Corporation of America (RCA), are all designed to collect data from several points of origin within a plant to a central processing point. Multiple transmitters, called "readers," or "input stations," are connected to a central unit (receiver) on a cable-sharing basis. Different systems permit the use of different media for input. In general input stations are obtainable that will read and transmit data punched into eight-channel tape, edge-punched cards, or tab cards. Data are received and compiled in the form of eight-channel tape. Figure 18-13 shows a schematic of the data-collection system.

In addition to reading alphabetical and numerical information from tape, edge-punched cards, or tab cards, the input stations may be used to enter manually variable numerical data for transmission.

Applications for plant-data-collection equipment include work-in-process accounting, machine-load forecasting, attendance recording (payroll), and receiving reporting as part of the procurement cycle.

Figure 18-14. Paper tape-to-card converter. Courtesy of Olivetti Underwood Corp.

Converters

A converter is a machine or implement for transforming information contained in one form of language medium into another form. Such equipment is of the following three types:

1. Tape to card
2. Tape to tape
3. Card to tape

Tape-to-Card Converters. Equipment in this category reads tape and transposes the coded information into tabulating cards. Information thus

Figure 18-15. Paper tape-to-magnetic tape converter. Courtesy of International Business Machines Corp.

converted can undergo further processing on equipment that makes use of punched cards as a medium. A typical converter of this type is shown in Figure 18-14.

Tape-to-Tape Converters. Machines are available that will change tape from one coded format to that of another; for example, five-channel tape may be converted into eight-channel tape or, conversely, eight-channel tape may be converted into five-channel tape.

Another type of tape-to-tape conversion for which equipment is available is that of transferring coded information from punched paper tape to magnetic tape (Figure 18-15). Equipment may be had for transferring information from magnetic tape to paper tape (Figure 18-16).

Card-to-Tape Converters. Information stored in punched cards can be converted to tape by means of card-to-tape converters (Figure 18-17). This is the reverse of the tape-to-card process. If need be the perforated tape can be forwarded to some other location and processed through a tape-to-card converter to transform the information back to punched card form.

THE RELATIONSHIP BETWEEN IDP EQUIPMENT AND COMPUTERS

Centralized data processing requires that information be gathered from widespread sources. In this respect units of IDP equipment are frequently used as auxiliary, or peripheral, equipment to go along with a central computer installation. In this capacity they really function in what is

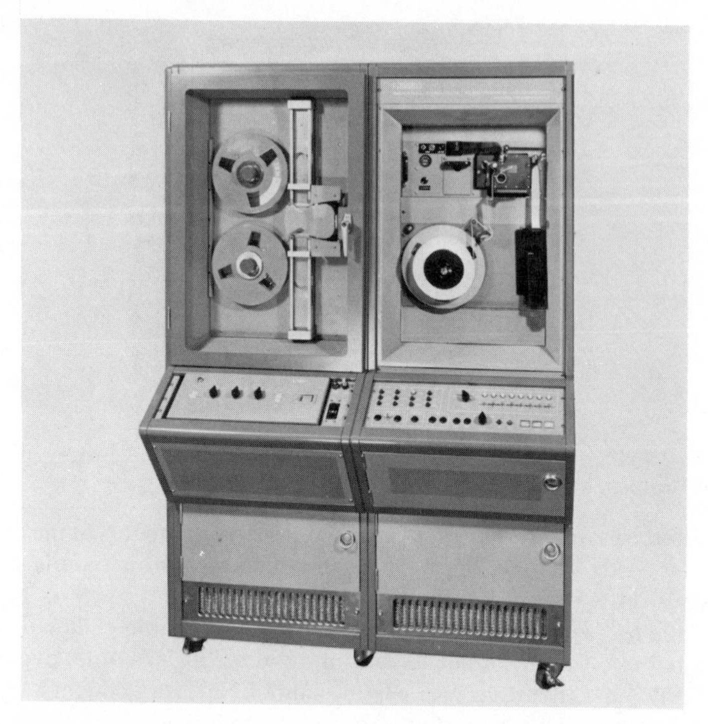

Figure 18-16. Magnetic tape-to-paper tape converter. Courtesy of General Dynamics Electronics.

Figure 18-17. Punched card-to-magnetic tape converter. Courtesy of General Dynamics Electronics.

known as off-line operations. When so used, the equipment itself is termed *off-line.*

Off-line equipment may be defined as equipment not connected directly to the main computer, but operating in conjunction with it. It is distinguished from on-line equipment, which is connected directly to the main computer by means of a conduit or cable.

Equipment of the IDP type provides a simple, accurate, and inexpensive means of capturing information at its point of origin. It is in general flexible and versatile, and may be used in combination with conventional tabulating equipment or with much faster, highly sophisticated equipment such as computers and associated devices. The common language may be converted to tabulating cards, magnetic tape, and punched paper tape with additional code channels to accommodate the native language of the high-speed data processors.

APPENDIX

OUTLINE OF AUTOMATED PROCUREMENT SYSTEM UTILIZING COMPUTYPER INSTALLATION AND THE "357" DATA-COLLECTION SYSTEM

Introduction

The following outline and accompanying schematic (Figure 18-18) relate the highlights of an entire procurement cycle for a manufacturing plant; that is, the procedures involving purchasing, receiving, and accounting. For convenience of understanding the reader is advised to relate the paragraph designations to those enclosed within hexagons on the schematic.

Contents

I. *Requisitioning*

A. For Standard Items

1. Originator will prepare a three-part Purchase Order (P.O.) Requisition to provide for the following distribution:

a. Purchasing
b. Accounts Payable
c. Originator

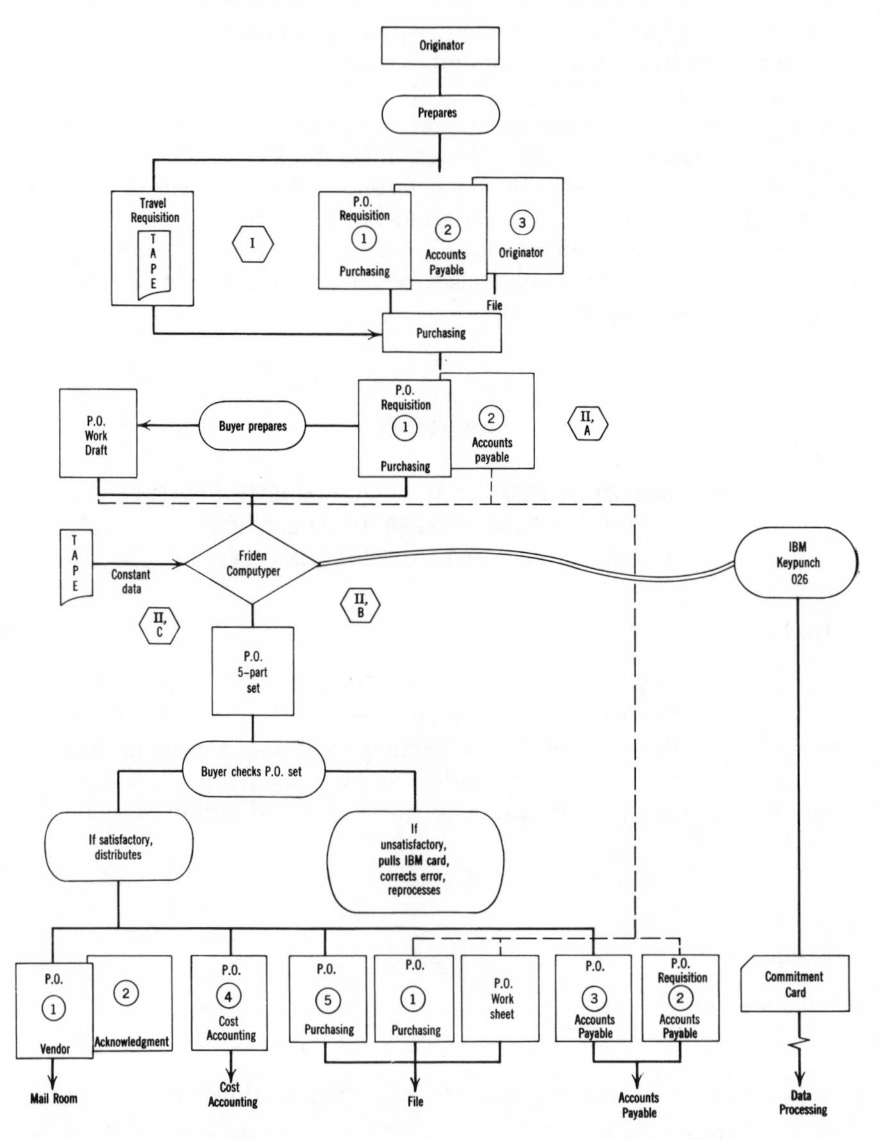

Figure 18-18. Outline of automated procurement system utilizing Collectdata and Computyper installations. Roman numerals and letters inside the hexagons refer to applicable paragraphs in the Appendix.

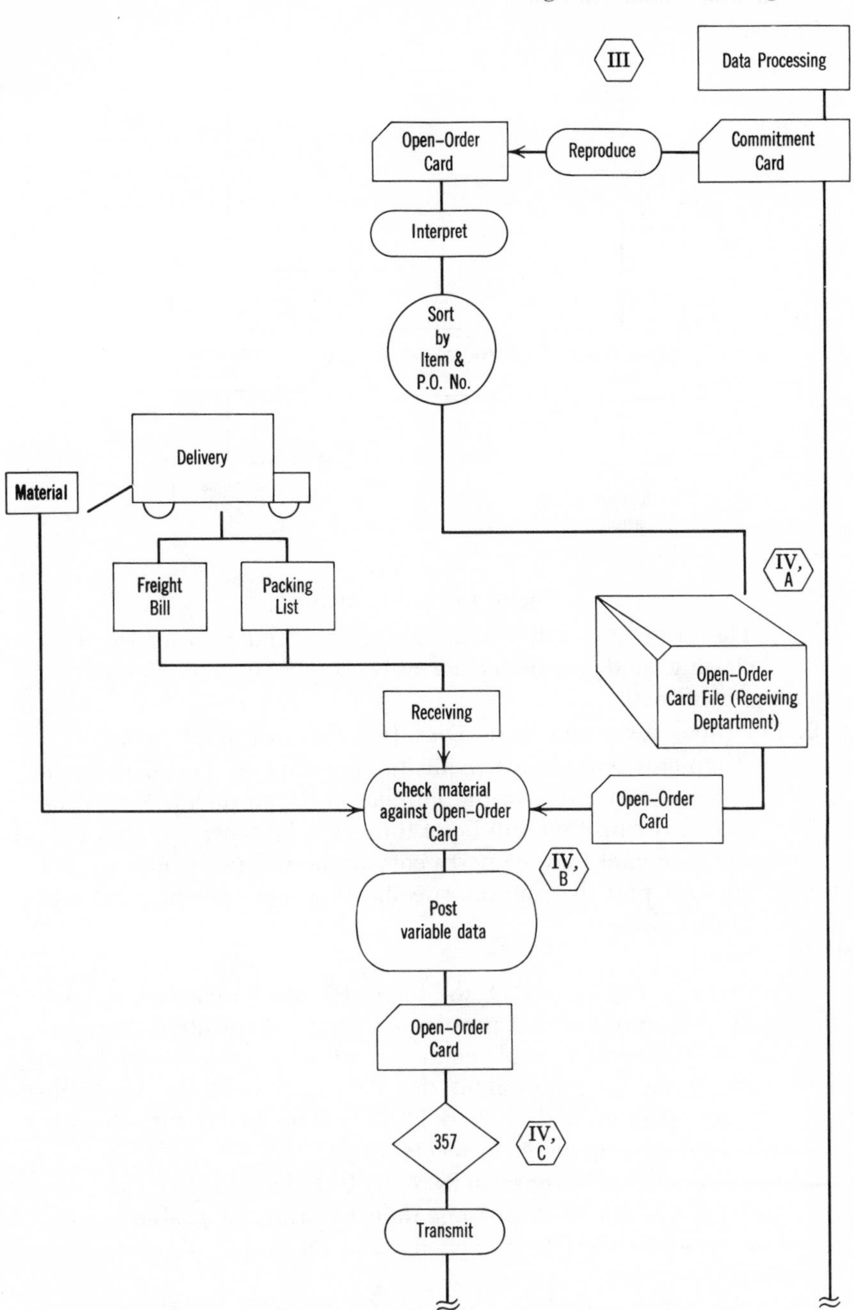

Figure 18-18. *Continued*

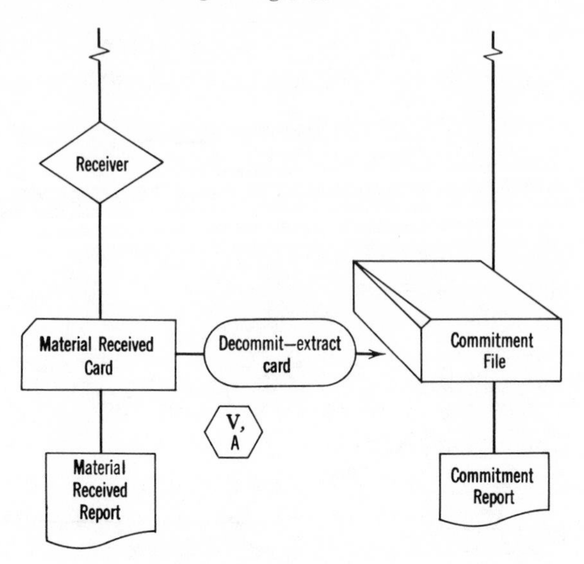

Figure 18-18. *Continued*

The originator will maintain one copy and forward the Purchasing and Accounts Payable copies to the Purchasing Department.

B. For Stock Items and Tools, Dies, Jigs, Fixtures, and Gages

1. Originator will record requisitioning data on Travel Requisition and forward it to the Purchasing Department. *Note:* The Travel Requisition will be of the envelope type, containing an eight-channel tape of pertinent constant data, including part number, part description, specification requirements, and any special clauses.

II. *Purchasing*

A. In order to initiate a P.O. the buyer (in the Purchasing Department) will complete that portion of the P.O. Requisition (Purchasing and Accounts Payable copies) reserved to record information enabling the preparation of the P.O. document by the Computyper operator. A P.O. Work Draft is to be used to furnish additional data requirements in special cases.

B. The operator will prepare a five-part P.O. document on the Computyper, which will at the same time, by means of a cable connection, produce IBM keypunched cards on an 026 Printing Card Punch.

1. The P.O. document will provide for the following distribution:
   a. Vendor
   b. Acknowledgment
   c. Accounts Payable

d. Cost Accounting
e. Purchasing

2. A keypunched IBM card will be automatically produced for each item on the P.O. These Commitment Cards are to be forwarded to Central Data Processing on a daily basis.
3. The buyer will be responsible for proofreading the P.O. prior to its release. If unsatisfactory, the buyer will void all P.O. copies, have the machine operator remove the Commitment Card(s), and he will provide for the processing of a satisfactory document.

C. Prepunched tapes containing coded information of a repetitive nature (vendors' names and addresses, part numbers, descriptions of items, specifications, and special clauses), plus codes to control positioning of the P.O. document, will provide a source of error-free input to the automatic writing machine.

1. Vendors' name and address tapes and those containing other constant information will be filed in a tub next to the operator.
2. In the case of items ordered by a Travel Requisition, the tapes containing the part numbers, item descriptions, and special clauses will be filed within the requisition itself.

III. *Central Data Processing*

A. From the Commitment Cards (Ref: Paragraph II, B, 2, above), Central Data Processing will automatically reproduce a deck of Open-Order Cards containing pertinent data for receiving operations.

B. The Open-Order Cards will be sorted by item number and P.O. number and forwarded to the Receiving Department.

IV. *Receiving*

A. The Open-Order Card will contain prepunched constant data (e.g., P.O. number, date of P.O., quantity of item ordered [Ref: paragraph III, A, above]), and will provide for the manual recording of partial as well as complete deliveries of each item.

B. Upon receipt of material the Receiving Department will extract the applicable Open-Order Card(s) from the file, check the material, and, if found satisfactory, manually record the variable data (e.g., quantity received, date received) on the card(s). It is to be noted that an Open-Order Card will be used until deliveries are completed against an item.

C. Open-Order Cards designating the receipt of material will be inserted in the Tab-Card Transmitter. The prepunched constant data will be processed automatically, and the recorded variable data will be keyed in on the transmitter. This will result in the production of a card by the Receiving unit in Central Data Processing.

V. *Central Data Processing*

A. The Material Received Cards will be used

1. To provide a periodic tabulation of materials received;
2. To decommit the Commitment File (paragraph II, B, 2, above), and thus furnish exact commitment figures;
3. For accounts-payable information;
4. To implement a vendor-evaluation program (Does the vendor meet delivery schedules? Are quantities satisfactorily fulfilled? Etc.).

part 7

Automatic Data Processing—Electronic

19

Computers—History, Development, and Terminology

Electronic data processing (EDP) refers to the processing of information via electronic computers and attendant equipment. The purpose of this chapter is to provide a background of knowledge about computers.

Essentially, computers are of two types: analogue (or analog) and digital. An analogue computer performs its operating functions by producing physical or electrical simulation of the magnitudes or dimensions of the various components in the problem to be solved. Weight scales, thermometers, and slide rules are familiar items that are examples of simple analogue computers. They *measure* "how much," rather than *count* "how many." A speedometer, another analogue computer with which we are very familiar, shows the speed of a vehicle by the physical movement of an indicator that corresponds to the rate of motion of the vehicle in miles per hour.

Industry utilizes analogue computers that are a great deal larger and more complex than the simple kinds noted above. The basic principle is the same, however. Accuracy is largely governed by the physical analogies employed and the dependability of the instruments for measuring that analogy.

Two features should be noted about analogue computers. The first is that each type must be designed for a particular application; each is essentially a special-purpose machine—for example, a slide rule cannot serve as a bathroom scale. The other feature is that analogue computers deliver answers in approximations, not in definitive values.

For business applications, and most scientific and industrial process control applications, there is the need for accuracy. The digital computer fulfills this requirement.

DIGITAL COMPUTERS

Whereas analogue computers in industry are used almost exclusively for scientific purposes, digital computers are employed for a multitude of purposes, both of a business and a scientific nature.

Digital computers differ significantly in speed and efficiency according to the make and design of the equipment, the ease of accessibility of data contained in the memory unit (where access time is a factor), and the particular kind of operation to be performed. Generally the requirements for handling scientific data are different from those for handling business data. Some computers will accommodate both kinds of processing with comparable facility, others will not. The reason is that in scientific applications the amount of input data is generally small in relation to the amount of data processing that takes place within the machine, and the output data can usually be expressed in short form. Business applications, by comparison, usually entail great amounts of input data, and, as with scientific applications, the internal processing that takes place may also be great. In addition, the product of these operations in terms of recorded data is normally of a considerable magnitude. Therefore, to meet the demands of business data processing, which usually involves a great many individual readings, it is necessary to employ high-speed input and output equipment. Otherwise, there would be a loss in overall data-processing speed.

EARLY HISTORY AND DEVELOPMENT OF COMPUTERS

Like many another scientific achievement, the modern electronic computer is the result of creative forces that took place over many thousands of years. Man's 10 fingers constituted the earliest digital computer!

Historians believe that man developed the decimal number system, which has 10 as its base, because of his 10 fingers. The word "decimal" itself is derived from the Latin word *decimus*, meaning "ten."

The Abacus

Man progressed to using stones or pebbles for counting. The ancient Egyptians devised a clay tablet with grooves in which they set pebbles in accordance with a counting scheme. Later on, using the Egyptian system as a basis, someone strung pebbles or beads on wires, affixed them to a frame, and the abacus was born. Invented two to three thousand years ago, this clever computing device was used extensively throughout ancient and medieval Europe, Egypt, and the Orient. Even today it is used in eastern Asiatic countries, particularly in China and Japan.

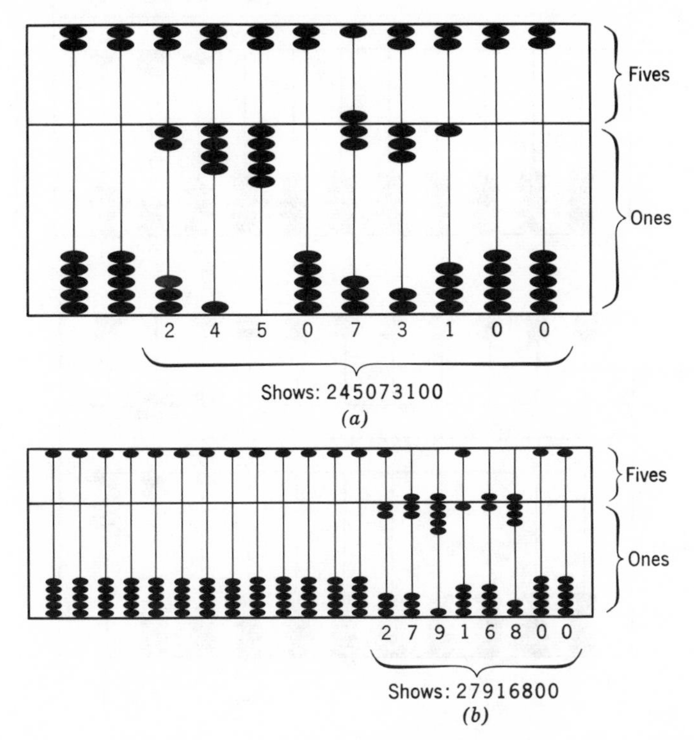

Figure 19-1. Two forms of the abacus: (*a*) the Chinese Sann Pan; (*b*) the Japanese Soroban.

The abacus holds more than a historical interest, because it employs a means of counting that has a direct bearing on computer arithmetic.

Only the number of rows of beads restricts the size of the numbers that can be represented on an abacus.

The rules for using the abacus are as follows:

1. Five beads on a line may be replaced by a single bead above.
2. Two beads in the space above may be replaced by a single bead on the row to the left.

Figure 19-2 shows how addition is performed with the Chinese abacus, which is a type of biquinary calculator. As we follow the three

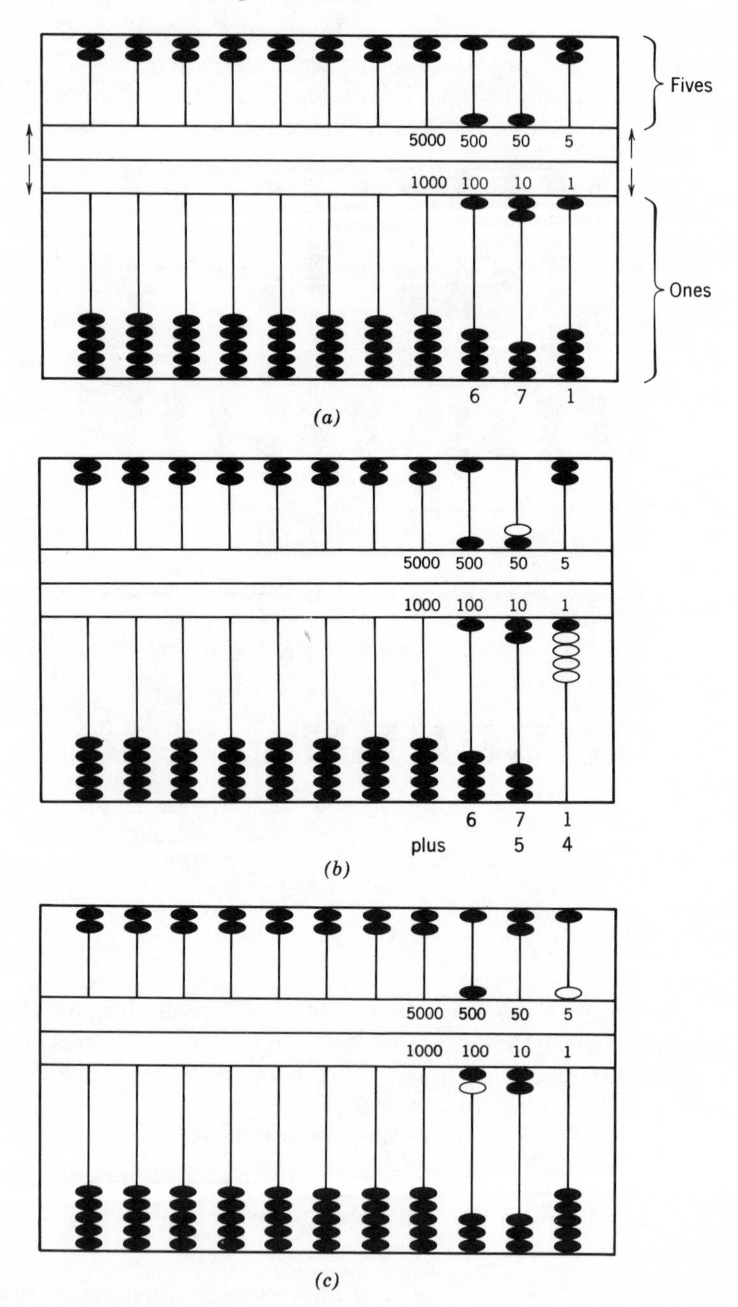

Figure 19-2. Adding 54 to 671 on the Chinese abacus: (*a*) first step: place beads to show 671; (*b*) second step: add beads to show 54—a single "50" bead and four "1" beads; (*c*) third step: replace the five "1s" by a "5" and the two "50s" by a single additional "100."

steps necessary to add 54 to 671 we see that much of the process is mental arithmetic.

Acquiring skill in the use of the abacus to perform rapid calculations requires much practice and patience. The instrument may be used to perform not only additions and subtractions but also multiplication and division. Individuals have mastered its use to a point where their speed at solving problems exceeds that of an efficient operator using a modern desk calculator of the electromechanical type.

Pascal's Counting Wheels

Blaise Pascal (1623–1662), the French philosopher and mathematician, invented the theory of probability, and constructed what is believed to be the first truly mechanical computing device (see Figure 19-3). Some of the basic ideas and mechanisms that he introduced are still being used in the manufacture of calculators today.

First Multiplying Calculator

The German scientist and mathematician Gottfried Wilhelm Leibnitz (1646–1716) invented the first calculating machine that would add, subtract, and multiply. The multiplication was performed by a series of additions. The machine itself was not built until 1694, but he had proposed the idea of multiplication by means of rapid additions as early as 1671.

Further Developments

Other early significant computer devices were Grillet's calculator (1689); Thomas' arithometer (1890), which could add, subtract, multiply, and divide; and the Edmundson Calculator (1885).

Dorr Eugene Felt (1862–1930), born in Beloit, Wisconsin, developed the first efficient key-driven adding machine. Felt patented his machine in 1886 and called it a comptometer. He is also credited with developing in 1888 the first key-driven calculator of the listing type—one that actually produced a printed output.

Calculators underwent numerous technological modifications and improvements in the years immediately following Felt's initial contributions. It was not until the 1920s, however, that a major advancement took place. Early calculators were hand-operated units, the more advanced type being operated by means of a hand lever or crank on the right side. In the 1920s, with the development of small electric motors, electrically operated calculators came into existence. The operator had only to strike the

Figure 19-3. Pascal's calculator. The first mechanical computer was probably this adding machine, designed in 1642 by Blaise Pascal. The machine adds when the wheels are turned with a stylus. Gears within the housing automatically "carry" numbers from one wheel to the next. (*a*) Decimal counting wheels; (*b*) Gears and mechanisms.

keys separately for the figures in each number to be acted upon and then to strike the appropriate instruction key, and the machine would perform the required calculation.

It was also in the 1920s that the punched-card method of data processing gained impetus. In 1928 the punched card was made to hold 80 columns of information, and by the early 1930s, an expanded line of machines were available for data processing. These machines could perform not only calculations but also the sophisticated routines required of an elaborate record-keeping program. The IBM 400 introduced in 1931 could process alphabetic information for business accounts. The IBM 407 in 1949 had the capability of printing 18,000 characters per minute as against 7000 for earlier machines.

As mentioned in Chapter 16, the Frenchman Jacquard had used his newly invented loom to prove the value of punched cards in the weaving of textile fabrics into elaborate and beautiful designs. Also, Charles Babbage worked on his "analytical engine" from 1833 to 1871. (See Figure 19-4.) His contemporaries dubbed the machine "Babbage's Folly," but, despite its lack of success, Babbage would probably have been successful had the machine technology of the day been as advanced as his ideas.

Both Jacquard and Babbage used mechanical devices with their punched cards but not until Hollerith applied the wonders of electricity to the processing of punched cards was the first truly electric data-processing machine born.

From the creation of the first crude abacus to the development of punched cards—the stage is set for new and still greater developments in computer technology.

The First High-Speed Computers

Before World War I manually operated office equipment provided the most advanced method of automatically handling data in the operation of a business. Soon after the war electromechanical machines came into extensive use. It was possible to perform more easily such operations as sorting, calculating, summarizing, and recording. However, these machines, including punched-card accounting machines, were far from being completely automatic. They still required the operator to perform many detailed functions.

A major step toward modern-day computers came from the recognition of compatibility between symbolic logic and electrical network theory. In 1938 Dr. Claude Shannon expressed this relationship in his famous paper, "A Symbolic Analysis of Relay and Switching Circuits." [1] It explained the basis for logical computer design and illustrated how to instruct such machines to perform logical as well as adding operations.

[1] This article appeared in the *Transactions of the American Institute of Electrical Engineers* and was based on his thesis for the M.S. degree at the Massachusetts Institute of Technology a year earlier.

Figure 19-4. A unit of the "Analytical Engine." Considered the forerunner of modern digital computers, the *analytical engine* was conceived about 1833 by the English mathematician and inventor Charles Babbage. Initially, in 1823, he started out to build a "difference engine" to compute the square of successive integers, but abandoned the undertaking some ten years later for the purpose of building a more complicated analytical engine which would perform any mathematical instructions fed into it. As planned, the machine had three components: a storage section, a calculator section, and a control section. Data could be fed into the machine by manual entry or by punched cards, like Jacquard's. Babbage spent a lifetime trying unsuccessfully to complete this machine. Still, despite his lack of success, his work is viewed as a significant contribution in the evolution of calculating machines. How proud he might be if he were to come back and see that crude machine of yesteryear developed in today's remarkable computers.

From the standpoint of history the first successful high-speed computer was built by the Bell Telephone Laboratories in 1940; it was an electrically operated special-purpose machine.

Four years later, in 1944, Dr. Howard Aiken, a professor at Harvard University, completed the first general-purpose automatic digital computer. Developed under the joint auspices of Harvard University and the IBM Corporation, this machine, commonly known as the Harvard Mark I, was more elaborately termed the Automatic Sequence Controlled Calculator. It was the type of computer that Babbage had envisoned more than 100 years earlier.

The Mark I was an electromechanical computer using mostly mechanical rather than electronic devices and relays instead of mechanical wheels for number storage.

In 1946 J. P. Eckert and J. W. Mauchley, two scientists at the Moore School of Electrical Engineering at the University of Pennsylvania, completed the first all-electronic digital computer. Known as the Eniac (ENIAC), for *E*lectronic *N*umerical *I*ntegrator *A*nd *C*omputer, this machine contained 20 registers in which numbers of 10 decimal digits could be stored. It used 18,000 vacuum tubes. Inputs and outputs were accomplished by punched cards, and it was designed to accommodate a prewired program.

The Eckert-Mauchley machine could process additions at the rate of 5000 per second and could carry out multiplications at a rate of 360 to 500 per second. Since then, addition speeds have climbed to more than 100,000 additions per second; and multiplication speeds, to more than 10,000 per second. Storage capacity has been increased to millions of registers.

The next important step forward came in 1949, when the *E*lectronic *D*iscrete *V*ariable *A*utomatic *C*omputer (EDVAC) was introduced. Built at the Moore School of the University of Pennsylvania, this machine was the first to embody the idea of storing the program within the computer itself. Up to this time the storage facilities of computers were used not for programs but solely to retain data and handle computations. Sequential instructions or programs were provided by punched cards, paper tape, or plug boards; thus the programs were stored outside of the computer proper. (The EDVAC used 5100 vacuum tubes and 12,000 solid-state diodes.)

Developed at about the same time as the EDVAC was another stored-program type computer: the *S*tandards *E*astern *A*utomatic *C*omputer (SEAC). This machine was built at the Bureau of Standards and put into operation in 1950.

Both the EDVAC and the SEAC were immediate forerunners of

the modern computer in the sense that they were the first to store both data and programs in their memory units.

Since 1950 the major improvements in computer design have been along four lines: greater speed, more storage capacity with greater flexibility, compactness through the miniaturization of piece parts and other components, and increased reliability.

Other computers of historical interest are the Whirlwind I (1951), developed by the Massachusetts Institute of Technology; the *Univ*ersal *A*utomatic *C*omputer (1951), known as UNIVAC, developed by Eckert and Mauchley; the IAS Computer (1952), developed by the Institute of Advanced Study in Princeton under the direction of Dr. John von Neumann; the IBM 701 (1951), and the IBM 650 (1954).

The family tree of computers is shown in Figure 19-5.

Figure 19-5. Family tree of computers. This illustration, presented through the courtesy of the National Science Foundation, is of historical interest. It should be pointed out, however, that the drawing was made in 1961, or thereabouts, and that the branching of the tree has increased greatly since that time.

EARLY HISTORY AND DEVELOPMENT OF THE INDUSTRY

The interval between the completion of the first general-purpose digital computer in 1944 and 1950 was a period of research and development by universities, government agencies, and business firms. The potentialities of the computer market were becoming evident, and manufacturers—especially those already in some phase of the data-processing industry—were seeking to get into this new and promising field. Among the first was the Remington Rand Corporation,[1] whose founders designed the Eniac. The Univac, which they later developed, was the first general-purpose electronic computer designed for business data processing.

To add a machine for scientific computations to its product line, the Remington Rand Corporation purchased Engineering Research Associates in 1952. Thus, dually equipped, they set out to market both machines with great gusto.

In 1951 the Remington Rand Corporation set up the first commercial computer installation—a Univac, delivered to the Bureau of the Census. Three years later, in January 1954, it delivered to General Electric Corporation the first large-scale electronic computer to process business data, the UNIVAC I.

Believing that the greatest market potential for computers was in the scientific rather than in business applications, the IBM Corporation passed up the opportunity to acquire the Eckert-Mauchley Corporation. In 1953, it concentrated its marketing on the 701, designed principally for scientific applications. Sometime later the 701 was modified to process business data and to compete with the Univac. The modification, the 702, was a failure. Subsequently, a large and expeditious program was instituted to replace the 701 with the 704, and the 702 with the 705—each changeover to occur by January 1956. In the interim the company planned its marketing strategy—to avoid the mistakes of the Remington Rand organization. The Remington Rand Corporation (the Sperry Rand Corporation as of June 1955) had neglected three vital areas: customer education, customer service, and the development of high-speed output equipment. The IBM Corporation adopted the marketing policy of holding off on the delivery of a machine until the customer was adequately educated and capable of making efficient use of the machine from the time of installation. This policy proved highly successful, and by the end of 1955 IBM had

[1] In 1955, the Sperry Corporation and the Remington Rand Corporation merged to form the Sperry Rand Corporation. The Remington Rand Division of the new Sperry Rand Corporation was to handle the company's computer business.

wrested the lead from Sperry Rand. At the midpoint of 1956 it had delivered 100 million dollars worth of its 700 series machines, compared to 70 million for Univac.

The Burroughs Corporation entered the computer field in 1956 when it acquired the Electro-Data Corporation and obtained the Datatron computer, which at the time provided strong competition for the IBM 650. Also, 1956 was the year in which the Radio Corporation of America sold its first Bizmac, after enormous investment in research and experimentation.

About the same time other firms were entering the field as follows: Bendix Corporation in 1955; General Precision, Inc. in 1956; the Minneapolis-Honeywell Regulator Corporation in 1957; the Philco Corporation in 1958; the Monroe Calculating Machine Company in 1958; the General Electric Corporation in 1959; and the Control Data Corporation, the National Cash Register Company, and the Packard Bell Computer Corporation in 1960.

Since 1960 many other firms have elected to manufacture computers, with an even greater number engaged in the making of peripheral and accessory equipment. Some of the early contenders have in the meantime bowed out of the digital-computer industry.

COMPUTERS—THE UNTHINKING THINKING MACHINES

In the early stages of marketing electronic computers, when they were first gaining acceptance and popularity, their wondrous technology and incredible speed captured the fancy of many. Writers were simply fascinated by the new subject matter and wrote about "giant brains," "thinking machines," and "educated robots." Among other attributes, computers were said to have the ability to select ideal marriage partners, design other and better computers, and write stories and poetry. Fact and fiction soon became entangled in a web of semantics.

Much as philosophers in the Middle Ages argued seriously over such issues as how many angels could sit comfortably on the head of a pin, there were those in our time who debated seriously about the thinking capabilities of computers. Both time and effort could have been saved if they had but defined their terms.

As the business potentialities of computers became more evident, practical businessmen of the "show me" school were not impressed by the fiction; they were seeking unadorned facts. Admittedly, the large-scale computers were engineering masterpieces, each a strange mixture of complexity and simplicity, but could they (or their progeny) possess true utility for carrying out the mundane activities of business? The answer, of course, was yes. The electronic computer with the superhuman "brain" was, after all, a near-relative of the desk calculator. It was capable of doing arithmetic functions and retrieving information with fantastic speed. Moreover, it could ascertain if a number was plus or minus, it could compare two numbers to determine if one was higher, lower, or equal to the other.

Credence was given to the concept that computers *think* by the use of such anthropomorphic terms as "decision-making" to describe the distinguishing capabilities just mentioned, "memory" to designate the storage area for new and processed data, and "machine language" to define a set of expressions used to convey information to the computer. It soon became clear, however, that computers do not think in the profound sense that people think. They cannot think in abstract terms, conceive original thoughts, feel emotional about anything, perform intuitively, or express feelings, courage, or ideals.

What they can do is operate with great speed, accuracy, and efficiency, but they must be told precisely and minutely what is to be done. In this respect they are more akin to a moron. Once properly instructed, however, in the space of a few seconds answers to problems that would otherwise take many man-months to perform are produced.

Certainly, many interesting machine-brain parallels have been drawn. We say that a computer *reads,* meaning that it accepts instructions from an input unit and places it in storage; that it *writes,* when it imparts information and records it on some output medium; that it has a *memory,* meaning that it can store data for use at a later time; and that it can perform basic arithmetic and logic functions.

The whole idea of imputing human intelligence to computers resulted in interesting but nonsensical presentations. Writings and speeches about "giant brains" and "thinking machines" did more harm than we know. Many companies found it difficult to gain acceptance of computers by their employees, especially middle-management personnel. The dignity of man seemed to be threatened by an increasing encroachment of technology; as in *Frankenstein,* by his own creation! Man is master of the machine, not the machine the master of man. A computer is a tool—a marvelous tool, to be sure—but still only a tool.

We welcome the idea that man will eventually be liberated from ignoble and wearisome toil—and free to perform more useful, enjoyable, and productive activities. The world could do with improvements, and eventually perhaps man will be able to achieve a society similar to Athenian life during the Golden Age of Greece. The ancient Greeks were free to devote themselves to politics, discussions in the market place, and intellectual or artistic pursuits—limited only by their individual taste and abilities—while slaves did the menial and tedious tasks. In a utopian and humanistic society of the future the computers could be the latter-day slaves, setting man free to develop his mind, and improve his material and social well-being.

COMPUTER LANGUAGE—A GLOSSARY

It may be well at this point to introduce a few special terms, terms that have developed in the computer field and are widely used. The following glossary is not an extensive one, but it will be of value in understanding the mechanics of computer operations.

Access time—The time interval required for a computer to locate data or an instruction word in its memory or storage facility and transfer it to its arithmetic unit. Also, the time interval required to transfer data from the arithmetic unit to the location in memory where it is to be stored.

Address—A designation that specifies where information is stored.

Binary—Constituted of two complementary alternatives; for example, on–off, yes–no.

Binary code—The representation of numbers, alphabetic characters, and special symbols by binary digits (0 and 1).

Binary-coded decimal—An encoding scheme for representing each decimal digit in binary notation.

Binary digit—A number, either 0 or 1, in the binary-coding scheme. Also, an extension of the word "bit": *b*inary dig*it*.

Bit—A unit of information represented by 0 or 1, a hole or no hole in a tape or card, or a spot or no spot on magnetic tape.

Block—A group of consecutive data (records, words, or characters) treated as a unit, particularly with respect to input and output. On paper and magnetic tape a block is a group of characters preceded and followed by a record gap.

Buffer—A storage device serving to compensate for differences in the speed of equipment or in the time of occurrence of events when transmitting information from one device to another, such as from an input device in which information is assembled from external or secondary storage and stored, ready for transfer to internal storage; or from an output device into which information is recorded from internal storage and held for transfer to external or secondary storage.

Field—A set of one or more characters that is treated as a unit of information.

Flip-flop—A circuit or device having two states. The circuit or device remaining in either state until the application of a signal causes it to change to the alternative state. The two states may be thought of as off and on, opened and closed, or binary digits 0 and 1.

Instruction—A set of characters that defines an operation to be performed by the computer. An instruction consists of two parts—an operation designation and one or more addresses.

Item—Two or more fields for specifying information about someone or something (e.g., a name and address).

Microsecond—A millionth of a second, or 0.00001 second.

Millisecond—A thousandth of a second, 1000 microseconds, or 0.001 second.

Nanosecond—A billionth of a second, or 0.0000009 second.

Off-line operation—Description of the functioning of certain items of peripheral equipment, physically unconnected and operating independently from the main processing units.

On-line operation—An operation in which peripheral equipment is directly connected to the main processing units. In such operations data are immediately introduced into the automatic-data-processing system as soon as it is originated.

Operand—Any of the quantities entering into or arising from an operation.

Program—A sequence of instructions that the digital computer must carry out in order to solve a problem.

Record—A collection of related words to be transferred between a peripheral unit and the main memory (see "Word").

Register—A data-storage unit with a specifically designated function and a fixed character capacity.

Software—The programs used to facilitate the utilization of digital-computer hardware. The term "hardware" refers to the physical components of the computer system.

Tab-card field—A set of one or more columns assigned for recording specific information.

Word—A combination of characters that is stored and handled as one unit; it is analogous to a field on a tab card.

Word length—The number of characters comprising a word.

20

Computer Arithmetic

NUMBERING SYSTEMS

Before the anatomy of computers is discussed some of their principal means of communication should be pointed out. Computers read and write in different *machine languages;* that is, the decimal system, the pure binary system, and various binary-decimal modifications such as the binary-coded decimal system, the binary-coded excess-3 system, and the biquinary-coded decimal system.

The desk calculator is an example of a computer that utilizes the decimal system. By adjusting a movable indicator figures may be entered, calculated, and answered as decimal numbers.

Large electronic computers that utilize the decimal system throughout must be able to distinguish among 10 distinctly different symbols or figures. They use the decimal system with the exclusion of other numbering techniques. Since they do not take advantage of the binary system, they require more intricate circuitry and parts that tend to have an adverse effect on computer reliability.

The major benefit of a decimal computer is eliminating the need for making binary-decimal conversions. However, this advantage is offset by functional and economic considerations.

The binary system and its various binary-decimal modifications are in keeping with the very nature of electronic devices—on or off, active or inactive, with no neutral position. Digital computers contain thousands of these devices, including relays, vacuum tubes, transistors, magnets, and magnetic cores, depending on the design of the computer. Because most computer devices do function on an on-or-off basis, the 0 and 1 symbols of the binary system can be easily applied to them. The digit 1 can represent that a circuit is *on;* the 0 can represent that a circuit is *off.*

Binary System

Because we are familiar with the decimal system, it is widely thought to be the simplest and easiest to use. Actually, the simplest system of numerical notation is the binary system, because it is based on the number 2, rather than the 10 decimal numbers (1 through 9, and 0). A mathematician, Gottfried Wilhelm von Leibnitz (1646–1716), thought that he had discovered the binary system, but there is reason to believe that it was known in China for at least 2000 years before Christ.

As previously stated, the binary number system is based on two symbols: 0 and 1. The 0 bit is referred to as *no-bit,* and the 1 bit is referred to as a *bit.* By assigning the symbols (0 and 1) to different positional, or place, values, all numbers (plus letters and special characters) can be represented by combinations of these two symbols. In the decimal system the place value increases from left to right, in a tenfold ratio.

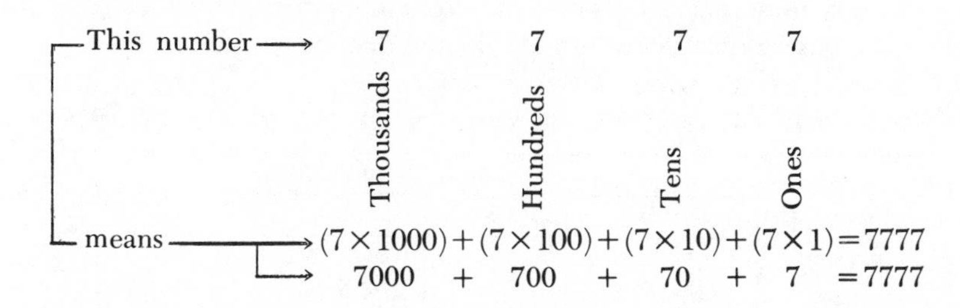

In the binary system the place value also increases from left to right, but in a twofold ratio.

This number → 1 1 1 1

Eights Fours Twos Ones

means → $(1 \times 8) + (1 \times 4) + (1 \times 2) + (1 \times 1) = 15$

$8 + 4 + 2 + 1 = 15$

As illustrated, the symbol 1 may be used to represent 1, 4, or 8, depending on its location, or placement.

To further illustrate the pure binary system, let us chart the place values and show an example of a binary number.

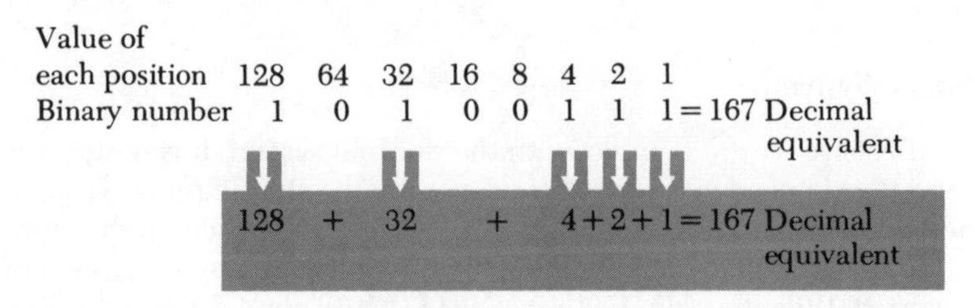

Another example is as follows:

Value of
each position 128 64 32 16 8 4 2 1
Binary number 0 0 0 1 0 1 1 0 = 22 Decimal equivalent

This number is read "one, zero, one, one, zero"

Binary numbers always have an equivalent decimal value. Table 20-1 gives the binary representation of selected decimal values.

Table 20-1. BINARY EQUIVALENTS OF DECIMAL VALUES

Decimal Value	Binary Equivalent					
	32s	16s	8s	4s	2s	1s
	2 × 2 × 2 × 2 × 2	2 × 2 × 2 × 2	2 × 2 × 2	2 × 2	2	1
	2^5	2^4	2^3	2^2	2^1	2^0
	32	16	8	4	2	1
0						0
1						1
2					1	0
3					1	1
4				1	0	0
5				1	0	1
6				1	1	0
7				1	1	1
8			1	0	0	0
9			1	0	0	1
10			1	0	1	0
11			1	0	1	1
12			1	1	0	0
13			1	1	0	1
14			1	1	1	0

Decimal Value	Binary Equivalent					
	32s	16s	8s	4s	2s	1s
	2 × 2 × 2 × 2 × 2	2 × 2 × 2 × 2	2 × 2 × 2	2 × 2	2	1
	2^5	2^4	2^3	2^2	2^1	2^0
	32	16	8	4	2	1
15			1	1	1	1
16		1	0	0	0	0
17		1	0	0	0	1
18		1	0	0	1	0
19		1	0	0	1	1
20		1	0	1	0	0
21		1	0	1	0	1
22		1	0	1	1	0
23		1	0	1	1	1
24		1	1	0	0	0
25		1	1	0	0	1
26		1	1	0	1	0
27		1	1	0	1	1
28		1	1	1	0	0
29		1	1	1	0	1
30		1	1	1	1	0
31		1	1	1	1	1
32	1	0	0	0	0	0
33	1	0	0	0	0	1
34	1	0	0	0	1	0
35	1	0	0	0	1	1
36	1	0	0	1	0	0
37	1	0	0	1	0	1
38	1	0	0	1	1	0
39	1	0	0	1	1	1

Binary Addition. In the binary system there are four rules of addition to remember. These are expressed as follows:

Binary Addition Table

1. $0+0=0$
2. $0+1=1$
3. $1+0=1$
4. $1+1=10$

Of these, the reasoning behind the first three is obvious, and only the fourth one needs explanation. As with the decimal system, it becomes necessary to carry a number when the largest digit that can be represented in a given position has been reached. In the decimal system the highest number is 9, but in the binary system the highest number is 1. The addition of two 1s in the binary system produces a result of 0, with a carry of 1 to the left:

```
   (1)          (carry)
  0001
  0001
  ----
  0010
```

Having familiarized ourselves with the four basic addition processes, let us perform two typical operations:

```
                       (111)        (carry)
(a) 1001           (b) 0011
    0110               0111
    ----               ----
    1111               1010
```

The first example does not involve carrying and is therefore relatively easy. In the second one, reading from right to left, the addition of 1 in the first column equals 0 with a carry of 1 to the column on the left. Since a 1 was carried into the second column, this means that three 1s must be added: $1+1=10+1=1$, with a carry to the next column. The carry into the third column is then added to the 1 already there, resulting in a sum of 0 with a carry to column four. Here, the 1 is added to the existing 0, producing a 1 with no further carry.

Binary Subtraction. As you learned in grade school, subtraction is the process of finding the difference between two numbers; it is the reverse of addition. There are four subtraction situations within the binary system, as follows:

Binary Subtraction Table

1. $0-0=0$
2. $1-0=1$
3. $1-1=0$
4. $0-1=1$

Here again the answers to the first three situations are evident. The last one involves a "take away," or "borrow." The basic technique for accomplishing this follows the same principles used in the decimal system; for example, to subtract 38 from 67, we cannot directly take the 8 from the

7 because the latter is larger. Hence we must move to the left and borrow 10 from the tens column—

```
              1
  67        ↗57
 -38   ➡   |-38
 ---        ---
            29
```

Essentially the same method is used within the binary system, the difference being that we borrow in multiples of two, rather than in tens:

```
Place values   2 1          2 1
               -----        -----
                             1 } Borrowed 2
               1 0    ➡    0 1 }
              -0 1         -0 1
              ----         ----
                            0 1
```

Now let us consider a more involved problem; namely, subtracting 0 0 0 1 from 0 1 0 0:

```
Place values   8 4 2 1          8 4 2 1
               -------          -------
                                     1 } Borrowed 2
               0 1 0 0    ➡    0 0 1 1 }
              -0 0 0 1         -0 0 0 1
              --------         --------
               0 0 1 1          0 0 1 1
```

In the above example we cannot subtract the 1 from 0, so we must take away from the nearest place to the left, which is the third column over with the placement value of 4. From here we borrowed the quantity of 4 and distributed it as required. The step-by-step approach for doing this called for putting a 1 in the second column (placement value of 2) and two 1s in the first column (placement value of 1 each); then the actual subtraction followed.

As with binary addition, subtraction within the binary number system has much in common with its parallel in the decimal system. A digital computer does not count as on fingers, but utilizes the addition table stored in its memory.

Subtraction by Complementing. You will recall that it is possible to subtract one number from another by the addition of complements. In decimal arithmetic, the tens complement of a number is the difference between that number and the next power of 10 above it. Accordingly the tens complement of 2 is 8 (10−2=8); the tens complement of 9 is 1 (10−9=1); the tens complement of 32 is 68 (100−32=68); and, the complement of 514 is 486 (1000−515=486). Basic complements in the decimal system are shown in Table 20-2.

Table 20-2. DERIVATION OF TENS COMPLEMENTS

Base		Decimal Number		Tens Complement
10	−	0	=	10
10	−	1	=	9
10	−	2	=	8
10	−	3	=	7
10	−	4	=	6
10	−	5	=	5
10	−	6	=	4
10	−	7	=	3
10	−	8	=	2
10	−	9	=	1

To subtract by means of complementary addition, add the complement of the subtrahend to the minuend. Example: Subtract 3 from 8.

Regular Subtraction		Subtraction by Complementary Addition
8	Minuend	8
−3	Subtrahend	+7
5		15

It will be observed that 8+7=5 with a 1 carry, and not just 5. This superfluous 1 in the carry position is ignored, and the correct answer—the number 5—is the result. The rule follows that a carry of 1 beyond the most significant digit is not to be counted when subtraction is carried out by the addition of complements.

Note that the tens complement of a number is equal to its nines complement plus 1. The tens complement can be calculated by taking the nines complement of every digit and adding the least significant digit. Suppose, for example, that we want to subtract 29,481 from 62,986:

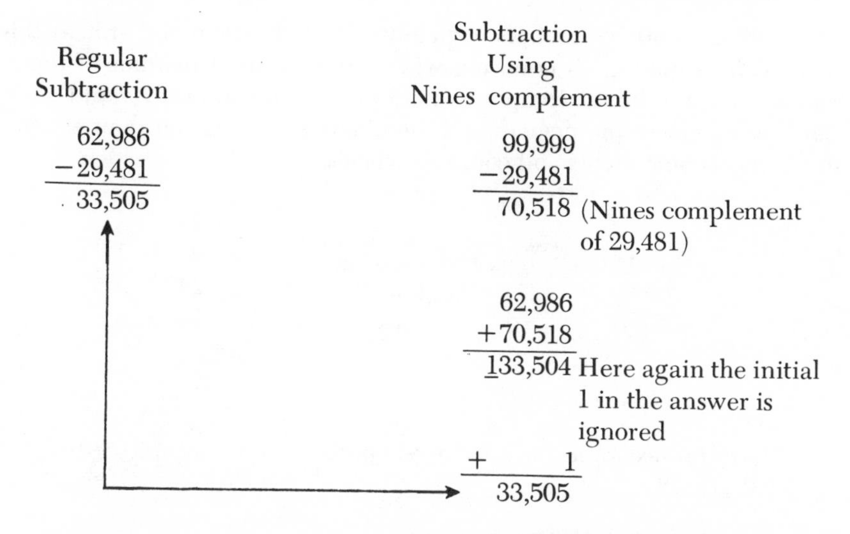

In the binary system the counterpart of the nines complement is the ones complement. We obtain the ones complement of a binary number by taking the difference between 1 and each of the digits of the number: $1-0=1$ and $1-1=0$. Complementing in the binary system therefore consists in transposing a 1 for a 0 and a 0 for a 1. Accordingly the complement of 1 1 0 1 0 is 0 0 1 0 1. To subtract by complementing in the binary system you add the ones complement plus 1 and ignore the initial 1, as in the decimal system. As an example consider the following problem:

Regular binary Subtraction

```
  110001=49
−  11011=27
  010110=22
```

Complementing

```
  110001=49
+  00100   (complement of 27)
  110101=21
+      1
   10110=22
```

Most digital computers carry out binary subtraction by adding the ones complement. The addition is performed by reference to the addition table stored in the memory, thus avoiding long drawn-out arithmetic processes.

Binary Multiplication. In any numbering system multiplication is essentially nothing more than repeated addition. It may be defined as the process of increasing a given number or quantity as many times as there are units in another number or quantity. A desk calculator utilizes this method by spinning its counting wheels the required number of times. Similarly, some digital computers perform multiplication by rapid additions, using electronic devices and the binary system rather than electromechanical components and counting wheels.

24	is the same as	24
× 5	adding	24
120	five 24s together	24
		24
		+24
		120

Here, for example, is how a computer might multiply 10001× 01000 (17×8):

Place values		Decimal
16 8 4 2 1		
1 0 0 0 1		17
1 0 0 0 1		17
1 0 0 0 1 0	←two times→	34
1 0 0 0 1		17
1 1 0 0 1 1	←three times→	51
1 0 0 0 1		17
1 0 0 0 1 0 0	←four times→	68
1 0 0 0 1		17
1 0 1 0 1 0 1	←five times→	85
1 0 0 0 1		17
1 1 0 0 1 1 0	←six times→	102
1 0 0 0 1		17
1 1 1 0 1 1 1	←seven times→	119
1 0 0 0 1		17
1 0 0 0 1 0 0 0	←eight times→	136
Place values 128 64 32 16 8 4 2 1		

Most computers are not designed to handle multiplication in this way, because it is a comparatively slow method. Instead they have a binary multiplication table stored in their memory—much the same as you and I have a decimal multiplication table stored in ours. This, of course, is another human parallel.

Multiplication within the binary system is carried out in a manner that corresponds to the decimal system. The rules are as follows:

Binary Multiplication Table

1. $0 \times 0 = 0$
2. $0 \times 1 = 0$
3. $1 \times 0 = 0$
4. $1 \times 1 = 1$

All conditions equal 0, except 1×1, which equals 1.

Example: Find the answer to 10001×01000 (or 17×8). The solution is as follows:

Place values	256	128	64	32	16	8	4	2	1
					1	0	0	0	1
				×	0	1	0	0	0
					0	0	0	0	0
				0	0	0	0	0	
			0	0	0	0	0		
		1	0	0	0	1			
	0	0	0	0	0				
	0	1	0	0	0	1	0	0	0

Binary Division. Multiplication has been described in terms of a series of repeated additions. Similarly, division is a series of repeated subtractions. It may be defined as the process of finding out how many times one number or quantity is contained in another number or quantity.

Dividing binarily requires the same basic steps that are used in dividing decimals into decimals. However, we multiply and subtract with the binary rules for these processes.

Example:

```
Place values  32 16 8 4 2 1
                     1 0 1
       1 1 1)1 0 0 0 1 1
              1 1 1
              -----
                 1 1 1
                 1 1 1
                 -----
```

```
   5
7)35
  35
  --
```

The same problem carried out in detail using subtraction is as follows:

Place values 32 16 8 4 2 1		
1 0 0 0 1 1		35
0 0 0 1 1 1		7
1 1 1 0 0	⟵one time⟶	28
0 0 1 1 1		7
1 0 1 0 1	⟵two times⟶	21
0 0 1 1 1		7
1 1 1 0	⟵three times⟶	14
0 1 1 1		7
1 1 1	⟵four times⟶	7
1 1 1		7
0 0 0	⟵five times⟶	0

A number of computers perform their division operations as a series of repeated subtractions.

Binary-Coded Decimal System

Retaining the place values of the binary number system, the basic format can be used in representing other than binary numbers; for example,

0		0000	22			0010	0010
1		0001	23			0010	0011
2		0010	24			0010	0100
3		0011	25			0010	0101
4		0100	↓				
5		0101					
6		0110	100		0001	0000	0000
7		0111	101		0001	0000	0001
8		1000	102		0001	0000	0010
9		1001	103		0001	0000	0011
10	0001	0000	104		0001	0000	0100
11	0001	0001	105		0001	0000	0101
12	0001	0010	↓				
13	0001	0011					
14	0001	0100	1000	0001	0000	0000	0000
15	0001	0101	1001	0001	0000	0000	0001
↓			1002	0001	0000	0000	0010
			1003	0001	0000	0000	0011
20	0010	0000	1004	0001	0000	0000	0100
21	0010	0001	1005	0001	0000	0000	0101

Figure 20-1. Binary-coded decimals.

combinations of bit positions can be designated as decimal numbers, alphanumeric characters, or special representations.

In the binary-coded decimal system each digit of the decimal system has its equivalent in binary form. Where in pure binary the number 14, for example, is written as one set of figures, 1 1 1 0, in binary-coded decimal it is written as two: 0 0 0 1   0 1 0 0. Figure 20-1 illustrates the development of such a system.

Excess-3 Code

The excess-3 code is a widely used variation of the binary-coded decimal system. In this scheme the binary value of a decimal 3, which is 0 0 1 1, is added to each numeral in binary to facilitate both carrying and complementing operations. Thus, 0 becomes binary 3 (0 0 1 1), 1 becomes binary 4 (0 1 0 0), and so on. One advantage to this scheme is that the addition of two numbers where the sum is greater than 10 produces a simultaneous "carry" in both the decimal and excess-3 systems. Another advantage is that complements can be obtained simply by changing each 0 to 1 and each 1 to 0.

Biquinary Code

The biquinary system of numerical notation, used on the Chinese abacus (the Saun Pan) and on some electronic digital computers, consists of two groups of digits—one of 2 (hence *bi*) and the other of 5 (hence *quinary*). The quinary group counts the five digits 0, 1, 2, 3, and 4, after which the binary part changes position, and then the quinary part counts over again. For comparison, biquinary notation and decimal notation are shown below:

Decimal Notation	Biquinary Notation
0	01 00001
1	01 00010
2	01 00100
3	01 01000
4	01 10000
5	10 00001
6	10 00010
7	10 00100
8	10 01000
9	10 10000

The biquinary system was used on a number of early computers that made extensive use of relays in their design. In this way it was possible to use only seven counters, rather than the ten that would be required with the decimal system.

NUMBER CONVERSIONS BY COMPUTERS

Although the use of the symbols 0 and 1 to represent quantities and characters is very efficient for internal computer operations, they are impractical for use on business media such as reports, records, statistical compilations, and the like. Under the pure binary system a $16.00 check would read 1 0 0 0 0. One can well imagine the difficulties that would arise if the business community were to attempt to use the binary system side by side with the decimal system for ordinary commercial transactions.

In the case of decimal computers there is no problem, as can be seen from the process shown in Figure 20-2.

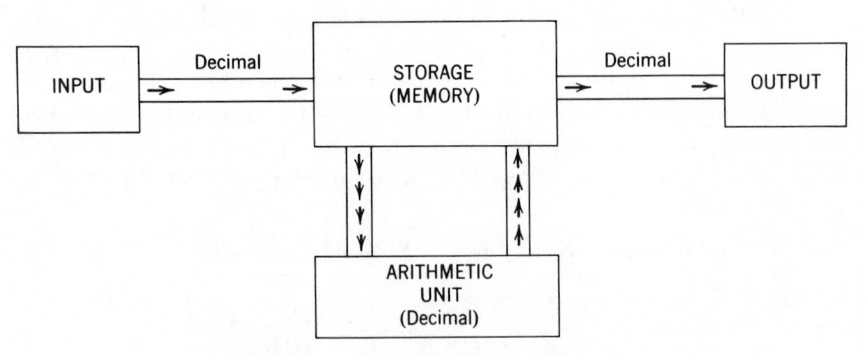

Figure 20-2. Flow of data in a decimal computer.

In binary digital computers, however, all input data are converted to binary notation. The arithmetic unit performs its operations in binary, after which the binary data are changed back to the decimal system of notation for output. The process is illustrated in Figure 20-3.

In the binary-coded decimal computer the binary system is used to encode each character. Conversion of data code by this method is rela-

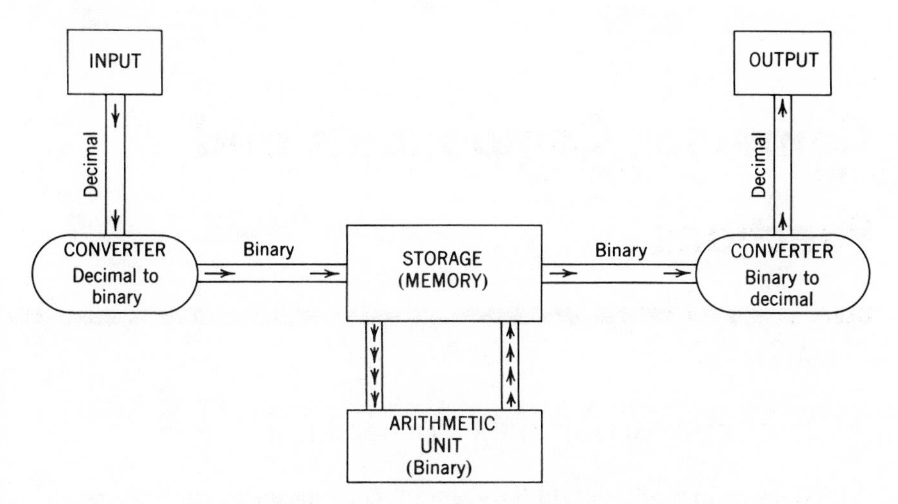

Figure 20-3. Flow of data in a binary digital computer.

tively simple, because a combination of binary numbers is interpreted and processed as a single digit or character. The process of conversion is shown in Figure 20-4.

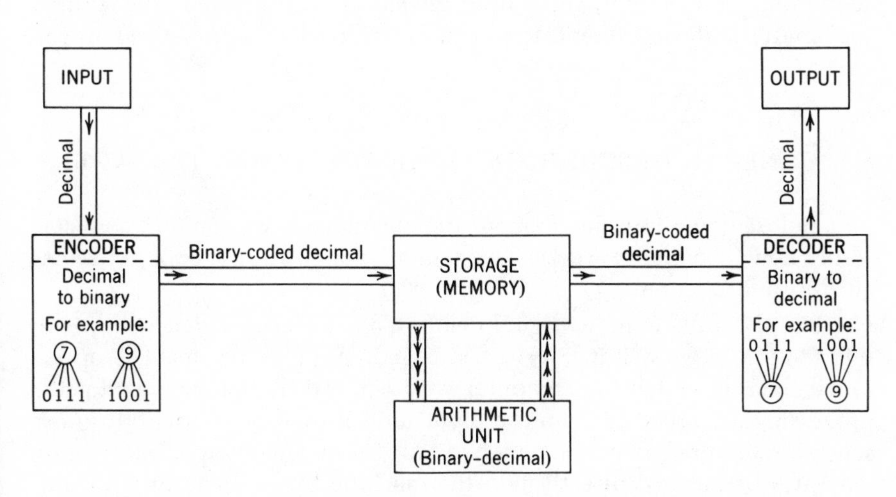

Figure 20-4. Flow of data in a binary-coded decimal computer.

21

Computer Components and Functions

THE COMPOSITION OF A COMPUTER

Makes and models of digital computers for business applications differ somewhat in outward appearance and in a considerable number of particulars, much as automobiles do. As with automobiles, which invariably have a frame, motor, wheels, body, and an electrical system, computers also have fundamental components. Figure 21-1 shows a block diagram of the major computer elements which comprise a computer system. Observe that there are five types of functional units: input devices, storage, arithmetic unit, control unit, and output devices. In combination, the storage, arithmetic, and control units form what is referred to as the central processor.

GENERAL DESCRIPTION OF HOW A COMPUTER WORKS

Before going into detail about the components we should know generally how a computer works. This will alleviate the danger of not seeing the forest for the trees.

Simply, an automatic digital computer is a machine that will accept information, process it at a very fast speed, and give results. The operational data on which the computer is to act and the instructions for the processing are received from the input unit in the form of digital signals sensed from punched cards, punched tape, or magnetic tape. Information may also be entered directly into the machine by an operator who activates keys similar to those found on a typewriter keyboard.

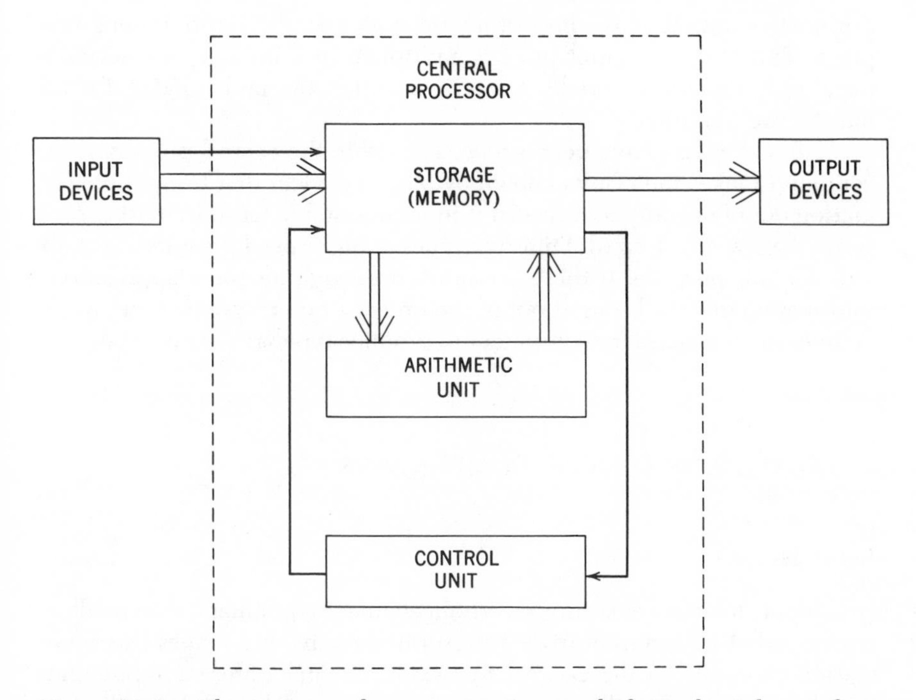

Figure 21-1. The anatomy of a computer. In capsule form, this is how it functions: The control unit directs the transfer of information, with necessary instructions, to designated locations in the high-speed storage unit. For processing data, the control unit receives an instruction from storage and interprets it. Then, in accordance with the instructions, the control unit selects the appropriate circuitry for executing these instructions, obtaining data from storage, carrying out the desired operation on the data in the arithmetic unit (that is, arithmetic and logical operations), and returning the results to the storage unit where they are held until required for further processing, or until they are transferred to an output device and onto an output medium.

Before the computer can begin to solve a problem, all of the relevant data and instructions must be stored some place in the computer system. This storage place is referred to as the *memory unit,* or simply *storage.* Both *data words* and *instruction words* are read into memory, containing characters and instructions, respectively, about the operations to be performed.[1] For processing the data words are taken out of memory and transmitted electronically, under the direction of the instructions, to the

[1] Here we have the *stored program* concept, peculiar to computers. The term describes the idea of storing within the computer's central processor a set of instructions specifying the precise sequence of coded steps that the computer is to follow in solving a problem or in processing data.

arithmetic unit. It is in the arithmetic unit that the computations take place. This is where a unit price is multiplied by a quantity, an accounts payable is debited or credited, and wages taxable under FICA for the quarter are computed.

After the data have been processed within the central processor, the computer makes the results known by output devices that bring the information out of memory and record it in some suitable form. In most cases a printed answer is desired. Different types of high-speed printers are available for this purpose. If these are not fast enough for some applications, information may be brought out of the processor on magnetic tape, which is processed on auxiliary equipment to produce typewritten material.

FUNCTIONS OF THE COMPONENTS

Input Devices

Input devices constitute electromechanical equipment that reads or senses coded information from a prescribed media and makes this information available to the control processor. Several different input units may be connected to the processor, the number and type depending on the design of the equipment and the applications involved.

The Central Processing Unit

The central processor—the heart of the system—controls, unifies, and integrates the entire computer system, and performs the actual arithmetic and logical operations on data. It consists of the following components:

1. An internal memory (storage unit);
2. An arithmetic unit;
3. A control unit; and
4. A console.

Storage. The memory unit stores both data and programs that the computer is to follow in machine language. Reading in a different program of instructions will enable the computer to process new kinds of data.

The memory consists of a large number of *locations* where the data and instructions are stored. Each location has a specially designated *address* where the information can be found.

To perform an arithmetic operation data are transferred from the memory to the arithmetic unit, where the calculation is performed. The answer is then transferred back into memory to be used later, perhaps for additional computations or printout.

Arithmetic Unit. This portion of the processor does the following:

1. Arithmetic operations—addition, subtraction, multiplication, and division.

2. Stores information temporarily, transfers information, and shifts information—such as from left to right in the same register, if required.

3. Logical operations—the comparing, selecting, matching, sorting, merging, etc. functions.

Control Unit. The control unit is sometimes compared to a telephone exchange. Just as there are connecting lines between all telephones serviced by a control exchange, so also in a computer all information-transfer paths are already in existence.

The telephone exchange provides for connecting one subscriber with another. Dialing causes a number of electrical impulses to travel from one place to the other, to connect and disconnect circuits, to ring the phone, and so on. At the exchange appropriate controls set up the connections for the flow of conversations. In the computer the carrying out of an instruction involves opening and closing switches (gates) and thus controlling the flow of data.

The control unit performs the following functions:

1. "Interprets" and executes the instructions stored in the memory unit. It follows commands such as "start" and the like.
2. Directs the sequence of operations within the system.
3. Controls the input-output units.
4. Performs automatic accuracy checks.

The Console. Often termed the "face" of the central processing unit, the console contains keys, switches, display lights, and signals through which the operator can monitor and control the system. By using these devices it is possible to—

1. Start and stop the computer.
2. If necessary, enter data manually by means of a keyboard. However, information is seldom entered directly into the computer in this way except during testing or special circumstances.
3. Visually determine the contents of certain internal registers.
4. Ascertain the status of internal electronic switches.
5. Select input-output devices.
6. Control the mode of operation so that when unusual conditions are

encountered, the computer will either come to a halt or indicate the condition and continue in operation.

7. Detect errors and reset the computer when the condition causes it to halt.

Output Devices

As mentioned earlier, the term "output device" describes the hardware used for extracting data from the central processor. As with input units, the number and type of output devices that are used depend on the applications involved. In general they serve two purposes:

1. To create records and reports in a language that people can understand.
2. To create new media that can be used to satisfy further automatic-data-processing needs.

KINDS OF INPUT DEVICES

This section presents the kinds of input devices and how they are used.

Magnetic-Tape Units

Magnetic-tape units are high-speed devices with the dual capability of input and output. As the name suggests, magnetic tape is used as the medium for introducing and extracting information from the central processor. Operating at great speed, it is capable of transferring data at a rate of 5000 to 100,000 characters per second.

If the input medium is magnetic tape, the usual practice is to have the data punched into cards and then transferred to tape by a converter. The reason for doing this is that magnetic-tape units read in data much faster than a punched-card device and therefore are more suitable for handling huge volumes of data.

Punched-Card Readers

Some computers use on-line punched-card readers. Cards are fed through the reading station; the transfer of information reaches speeds of up to 1000 cards per minute (or 80,000 characters per minute).

Cards may be read in either an alphanumeric mode or in a binary mode. The alphanumeric mode uses the standard IBM card code, which permits one character per card column. Two methods of recording information in the binary mode are the *row binary format* and the *column binary format.*

Figure 21-2 shows a card in the row binary format. Note that only 72 columns of the 80-column card are used and that these are divided in half. The left half of the card consists of columns 1 through 36; and the right half, of columns 37 through 72. Each of the 24 half-rows may be treated as a series of 36 binary digits or as a word. As many as 24 words may be contained on one card.

For purposes of identification the half-rows are referred to by numbers; for example, the row with the encircled 3 is known as *three-row left,* and the one with the encircled 8 is called an *eight-row right.* Where applicable, each punched hole in the card is treated as binary 1; the absence of a hole indicates binary 0. In arithmetic applications the farthest position to the left in each row is reserved to indicate a plus or minus sign. No hole means a *plus,* whereas the presence of a hole indicates a *minus.*

A conventional card-reading station is made up of a lamp and 12 photoelectric cells. During the passage of a card between the lamp and the photoelectric cells, the punched holes are sensed. Each card is read column by column, from left to right, and the data are transferred into the memory unit.

In the column binary format, the information is arranged parallel along the card. Each column is capable of containing 12 information bits, and three columns are used to designate a word. Hence 36 bits comprise a word. Figure 21-3 shows such a card.

Under either card format the largest binary number that can be recorded is 36 successive 1s, which has the decimal equivalent of 8,589,934, 591.

Paper-Tape Readers

As with punched-card readers, on-line paper-tape readers sense holes in paper, and the electrical impulses thus produced transmit the information to the central processor. Reader speeds vary, with typical rates ranging between 200 and 1000 characters per second.

Character-Recognition Units

Devices designed for character recognition fall into two broad classifications; namely, magnetic character readers and optical character read-

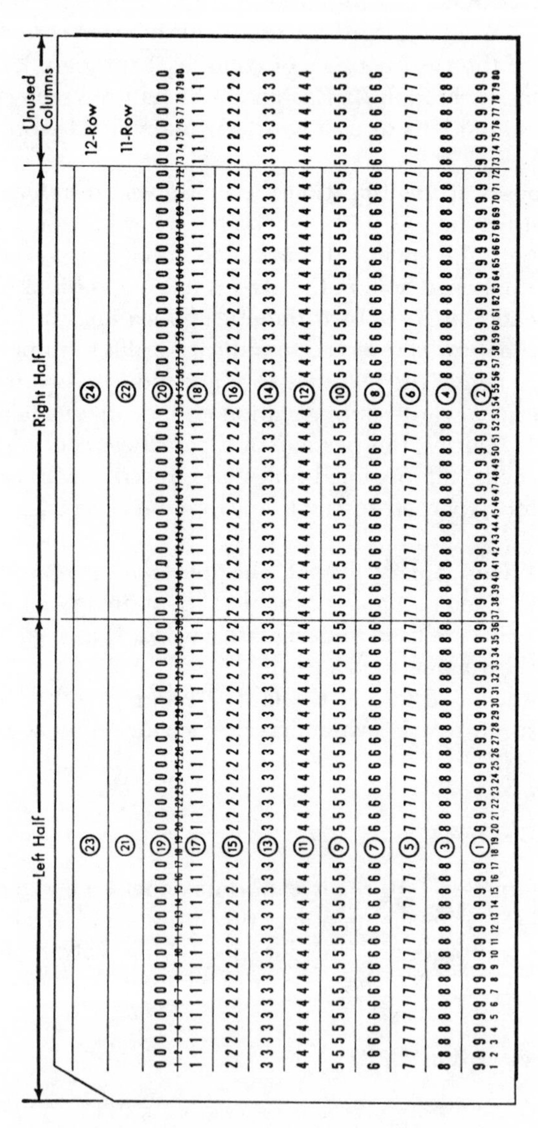

Figure 21-2. Row binary format.

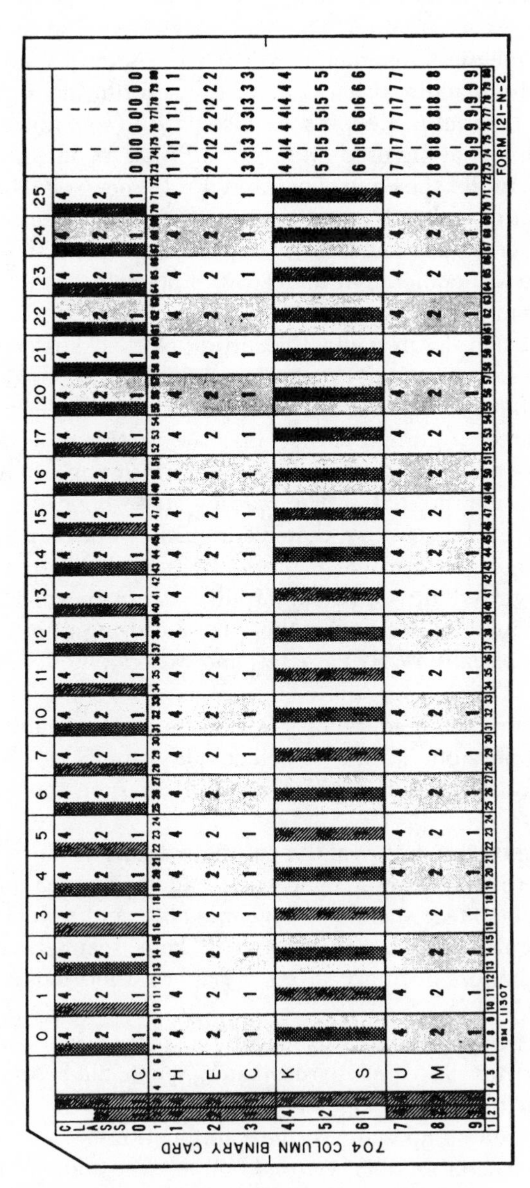

Figure 21-3. Column binary format.

ers. Both offer the advantage that the characters can be both machine-read and man-read.

Magnetic Character Readers. These machines can read checks and documents inscribed with the magnetic ink characters approved by the American Bankers Association. The reader is designed so that it senses the shape of each magnetic character as it passes under a read head. When used *on line,* the machine is capable of transmitting data directly to the memory of the computer. Readers are also used to sort documents, and in this capacity they may be used *on line* or *off line.* As on-line units, sorter-readers may be under the control of the central processor. The information read by the reader is thus transmitted to the central processor, which will automatically send back another set of pulses directing the sorter-reader on where to file a document within its row of pockets.

One of the important applications of the sorter-reader is on an off-line operation; for example, in banks the sorting of checks is commonly done independent of the computer.

Optical Character Readers. Machines of this classification can read characters printed in normal carbon inks by electric typewriters, adding machines, high-speed printers, and other imprinting mechanisms, such as those used with charge plates. The reading station consists of an intense light source and a lens system that distinguishes between reflected light patterns. The light pattern that is produced is read as a quantity of small dots and is converted into electrical impulses to create a character pattern. Within the circuitry of the reader there is a built-in pattern to match the pattern of the optically read character, and, when matched, the character is recorded.

Optical character readers vary widely in their capabilities. Certain units will read only one kind of typeface. Other units can be adjusted to read various intermixed and different kinds of type fonts. (See Figure 21-4.)

Depending on the design of the equipment, one or more lines of information can be read at a time. Some units, called *page scanners,* as distinguished from *document scanners,* have movable spot readers that permit them to read entire pages at high speeds. Page scanners are especially constructed to handle ordinary sheets of paper rather than cards or heavy paper stock.

Optical character readers also have the capacity to *mark-read;* that is, to sense ordinary pen and pencil recordings when made to specification. This feature enables source documents so marked to be fed directly into the machine—opening up many systems applications.

Optical character readers are used both *on line* and *off line.*

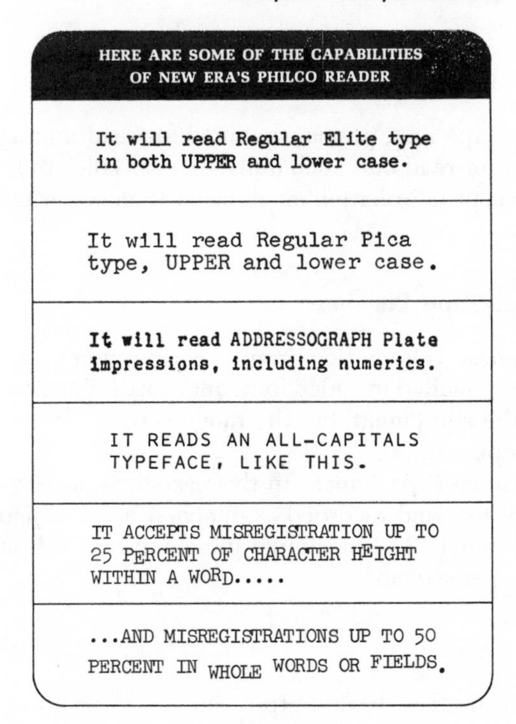

Figure 21-4. Some of the capabilities of an optical character reader.

Buffer Storage Units

The central processor can ordinarily manipulate data at a faster rate than input units can transmit them. To overcome the difference in speed, buffer storage units are used. A *buffer* may be an external storage device between the input unit and processor storage. In this kind of arrangement it is used as temporary storage for assembling data for transfer to internal storage. In some computers part of the main storage unit is used for buffering.

KINDS OF OUTPUT DEVICES

Most computing systems use one or more of the following kinds of output devices.

Magnetic-Tape Units

A magnetic-tape unit, previously noted for read-in, may also be used interchangeably for read-out. Such units are very efficient for this purpose because they accept information much faster than either punched tape or cards.

Card and Paper-Tape Punches

Card Punches. Information from the computer's memory may be brought out and punched in coded form onto cards. Punching speeds vary, depending on the equipment, but the range is from 100 to a maximum of about 300 cards per minute.

On-Line Paper-Tape Punch. In this case the data are received from the computer system and recorded as punched holes in paper tape by an automatic tape punch. The maximum punching rate is from 20 to 500 or more characters per second.

High-Speed Printers

Printers are used as basic output devices to produce characters in printed or typewritten form. Speeds of printing range from 10 to 2000 characters per second. When used on line, they save an intermediate step by converting the output medium to a visual record of data. This should be done with the realization that the output rate is still well below the speeds at which the data can be manipulated by the central processor.

Special Types of Output Devices

Another type of output device is the electronic character display tube, something like the television picture tube, which may be used in conjunction with printing. The *Charactron tube* receives signals from the computer, deflects the beam of electrons through the proper matrix characters, and forms a page of information on the tube. When the information has been fully read out, the information on the screen may be captured in permanent form, either by the Xerox photocopying process or by photographing it. This method of output operates at a rate of speed that is comparable to computer processing speeds.

Another output device of interest is one that permits the computer to "talk" to people. This is accomplished by the activation of prerecorded verbal messages.

Buffer Storage Units

Just as a buffer may serve as a temporary storage device between the input unit and processor storage, so a buffer may serve to compensate for the difference in speed between the central processing unit and ouput. Output buffering works in reverse: the data are received out of storage by the buffer at electronic speed and then transferred to the output unit at a speed that it will accept. As with input buffering, in some computers part of the main storage unit may be used for output buffering.

PERIPHERAL EQUIPMENT—INPUT-PREPARATION DEVICES

Peripheral, or off-line, units provide a method by which many operations can be performed in conjunction with the computer but not directly as a part of it. One advantage of such equipment is to unburden the computer of many time-consuming tasks, thus permitting it to perform high-speed data processing. Another advantage, mentioned in Chapter 23, is that certain of this equipment may be used to obtain data at its point of origin.

Card Punches. Punched cards, as already noted, are a major form of input in many data-processing installations. The cards can be created by a card punch such as the IBM 026, a reproducing punch, or a tape-to-card converter.

Sorters. Computers are not especially suitable for sorting. To sort a large number of records into sequence it is frequently more economical to use a conventional card sorter on an off-line basis.

Paper-Tape Punches. Tape-generating equipment was covered in the preceding chapter. It includes electric typewriters, adding machines, calculators, cash-register machines, bookkeeping machines, and card-to-tape converters. Machines in each of these catagories can provide tape for input to a computer.

Character-Recognition Units. As pointed out in the discussion on input devices, both magnetic character readers and optical character readers are often used *off line.*

Converters. By their nature converters fall within the classification of peripheral equipment. This includes paper-tape–to–card converters, paper-tape–to–magnetic-tape converters, punched-card–to–magnetic-tape converters, and magnetic-tape–to–paper-tape converters.

Printers. These off-line–on-line devices can be operated from various data media, such as punched cards, paper tape, and magnetic tape. The print unit may consist of a number of type bars, type wheels, or a print chain. Certain units of the high-speed category are capable of printing 130 positions on one line at a speed in excess of 1200 lines per minute.

Magnetic-Tape Writers. Manually operated keyboard machines are available that will record data onto magnetic tape in much the same way as a card punch records data on a card. Such a machine is illustrated in Figure 21-5. By means of the keyboard it is possible to transcribe information from source documents directly onto magnetic tape. A switch setting provides operation in any of three modes—original recording, verification, or searching.

Figure 21-5. Magnetic tape data recorder.

Tag and Ticket Readers. Representative applications of tag and ticket readers are in inventory control, interdepartmental billing, work-in-process control, and finished-goods control. Retailing concerns are widespread users of these machines, especially in conjunction with a merchandise-control system. In a typical situation special devices are used to prepunch and preprint tickets en masse with such information as item identification, price, style, color, and so on. (See Figure 21-6.) The tickets may be processed in any of the ways shown in Figure 21-7.

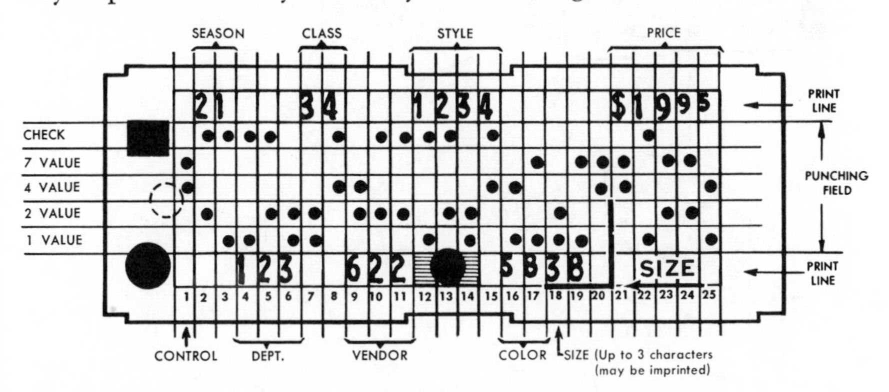

Figure 21-6. Representative coding format for a print-punch ticket. Source: *Print-Punch.*

INPUT MEDIA

Data may be fed into a computer by one or more of different kinds of input media, and the kind used in any particular case depends on the design and composition of the computer involved. The three most commonly used input media are punched cards, paper tape, and magnetic tape. Since the design and general make-up of both punched cards and paper tape have been treated in detail elsewhere, only the last one mentioned is in need of elaboration.

Magnetic Tape

In construction magnetic tape is a ribbon of paper, metal, or plastic coated or impregnated with a material that can be magnetized. Information is recorded on the tape by a combination of magnetized spots (bits). The actual recording process is similar to audio-recording methods. A significant difference, however, is that tape recorders register information

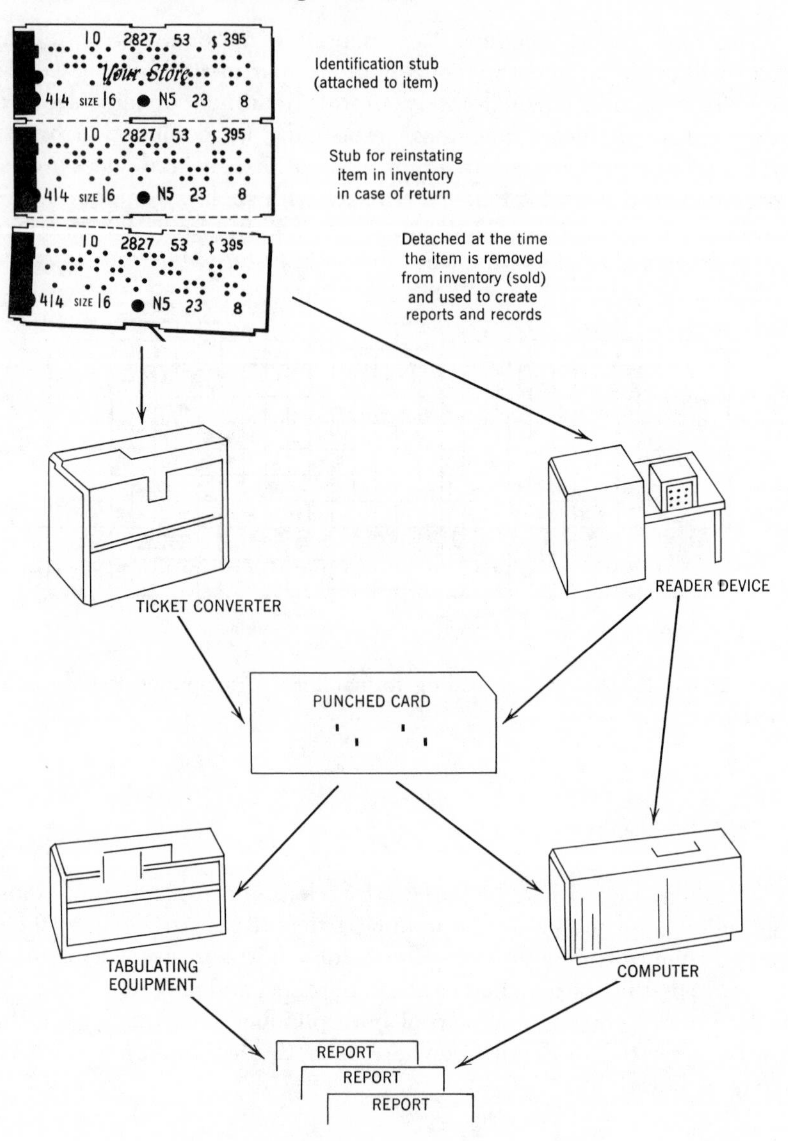

Figure 21-7. Alternative routines for processing tickets.

continuously, whereas for computer applications the information is registered discretely—that is, bit by bit.

In addition to speed, magnetic tape offers the following advantages:

1. It provides compact storage for large amounts of data.
2. The tape is not permanently altered as in the case of perforated

paper tape; the recorded information can be automatically erased, and the tape can be reused innumerable times.

3. A mistake can be set right by erasing the error and recording the correct information.

Tape may be ½, ¾, or 1 inch wide. The width has a definite relationship to the number of parallel channels or tracks: ½-inch tape commonly has seven parallel tracks; 1-inch tape is available with 10-track capacity.

Two main modes are used to record data on magnetic tape—binary coding or binary-coded decimal. The design of the computer determines which mode of recording is used.

Binary Coding on Magnetic Tape. Figure 21-8 illustrates binary recording on seven-channel tape. With an odd-parity system, the sum of the

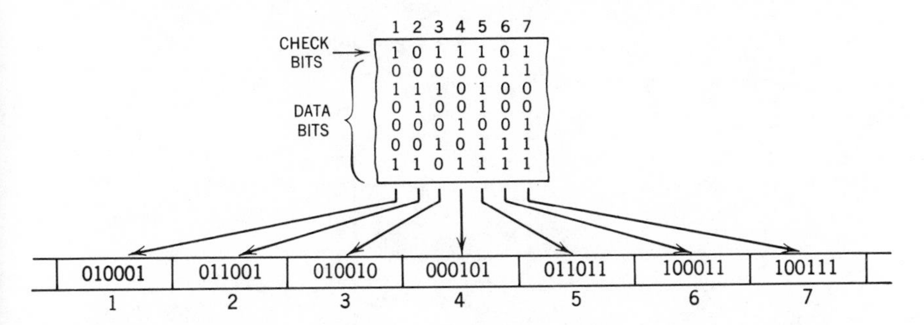

Figure 21-8. Binary notation on seven-track magnetic tape.

total number of pulses generated by the bits within each column is counted; the result must add up to an odd number of pulses, or a transcription error is indicated. Note in the illustration that the 1s in each column add up to 3, 3, 3, 3, 5, 3, and 5, respectively. Certain equipment is designed

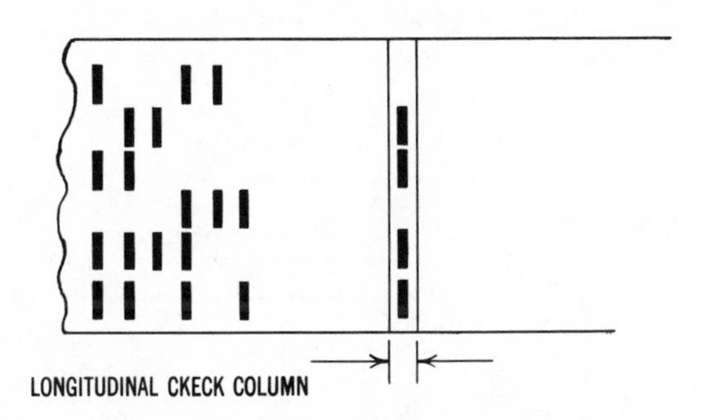

Figure 21-9. Horizontal validity checking.

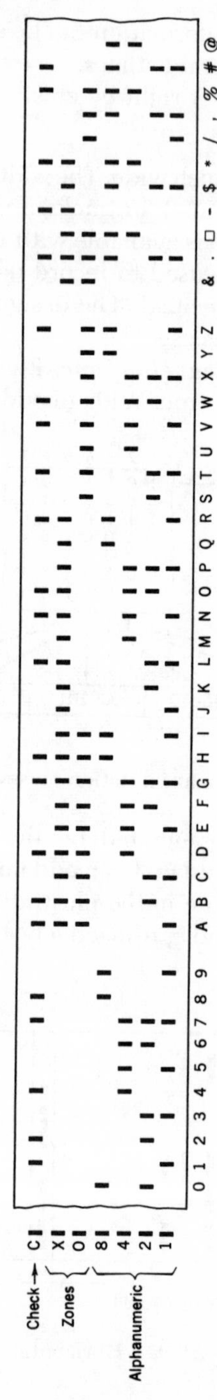

Figure 21-10. Seven-track magnetic tape showing BCD coding.

for even-parity checking. In either case, verification by individual columns is known as vertical parity checking.

In addition, equipment may provide for longitudinal parity checking. A longitudinal (lengthwise) record check is made by verifying the total number of bits in each horizontal track of a record block, including the check track (Figure 21-9); they must total to either an odd or even number, depending on the system, or an error is indicated. This is how it is accomplished. As a record block is being recorded on tape, an odd or even designation of the number of bits in each track is made. If the equipment is designed for odd-parity checking, a bit is recorded at the end of the record block for each track with an even number of bits. Under this method the bits along each track must total to an odd number or the odd-bit rule is violated. Equipment utilizing the even-parity bit method employ the same basic principle with an even count being required, and with either method the gain or loss of a single bit indicates the presence of an error.

Binary-Coded Decimal Coding on Magnetic Tape. Decimal numbers, letters of the alphabet, and a quantity of special characters may be designated on magnetic tape by using binary-coded decimal (BCD) coding. Figure 21-10 shows such an arrangement. In this mode also, the bits of information are given vertical and/or parity checks.

MEANS OF REPRESENTING DATA

It is one thing to represent data on tape or cards, but there must also be some way of representing data within the computer. This is done by electronic indicators or devices such as electrical pulses, relays, vacuum tubes, transistors, and magnetic cores. The selection of piece parts and components depends on design considerations and need not be restricted to any particular type.

Two distinct modes of electrical pulses are used to transfer and store data within computers. One is the *serial* mode, which derives its name from the fact that all the pulses follow a continuous course, one in back of another in straightforward succession. The following illustration shows how electrical pulses represent data in serial fashion:

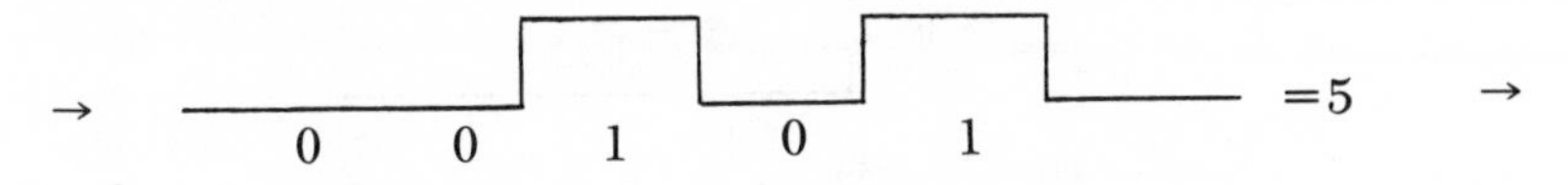

Movement in this form is also referred to as a pulse train.

The other mode is called *parallel.* In this mode the movement of data takes place simultaneously, but along different courses or wires.

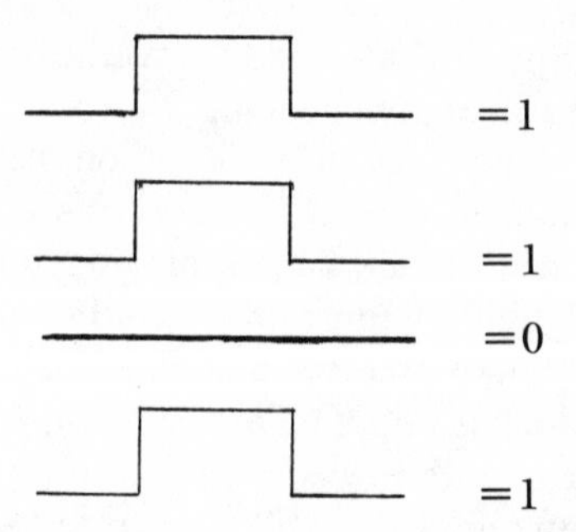

Early computers of the electromechanical type used switches and relays to perform their computations. Turning to Figure 21-11, we can see, in an elementary way, how these devices are able to compute the binaries: 1 + 1 = 10.

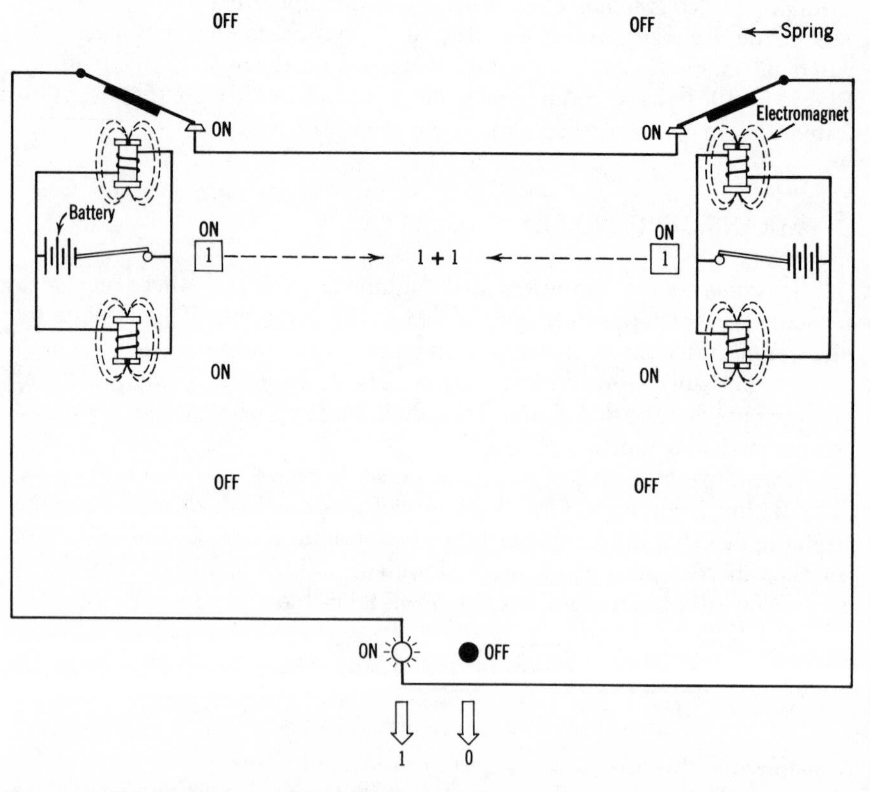

Figure 21-11. Electromagnetic relay schematic showing the computation of 1+1=10.

Vacuum tubes as well as small semiconductor devices such as transistors, are another means of representing data, as shown in Figure 21-12.

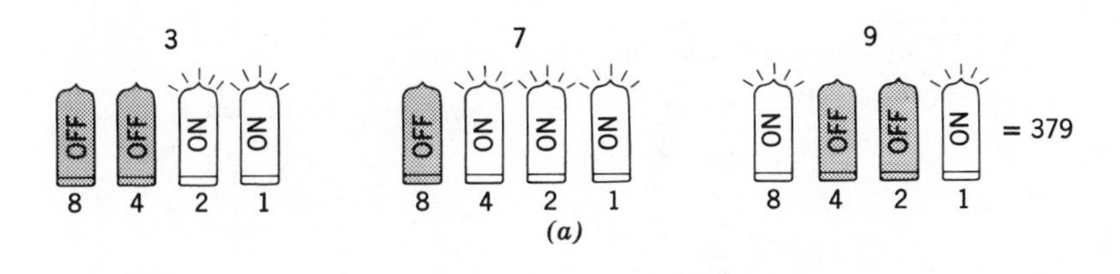

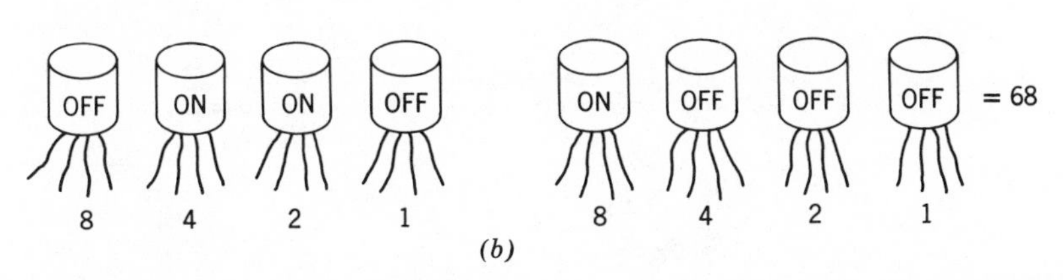

Figure 21-12. Elements of electronic circuits used to represent binary numbers; (*a*) vacuum tubes; (*b*) transistors.

STORAGE OF DATA

Based on location, the storage of data for a computer is of two types—*external storage* and *internal storage*. External storage is accomplished outside of the computer proper with the data being contained on cards or tape. Such storage is commonly used for long-term duration.

On the basis of accessibility internal storage is classified into two groupings—*main storage* and *secondary*, or *auxiliary*, *storage*. Of necessity all automatic computers have the former type as an integral part of their make-up; auxiliary storage is used where additional storage capacity is required. The difference between the two types of storage is shown in Figure 21-13. It will be observed from the diagram that the main storage unit of the central processor is the immediate-access storage facility for the

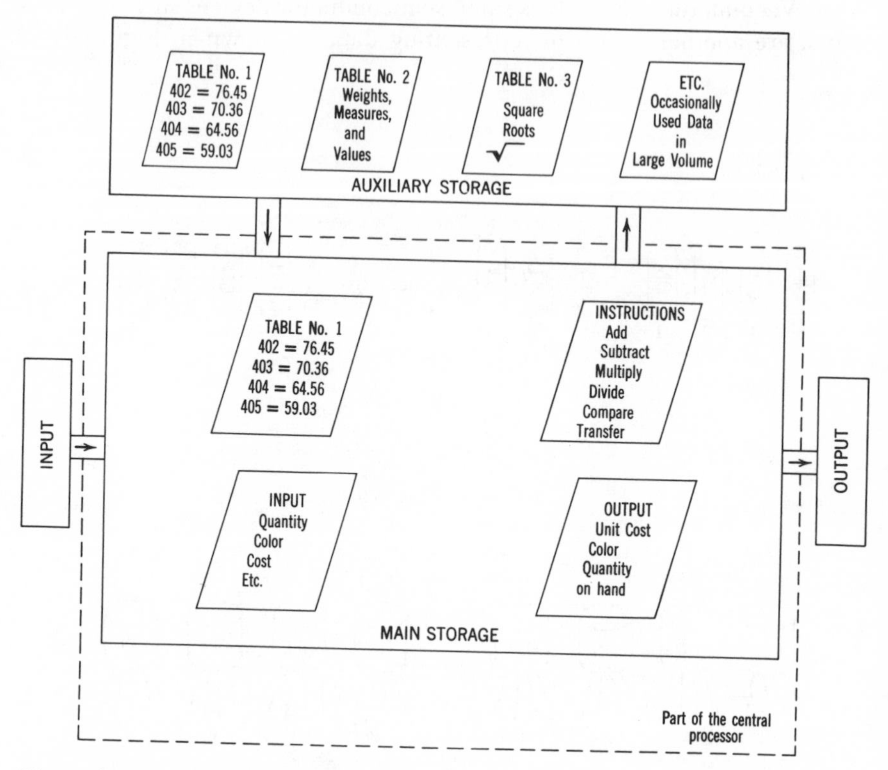

Figure 21-13. Diagrammatic portrayal of main and secondary storage facilities.

entire data-processing system. In this case the capacity of main storage is supplemented by an auxiliary memory. The auxiliary memory is not directly accessible to the arithmetic unit or to input-output devices. Data destined for entry or removal from auxiliary storage must be routed through main storage.

It takes a computer a certain length of time to locate and transfer data to or from their storage position. This interval is known as *access time*, and it is typically so brief as to be measured in milliseconds or even fractions thereof.

Storage facilities are said to be *random access* when conditions are such that the next location from which data are to be extracted (or to which they are to be delivered) is independent of the previous one. The opposite of random access is *sequential access*, and this is exemplified by the tape method of storing data. To locate any particular item of information on a reel of tape might necessitate running the tape from one end to the other. Although the tape moves at very high speed, it still is a relatively slow process compared with the capabilities of a processor; for example, a drum-type memory is designed so that all portions of it can be reached immediately. The approach is similar to the manner in which manual files are searched by individuals to obtain specific information. Instead of perusing the entire file, we normally pass over the folders that are not pertinent and go directly to the one in which we are interested. Similarly, random-access devices can skip over large amounts of irrelevant information and quickly retrieve only that which is sought.

Internal Storage of Data

The means of internal storage in computers take various forms. These include the following:

1. Mercury storage
2. Magnetic cards
3. Magnetic rods
4. Magnetic drums
5. Magnetic disks
6. Magnetic cores

The last three are widely used, and an explanation of them will enlarge the understanding of the *how* and *why* of storage.

Magnetic-Drum Storage. A magnetic drum is a cylinder capable of rotating rapidly, the surface of which is coated with a material that can be magnetized and thus used to store information in the form of magnetized spots on the curved surface. Each drum possesses a definite number of storage locations, the exact capacity depending on the diameter of the drum and its width. Drums range in size from 4 to 15 inches in diameter, and rotate at a speed of 1200 to 17,000 revolutions per minute. They may contain between 500 and 180,000 characters. The schematic of one arrangement of data on a drum is shown in Figure 21-14.

Individual read-write heads are usually employed for each track for reading, recording, or erasing magnetized spots of information. These heads are situated very close to the circumference of the drum. Once information has been stored on the drum it can be retained indefinitely. It may be read from the drum over and over again whenever necessary—or replaced with different information. The recording of new data in place of the old automatically erases the latter. Magnetic drums provide a high-density type of storage, and any data on the surface of the drum can be extracted within a few milliseconds.

Magnetic-Disk Storage. Another type of EDP system utilizes magnetic disks for storing information. These disks resemble an orderly

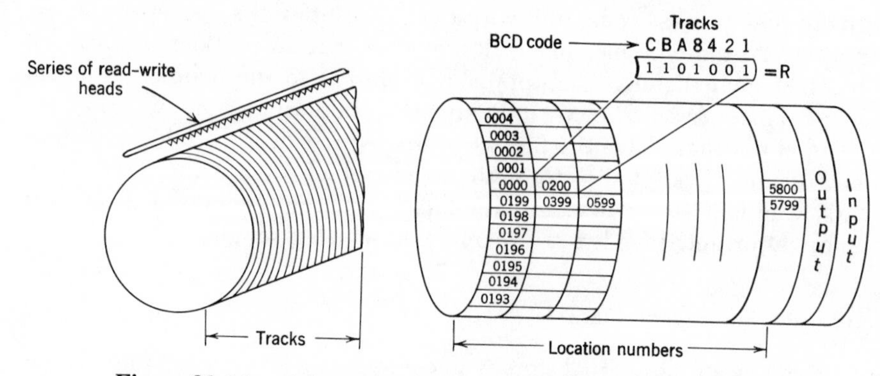

Figure 21-14. Schematic of one arrangement of data on a drum.

arrangement of phonograph records, stacked one next to the other in a series, with a slight space between them.

Like the surface of a drum, each disk is coated with a magnetic material, thus making it possible to store information in the form of magnetized spots. The space between the disks provides room for an access arm with read-write heads so that information may be recorded or extracted, as desired. Design engineers have fashioned units of this type in a variety of ways. Some are constructed with only one access arm that is activated to travel to the selected disk and then to the selected track on that disk; others have stationary arms—one for each disk surface; and still another design has single arms for each of two adjacent disk surfaces. Access time is, of course, affected by the particular design employed. In all cases the arm reads, writes, or erases while the disk rotates.

Just as the surface of a drum can be used repeatedly, so can the surface of a magnetic disk. When new information is stored on a disk, the old information is automatically erased. In this case also the recorded data may be read again and again, as needed. The data will remain on the disk until they are written over or purposefully removed.

Makes and models of computers that utilize disk-type memories differ with respect to the organization of data on disks. Typically, however, characters are read in and read out in bit-serial fashion. The bits making up the characters are stored in concentric tracks around the disk. (See Figure 21-15) Some disks are designed to store data on several hundred tracks on each surface. The data capacity of a single disk is commonly between 100,000 to 200,000 characters. Certain disk-storage units can retain data in excess of five billion characters.

Disk-type storage is used as the main storage unit of many processors. In some computer installations it provides supplemental storage capacity

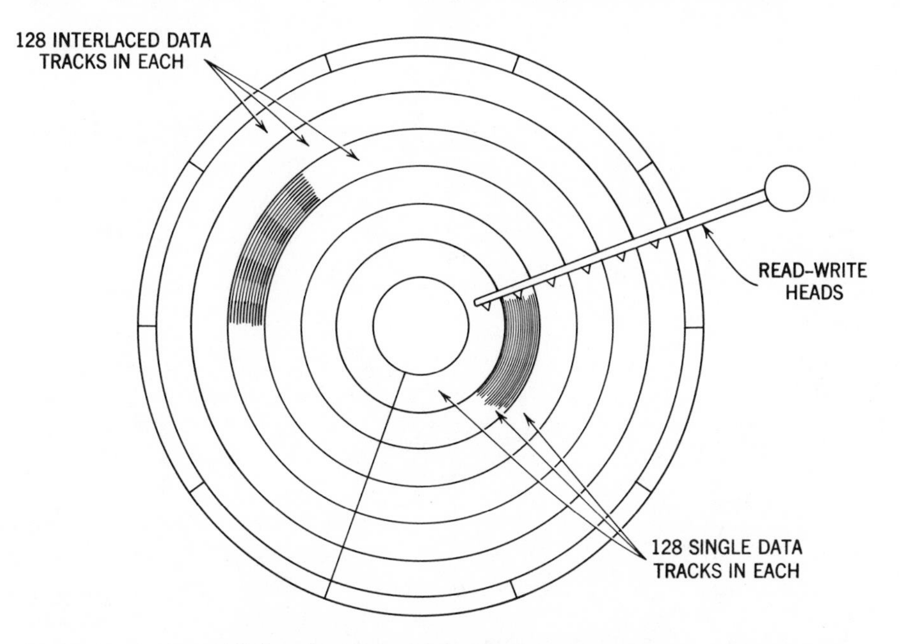

Figure 21-15. Data recording on a disk.

in addition to the main high-speed memory unit. In general disk-storage devices are slower than drum devices, but they do offer much greater capacity.

Magnetic-Core Storage. This type of memory unit provides access to data much faster than either drum- or disk-type memory units, but it is a relatively expensive medium.

A core memory is a network of tiny, doughnut-shaped circlets of ferromagnetic material that can be magnetized to retain information indefinitely. Each core is located at the intersection of two wires, as shown in Figure 21-16, and may be magnetized in either of two directions—clockwise or counterclockwise. Accordingly the direction of the magnetism can be used to represent alternative conditions such as 0 or 1, plus or minus, and on or off. A specific amount of current is required to magnetize a core. By reversing the direction of the current the existing magnetic state is changed to its opposite condition. Whatever the state—clockwise or counterclockwise—when the current is removed the core remains magnetized until some action is taken to change the situation.

Obviously, some method is needed for selecting the individual cores to be magnetized. For any given core this is accomplished by sending a half charge of electricity over each wire. (See Figure 21-17.) Thus only the core at the intersection of the wires receives a full charge and is magne-

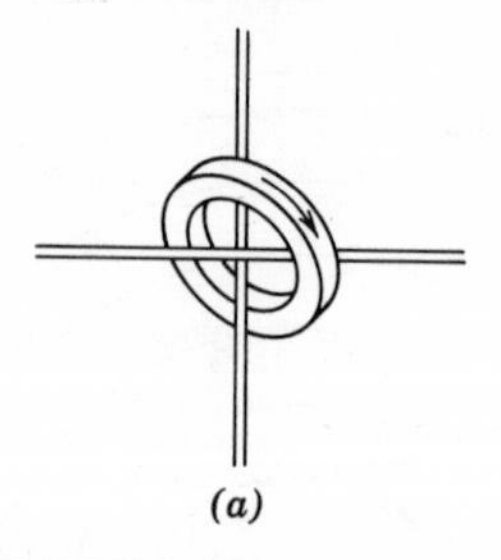

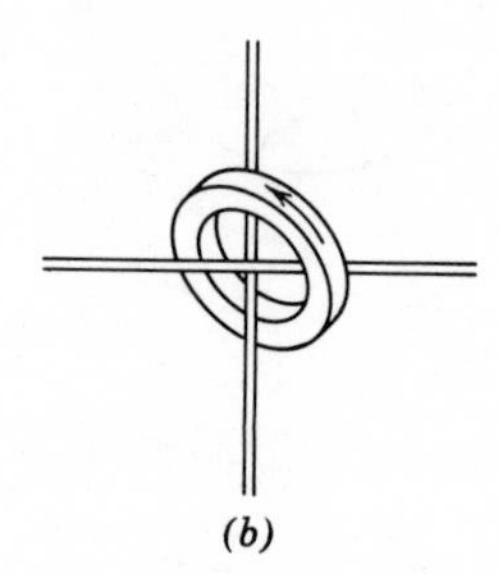

Figure 21-16. Magnetic cores showing (*a*) clockwise polarity and (*b*) counter-clockwise polarity.

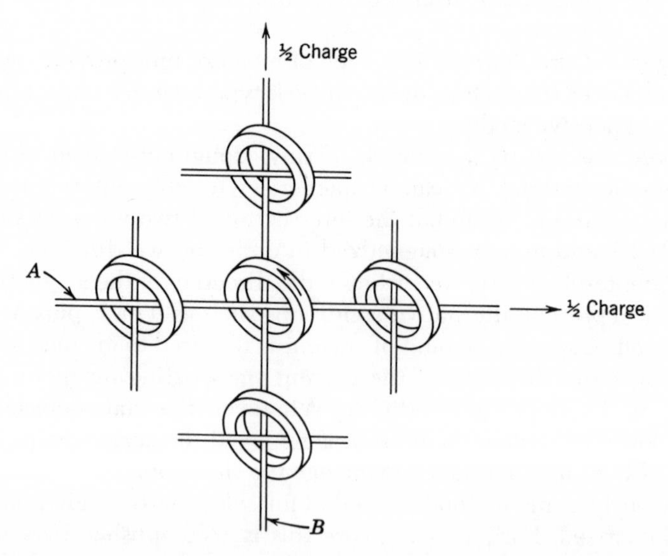

Figure 21-17. Selection of a core. One-half a charge of electricity through wires *A* and *B* will reverse the magnetic state of only the core at their place of crossing.

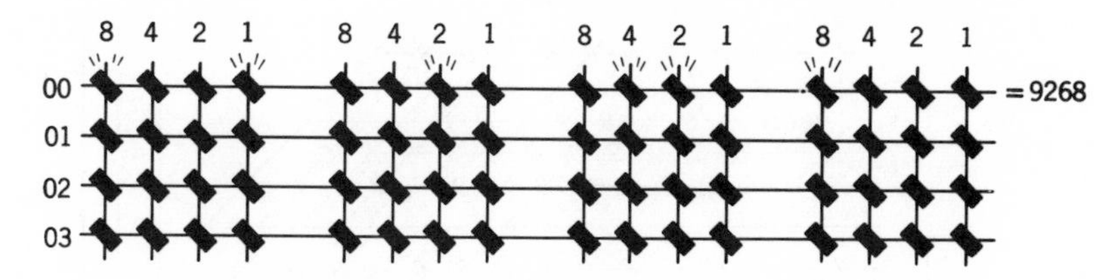

Figure 21-18. Representing data by means of cores.

tized. In this way, none of the other cores—the cores along either of the affected rows—will be switched from one state to the other.

Figure 21-18 shows how these tiny cores, about the size of the head of a pin, can represent data.

Figures 21-16 and 21-17 do not show a third wire, which in actual cases threads all cores for sensing. (See Figure 21-19.) Information is thus called out of memory by means of this third wire.

In some computers memory-core reading is destructive, because in the process, 0s remain 0s but 1s are converted to 0s. These computers, however, are designed to regenerate the original information in order to permit its retention. Entire portions of a memory unit may be cleared at any desired time.

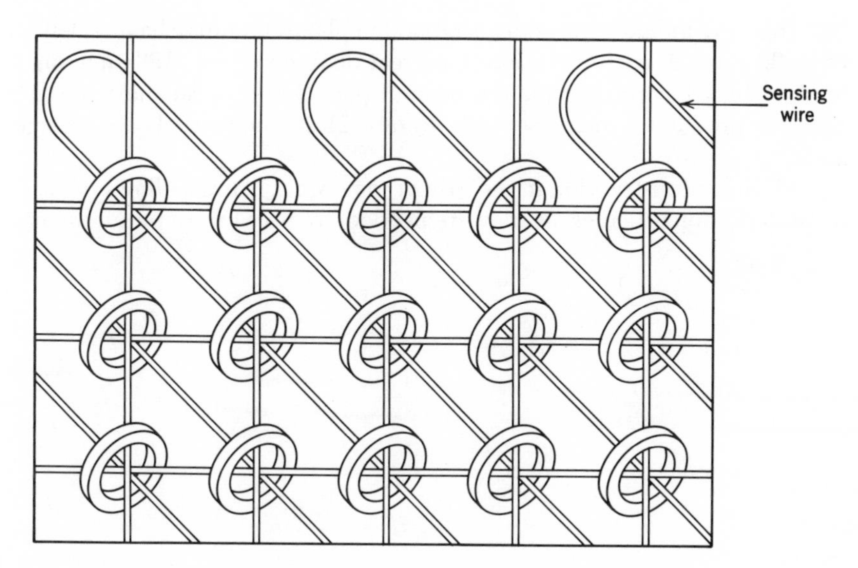

Figure 21-19. Sensing wire in core network.

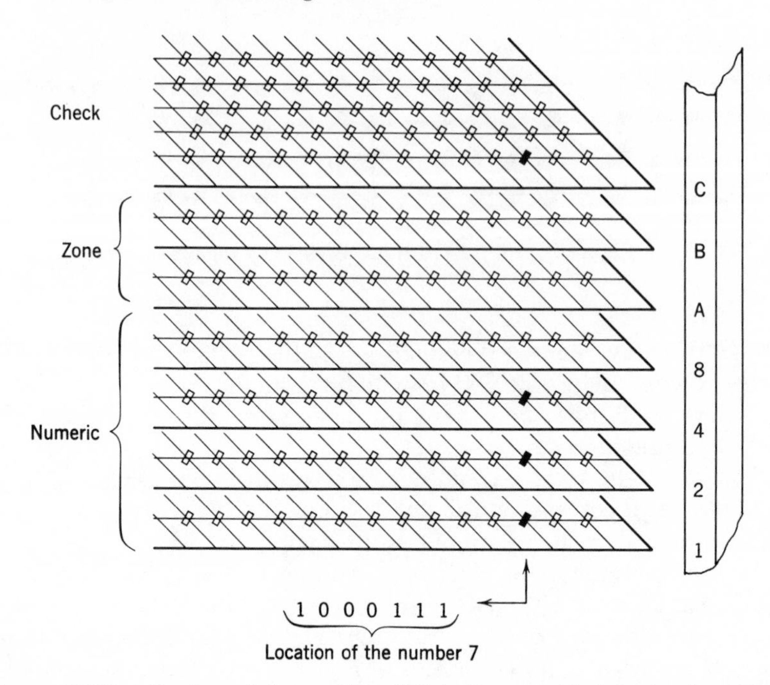

Figure 21-20. Storing of the number "7" in binary-coded decimal representation.

In a core memory cores are arranged in planes that may have as few as 25 or 30 cores in each direction or a great deal more (e.g., 128—making a total of 16,384 cores per plane). Several planes are placed next to each other to produce a memory stack. Figure 21-20 shows such an arrangement.

Modular construction in the case of some storage units allows the user to add additional planes, as they are needed, to meet future requirements.

22

An Introduction to Programming

Chapter 21 was concerned primarily with computer components and their functions. However, the equipment, the *hardware,* is only the means for carrying out operations. The hardware must be instructed as to the operations to be followed. This is accomplished through the medium of programs, collectively called *software.*

Since a computer must be instructed to perform each basic element of processing, a program is essential to its operation. The program controls the entire flow of data and consists of a precise sequence of coded instructions, so that each individual operation depends on another and requires no external control to complete the sequence.

Each operation of a computer is governed by a unit of specific information located in main storage, called a program step. If data are to be processed, the instructions—the program steps—direct the computer to read the data from the input medium, and convey them to specified processing areas for addition, subtraction, multiplication, division, shifting, or whatever action is indicated. If some unit of equipment is to be activated (e.g., a magnetic-tape unit), the instructions designate the particular unit and the required operations.

Instructions consist of two parts: (*a*) a command part that directs the computer to add, subtract, compare, shift data, and so forth, and (*b*) one or more addresses that designate the location of the information or device that is needed to carry out the specified command. Based on the number of addresses, computers are said to be single-address or multiaddress machines. The format of an *instruction word* is determined by what the computer is designed to accept. Representative instruction formats for four-, three-, two-, and one-address computers are given in Figure 22-1. The term "operand address" designates the location in which a quantity or set of characters is to be stored. A result address identifies a location in which a result is to be stored.

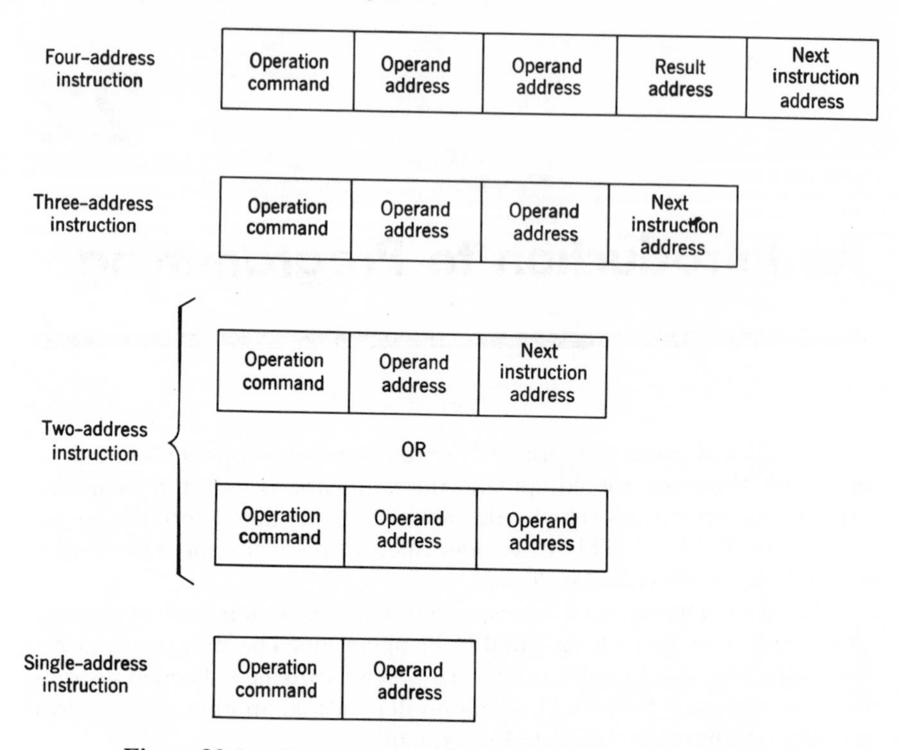

Figure 22-1. Instruction word formats used in digital computers.

INSTRUCTION FORMATS

Four-Address Instruction

Of the four basic types of addresses shown in Figure 22-1, the first comes closest to handling an arithmetic operation in a *step-by-step* fashion. This type of instruction contains the operation command code, the addresses of both operands, the address where the result is to be stored, and the address of the next instruction. The operands (quantities) are routed to the appropriate registers in the arithmetic unit, the operation code is decoded, the operation is performed, the result is stored in still another register, and subsequently a new instruction is located and acted upon.

A number of the early computers were built to accept four-address instructions. Although more versatile than machines with fewer addresses, four-address machines are more wasteful of programming time and result in slower operating speeds. Consider, for example, a simple type of instruction wherein a *whole* number is to be taken from storage and added to another number already in the arithmetic unit. Regardless of the number of address instructions, it is only necessary to specify one operand and the instruction command. Yet, where a four-address format is concerned, the remainder of the instruction must be filled in with zeros.

Three-Address Instruction

With the exception that no address is specified where the result is to be stored, the form of the three-address instruction is the same as that of the four-address type. Depending on the design of the computer, the result is always stored in the same place or another instruction governs where the result is to be placed. Only a limited number of computers are three-address machines.

Two-Address Instruction

As shown in Figure 22-1, there are two types of two-address formats. The one shown at the top is widely employed. In this programming system the instruction word consists of an operation command code, the address of one operand, and the address of the next instruction. For processing, the operand on which the operation is to be performed is selected from storage and routed to the proper arithmetic register, where the operation is decoded and performed; after this the address of the next instruction is located. Typically the new instruction contains the address of the second operand, the code for the operation to be performed on it, and the address of the third instruction. The third instruction will tell the machine where the result is to be stored.

The other type of two-address instruction shown in Figure 22-1 provides for two operands and a code for an operation. Observe that the format makes no provision for the address of the next instruction or the address where the result is to be stored. It is necessary therefore to employ at least one additional instruction to tell where the result is to be stored. The next instruction, by means of the control unit, is *automatically* selected in consecutive order from storage.

Single-Address Instruction

Most modern computers use the single-address type of instruction. Each instruction contains only an operation command code and one address, the operand address. Although such a format appears to be restrictive, it is not; indeed, it provides for faster processing since many ordinary operations require only one operand, a second one being already contained in the accumulator register of the arithmetic unit; for example, to execute a series of additions, the first operand (number) is selected from storage, routed to the arithmetic unit, and added to the zeros in the arithmetic register, which has previously been cleared. The second operand (or number) is then routed, in a similar manner, and added to the first, already in the accumulator. This process is repeated for each consecutive addition. Accordingly only one operation command code and one individual address instruction is needed for each successive addition. When the sum is finally developed in the arithmetic register, another instruction is needed to tell where the result is to be placed in the memory. This new instruction (as in the case of the two-address instruction with the two operand addresses) is *automatically* selected from storage on a sequential basis.

TYPES OF INSTRUCTIONS

Instructions may be classified into the following five basic types:

1. *Input-output instructions.* These are concerned with the transfer of instructions and data from an input medium (such as cards, paper tape, or magnetic tape) to the central processor; the transfer of results to the output medium; and other input-output operations.

2. *Arithmetic instructions.* These instructions command the computer to add, subtract, multiply, divide, compare, etc.

3. *Transfer instructions.* These provide for the transfer of data and instruction words from one storage location to another.

4. *Branch instructions.* Branch instructions are of two types—unconditional and conditional. An unconditional branch instruction tells the computer to start a new instruction or make a loop. A conditional branch instruction tells the computer to choose between alternative conditions; for example, such an instruction might say, "If the number in the accumulator is negative, branch to address 047; if it is positive, use the next control address as usual." (See Figure 22-2.)

5. *Miscellaneous instructions.* Within this category are instructions such as "halt," which stops computer operations, and "clear," which empties the contents of the designated operation register.

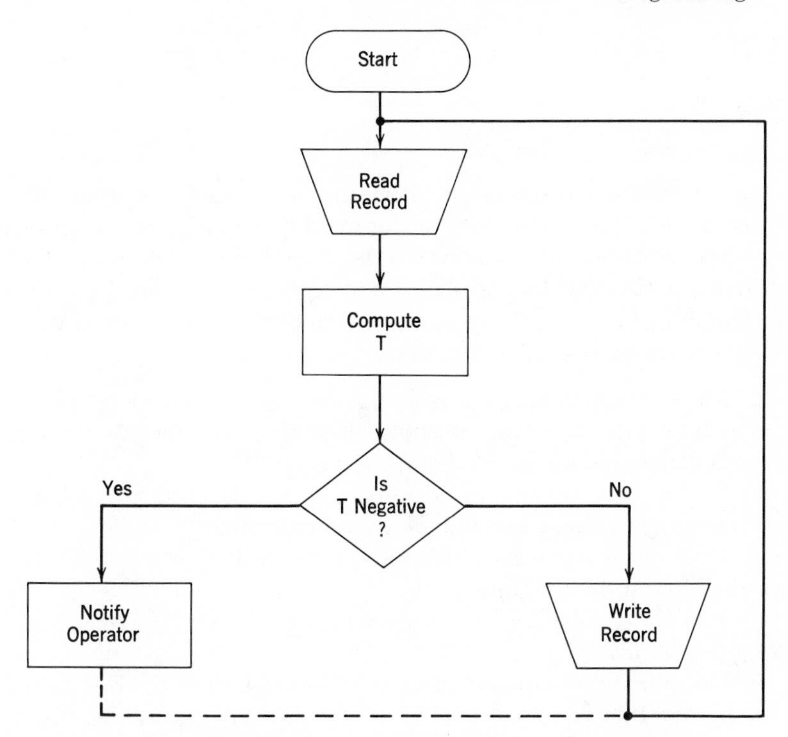

Figure 22-2. Conditional transfer. Source: IBM.

STORAGE OF INSTRUCTIONS AND DATA

Instructions are stored in the same manner as data. To provide for their entry into the computer, they are recorded on an input medium such as cards or tape. The term "loading" is used to describe the process of reading the instructions from the input medium into their designated locations in the memory unit. After the program has been stored internally, execution of the operations begins with the first instruction or by manually setting the machine for the selection. The first instruction is then executed. The computer subsequently continues to locate and execute instructions in the sequence specified by the program until the job is completed or the computer is instructed to stop.

Since instructions (programs) are recorded on cards or tape, they may be maintained external to the computer. When a particular job is finished, the computer can be set to perform a different task by simply changing the program.

PROGRAM DEVELOPMENT

In developing a program the usual practice is for the systems analyst to provide the programmer with one or more flow diagrams, the input and output specifications, and other material including a fairly precise set of requirements that can be translated into machine language by means of programming. After acquiring a knowledge of the requirements the programmer performs the following tasks:

1. Block diagramming—portraying the logic required by the computer to transform input into output. These diagrams usually undergo a series of changes before the final one is developed.
2. Coding—the process of converting a detailed block diagram into the machine instructions that will be stored in memory.
3. Desk checking—visual checking of the coding and preparation of input data to test the program.
4. Testing–tests should be made with sample data under actual operating conditions.
5. Documentation—preparation of a description of the program and its operation. Two purposes are served by this: (*a*) it provides the computer operators with the instructions to run the program, and (*b*) it enables one programmer to understand, review, and revise a program written by another programmer.

OPERATION OF A HYPOTHETICAL COMPUTER

To illustrate how computers are programmed and how they operate, it is useful to consider the processing of data on a hypothetical machine. Inasmuch as the basic code that a computer can accept, interpret, and execute varies with different makes and models, the explanation that follows is general rather than specific. Any similarity to a particular machine is not intended. It should also be pointed out that this is a highly simplified version; modern computers are considerably more complex and involved than the illustration indicates.

For our illustration let us assume that the following program has been prepared, recorded on cards, and is ready to be loaded into a magnetic-drum memory. It will be observed that this program has been developed for a machine utilizing single addresses. Let us further assume that the problem is to add two quantities and subtract a third. The processes involved can be visualized, in step-by-step fashion, from the schematics presented below. Note, however, that although decimals are used to conveniently show the storage of instructions and data in a computer, when actually lodged internally the instructions and data would be in binary notation.

Instruction		
Operation Code	Storage Address	Meaning
10	100	Clear and add to the accumulator the quantity stored in memory location 100.
15	101	Add to the accumulator the quantity stored in memory location 101.
20	200	Subtract quantity in location 200 from accumulator.
25	204	Store results in memory location 204.
32	204	Read out of memory location 204; print the results. Return for next set of input values.

First step: Read the above program into memory locations 000 through 004.

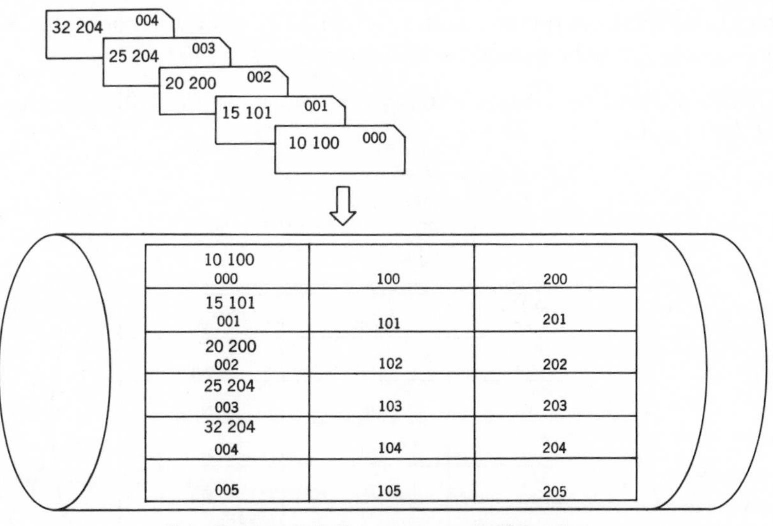

Note: For demonstration purposes only 18 locations are shown. Actually computer memories may contain many thousands of storage locations.

Second Step: Read the quantity 17 into memory location 100 from the first data card.

17 100

10 100 000	17 100	200
15 101 001	101	201
20 200 002	102	202
25 204 003	103	203
32 204 004	104	204
005	105	205

Accumulator
17
Arithmetic unit

Execution: First instruction causes the quantity stored in computer memory location 100 to be placed in the accumulator.

Third step: Read the quantity 162 into memory location 101 from the second data card.

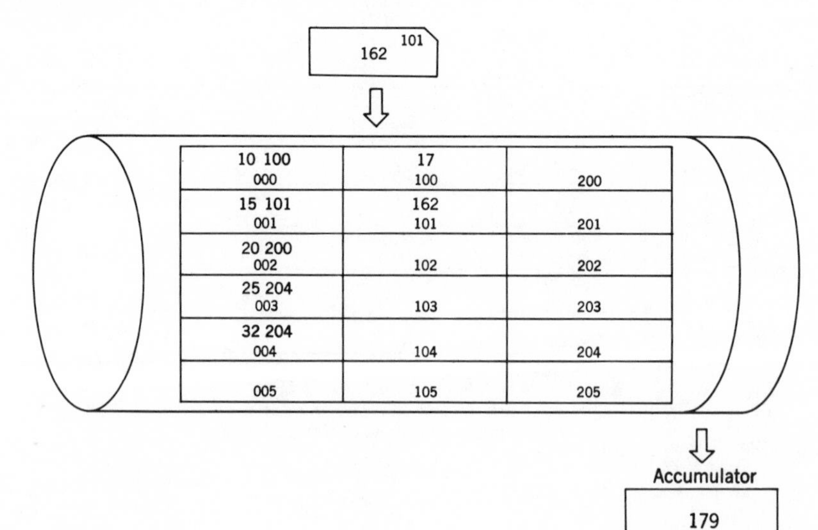

Execution: The second instruction, to which the computer automatically proceeds after carrying out the execution cycle of the first instruction, causes the quantity 162 to be added to the 17, making a total of 179.

Fourth step: Read the quantity 45 into memory location 200 from the third data card.

45 200

10 100 000	17 100	45 200
15 101 001	162 101	201
20 200 002	102	202
25 204 003	103	203
32 204 004	104	204
005	105	205

Accumulator

134

Arithmetic unit

Execution: The third instruction causes the quantity 45 to be subtracted from the 179, making a total of 134.

Fifth step: Store result in memory location 204.
Execution: The fourth instruction causes the result of the computations to be placed in memory location 204.

10 100 000	17 100	45 200
15 101 001	162 101	201
20 200 002	102	202
25 204 003	103	203
32 204 004	104	204
005	105	205

Accumulator

134

Arithmetic unit

Sixth step: Read out of memory location 204; print the results.

10 101 000	17 100	45 200
15 101 001	162 101	201
20 200 002	102	202
25 204 003	103	203
32 204 004	104	134 204
005	105	205

Execution: The fifth instruction causes the result to be read out of memory location 204 and printed on a form.

AUTOMATIC PROGRAMMING

Programming the early electronic computers was a long, expensive, and painstaking task, because instructions had to be given in special code or in actual machine language. To gain some idea of the enormity of the undertaking, consider the typical 34-bit instruction-word format shown in Figure 22-3. It will be seen that, in addition to the parity bit, the instruction word is made up of two principal parts—a code for the *operation* and coding for the *addresses*. An input that is typical for a machine holding such an instruction is shown in Figure 22-4.

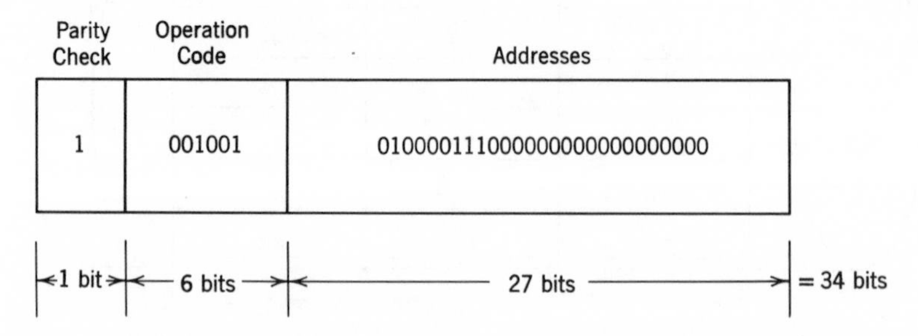

Figure 22-3. Instruction word in machine language.

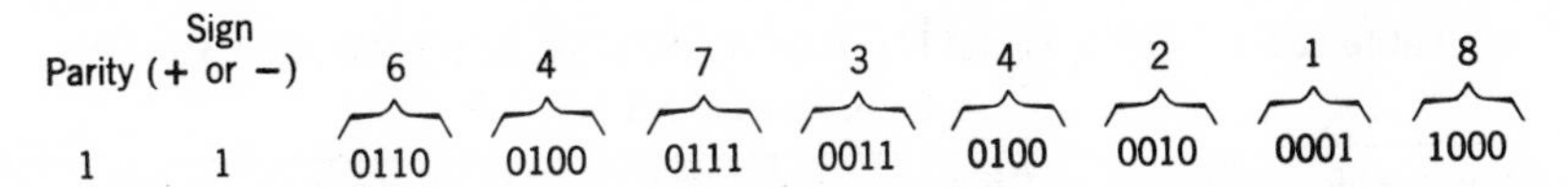

Figure 22-4. Typical arithmetic word in machine language (binary-coded decimal).

Many of the difficulties in writing programs were eliminated with the development of computers that can automatically encode and decode decimals and characters. These machines, as we have already seen in Chapter 20, directly accept Arabic numbers and convert them to binary notation. This led to the use of alphanumeric coding schemes, which are easier to work with and understand than the basic language of the machine. Three of these schemes are widely employed: the Hollerith code, the Powers code, and *specific* coding, wherein a letter or number is converted to binary notation without reference to another kind of coding scheme. However, it is worth emphasizing that, whatever scheme is employed, the numbers, letters, and special characters are eventually converted to binary notation—the only machine language a digital computer, using the form 0 and 1, is designed to understand.

Table 22-1 shows the Hollerith code with binary-notation equivalents. With any three of the coding schemes mentioned, the general form of an instruction might look as follows:

Instruction		
Operation Code	Storage Address	Meaning
20	130	Read 34 into 130

or

Instruction			
Operation Code	Storage Address	Storage Address	Meaning
S	132	194	Subtract the number at the 194 address from the 132 address and leave result at 194 address

In the past writing instructions and coding input data constituted a major bottleneck in computer usage. Programmers had to commit to memory or be continually looking up the operation code characters—a sizable undertaking when you stop to think that over a 100 operation com-

Table 22-1. HOLLERITH CODE WITH BINARY-NOTATION EQUIVALENTS

Character	Hollerith Code	Binary Notation			Character	Hollerith Code	Binary Notation		
		C	BA	8421			C	BA	8421
0	00	1	00	1010	O	11–6	0	10	0110
1	11	0	00	0001	P	11–7	1	10	0111
2	22	0	00	0010	Q	11–8	1	10	1000
3	33	1	00	0011	R	11–9	0	10	1001
4	44	0	00	0100	S	0–2	1	01	0010
5	55	1	00	0101	T	0–3	0	01	0011
6	66	1	00	0110	U	0–4	1	01	0100
7	77	0	00	0111	V	0–5	0	01	0101
8	88	0	00	1000	W	0–6	0	01	0110
9	99	1	00	1001	X	0–7	1	01	0111
A	12–1	0	11	0001	Y	0–8	1	01	1000
B	12–2	0	11	0010	Z	0–9	0	01	1001
C	12–3	1	11	0011	&	12	1	11	0000
D	12–4	0	11	0100	.	12–3–8	0	11	1011
E	12–5	1	11	0101	□	12–4–8	1	11	1100
F	12–6	1	11	0110	—	11	0	10	0000
G	12–7	0	11	0111	$	11–3–8	1	10	1011
H	12–8	0	11	1000	*	11–4–8	0	10	1100
I	12–9	1	11	1001	/	0–1	1	01	0001
J	11–1	1	10	0001	,	0–3–8	1	01	1011
K	11–2	1	10	0010	%	0–4–8	0	01	1100
L	11–3	0	10	0011	#	3–8	0	00	1011
M	11–4	1	10	0100	@	4–8	1	00	1100
N	11–5	0	10	0101	blank		1	00	0000

The above table shows the binary coding of the characters in a standard IBM 1401 system. This "seven-bit alphameric" is in the binary-coded decimal concept. The four bits on the right (8, 4, 2, and 1) are called the numerical bits, since they correspond in an approximate way to the Hollerith code. The two to the left (B and A) are called zone bits for the same reason; the seventh bit (C) is used for parity checking and is known as the *check bit.*

Note how the decimal digits and letters of the alphabet are expressed in terms of various zones plus numerical bits. All the decimal digits have representations in which the zone bits are both 0, corresponding to the fact that their Hollerith code representations have no zone designations. The letters A through I have zone bits that are 1 and 1, corresponding to a Hollerith-code zone designation of 12. In like manner, the coding of the letters J through R have zone bits of 10, corresponding to an 11 zone designation; and the letters S through Z have zone bits of 01, corresponding to a zone designation of 0.

mand codes may be involved. Programmers can readily learn the more frequently used codes since they are repetitive, but the operation codes and operand addresses—together—are capable of many combinations and are different for every program that is prepared. Thus, for example, when the operation code 15 is given, meaning to multiply, not only must the language code 15 be indicated but also the memory address or location of the number to be multiplied—a fact already explained in the previous section. Keeping track of hundreds or thousands of such instructions and knowing which locations remain unused requires much time and effort; moreover, working with these codes is cumbersome, leading to errors that are often difficult to find and correct. Automatic programming languages, languages that are problem oriented rather than machine oriented, were therefore devised and computer programs were developed to translate programs in the automatic languages into the machine codes of specific computers.

These automatic languages are so called because the computer itself is used to develop coded programs. Because of this fact computers are sometimes said to be capable of "writing their own programs." Strictly speaking, however, computers do not program themselves. Someone must initially prepare a program of instructions that enables computers to carry out the automatic coding.

A main characteristic of automatic programming is that symbols are used in place of *absolute codes;* that is, in lieu of specific computer codes. After the preparation of the entire program in symbolic language the symbols are translated into absolute coding by means of a *processor.* The processor is itself a program, not hardware, and must provide for conversion from a particular problem-oriented programming language to a particular basic computer machine language. The program, as written in symbolic language by a programmer, is called a *source program.* The computer-generated program, the output after translation, is known as an *object program.* This object program is then read into memory so as to direct the actions of the computer in its processing of data.

Inasmuch as processors convert from an original programming language into a particular final computer machine language, it follows that there are many different kinds of processors. In general, however, they are spoken of as being either "low level" or "high level." A low-level processor is one that carries out the translation from a source program to an object program on a more or less one-for-one basis, meaning that for every source-language instruction usually only one machine-language instruction is created. Both "assemblers" and "interpreters" are examples of relatively low level processors.

An assembler is a program that converts a source program in

mnemonic instructions and symbolic addresses into a machine-language object program. As an example of a symbolic assembly-programming system, consider the following instructions:

Instruction		
Operation Code	Storage Address	Meaning
CLA	A	Clear and add to the accumulator the quantity stored in memory location A.
SUB	B	Subtract in the accumulator the quantity stored in location B.
STO	C	Store the contents in the accumulator in location C.

In converting from symbolic language to machine language the processor substitutes machine operation codes for the operation abbreviations CLA, SUB, and STO. The processor also automatically substitutes machine operation codes for the sumbols A, B, and C. It should be realized, of course, that the processor program must be in storage and that the storage locations were previously assigned. Accordingly the machine language produced by the processor might take the following form:

Instruction	
Operation Code	Storage Address
003	10
004	11
005	20

In the preceding example the processor produces only one instruction in machine coding for one instruction written in symbolic language. This is what takes place when the processor requires the programmer to adhere closely to machine-language statements. To increase the effectiveness of programming systems, however, most symbolic languages make use of *macro instructions*. A macro instruction is a single instruction that, when translated, causes a series of machine-language instructions to be produced.

An interpreter, the other type of processor mentioned, is a processor that translates a program into machine language and immediately performs the indicated operations. With an interpreter no object program is needed since the source program is also the object program. By reason of this feature interpreters are frequently referred to as "load and go" processors.

As previously mentioned, assemblers and interpreters are relatively low level processors. The more powerful high-level processors, called *compilers*, can create a large number of machine instructions from a single source-language command. COBOL (Common Business Oriented Language), FORTRAN (Formula Translation), and ALGOL (algebraic-oriented language) are examples of such compilers. A typical compiler system is shown in Figure 22-5.

The steps for using a compiler are as follows:

1. Write the instructions in the symbolic language. This is known as a *source program.*

2. Record the source program on an input medium (e.g., cards or tape).

3. Load the processor (the compiler) into the computer's memory and start execution of it.

4. In the execution of the processor the source program is read, one instruction at a time.

5. Instruction by instruction, the processor converts the source pro-

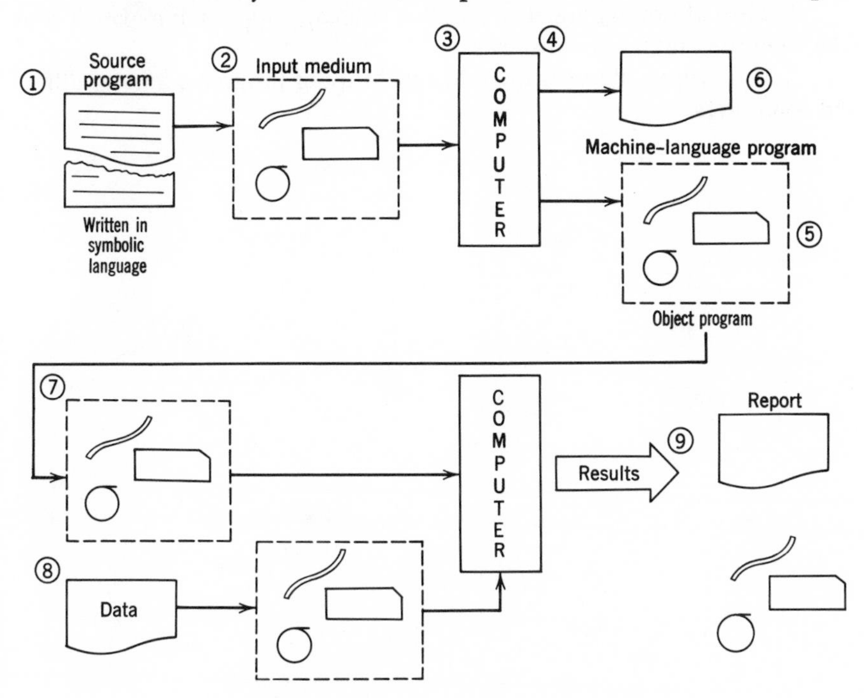

Figure 22-5. Typical compiler system.

gram into an object program. The object program is produced on an output medium such as cards or tape.

6. If desired, obtain an *assembly* listing that shows both the original source program and the final abolute object program.
7. The object program is read into memory.
8. Data are run through the machine.
9. Results are obtained.

ADVANTAGES OF AUTOMATIC PROGRAMS

The advantages of automatic programs include the following:

1. Ease of coding. No special knowledge of machine codes is required. The necessary language code and rules can be learned in a short period of time.
2. Reduction of errors. Many of the error-prone situations associated with manual programming are handled automatically by the compiler; for example, repetitive processes, or "loops"; assignment of available storage locations; input-output processing.
3. Ease of debugging. The English-language approach makes it simpler to locate and rectify mistakes.
4. Uniformity. Permits interchangeability of programs among different computers.

23

EDP Feasibility and Preparation Studies

A decision on whether or not to install EDP equipment should be based on a preliminary study, called a feasibility or justification study, that sets forth considerations essential to a sound judgment on the question. Such studies are normally time consuming and can in themselves be expensive. In the interest of saving time and money some companies have waived the preliminary study phase or settled for a hasty, incomplete evaluation. Where this has been done, experience has shown the results to be generally less than satisfactory.

The feasibility study is logically the first step in considering the acquisition of an electronic computer. The substantial cost of installing EDP equipment requires assurance that the basis for applying EDP resources to existing operations is essentially sound; that definite advantages will result; and that the company has, or will have, the personnel to operate the installation successfully. The feasibility study must be planned and conducted so as to determine the impact on the operations of the firm and to prepare the organization for the installation and operations of the computer.

The pioneer users of EDP equipment were lured by three enticing sirens—labor-saving, speed, and hardware. Managers sought these blindly. They were aware of how a number of functions would be eliminated but failed to recognize the extent to which new functions would be created. Moreover, they neglected to reckon the cost in relation to other factors, such as giving the computer a sufficient volume of work in terms of an efficient operating load. Thus the early economic outcome of computers frequently was found to be negative.

The management considering installation of EDP equipment today is fortunate. Potential users benefit from the experiences of the past, survey the requirements, determine the need, and come up with a practical plan that will deliver results.

SELECTION OF PERSONNEL TO CONDUCT STUDY

One individual or a study team of from two to six persons may be assigned to conduct the feasibility study. The number of persons engaged in the activity will depend on factors such as the number of systems to be studied for possible computer application, the number of departments involved, budget and time limitations, and the extent to which the subject-matter knowledge is concentrated.

The composition of the study team is important and varies with the particular situation. Some teams are made up entirely of systems department personnel; some also include representatives from major operating departments such as marketing, manufacturing, and industrial engineering; and others involve a combination of outside consultants and company personnel.

The man chosen to direct the study team must be extremely capable, because he has a heavy responsibility. Sizable expenditures of time and money, and at times the administrative well-being of the company, may be at stake. He must be a manager in every respect. His duties will include overseeing the work of the members of the team, establishing schedules and checking progress, and determining which problems require top-management decisions and which can be resolved without needless demands on executive time.

STATEMENT OF OBJECTIVES

Prior to proceeding with a feasibility study, sound practice calls for the determining of objectives to guide the study in relation to specific data-automation goals. Each objective should be carefully considered and its value carefully weighed. Doing so helps to prevent misconceptions and going off on tangents. If, for example, the prime objective of management is economy, then it ought to be determined as soon as possible whether economies are possible.

The following list is a summary of typical objectives that motivate a company to employ EDP equipment:

1. To reduce clerical costs.
2. To obtain management information that exceeds in operational

usefulness, accuracy, and timeliness the information that could be obtained by other means.

3. To acquire needed data that is not otherwise obtainable.

4. To permit the development of a minimum number of information systems to serve a maximum number of purposes.

5. To improve a competitive position; it may be desired to gain an advantage or simply to eliminate a disadvantage.

When the objectives have been established, the stage is set for the preliminary fact-gathering that is characteristic of all systems studies. A critical examination of the operations may disclose that methods other than electronic data processing may be effective. Systems redesign (through eliminating duplication, rearranging or abolishing certain processes, changing reporting requirements, redesigning forms, and the possible use of some mechanical device) may make a significant difference in the type or size of EDP equipment required; if EDP equipment is needed at all.

ASCERTAINING SYSTEM REQUIREMENTS

If EDP equipment is decided upon, the exact requirements must be established. On the basis of the systems studied and a knowledge of the kinds of information management needs, the next step is to formulate the requirements that must be fulfilled by the computer. Each system selected for automation must be considered to identify the character and volume of each type of input, of each type of output, and the processing requirements of the computer system. This information serves the following three basic purposes:

1. It provides facts and data that are necessary for soliciting proposals from equipment suppliers.

2. It serves to determine the machine capacity that will be needed (scale of computer required).

3. It forms the basis for developing cost-benefit relationships.

SOLICITING EQUIPMENT PROPOSALS

Once the system requirements are known, a practical approach is to become familiar with the equipment of the various suppliers. This is accomplished by contacting the venders' representatives, observing installations at showrooms or in user companies, and compiling literature on the subject.

Next the suppliers who have suitable equipment are requested to submit equipment proposals. The complying companies should be given written descriptions of the requirements. These written requirements, or specifications, are usually presented to the suppliers in the form of a brochure, setting forth information about the applications to be converted and including flow diagrams, rough drafts of forms, and work volume estimates.

Preparing Specifications

Equipment suppliers can prepare and present useful proposals only if the prospective customer provides proper information. The specifications must be uniform and prepared so that all proposals will contain comparable information. Typical of what is to be obtained from each supplier is the information covered by the following outline:

A. General
   1. Description of equipment—identification by make, model, quantity of components, and accessory equipment.
   2. Reliability of equipment.
   3. Availability and adequacy of maintenance services.
   4. Training and programming assistance.
   5. Ability to meet required delivery schedules, both for hardware and software.

B. Costs
   1. Equipment—detailed breakdown.
   2. Comparative costs:
      a. Lease method;
      b. Lease-with-option-to-buy method.
   3. EDP equipment operating and administrative personnel.
   4. Supply items.
   5. Site requirements—space, electrical power, air conditioning, etc.
   6. Conversion from existing data-processing systems.

C. Capabilities of equipment
   1. Kinds of input (punched card, magnetic tape, or paper tape).
   2. Sufficiency of storage and processing capacity.
   3. Adequacy of data handling.

4. Time cycles (e.g., to process a document or handle an inquiry).
5. Production of output in required form (printed copy, punched card, magnetic tape, or paper tape).
6. Degree of accuracy and freedom from malfunction.
7. Special features.
8. Adequacy of controls.
9. Potential for expansion to accommodate future growth.

Preparation of equipment specifications does not infer that new systems have been designed fully. Such an undertaking would go beyond the scope and depth of a feasibility study, increase time and manpower requirements, and defeat the overall purpose. At this point the objective is to postulate systems in general terms that can be applied to any capable equipment. Actually, ultimate systems cannot be designed completely until specific equipment is selected.

SELECTION OF EQUIPMENT

Evaluating suppliers' proposals is a complicated and difficult task. Guidelines for the selection are as follows:

1. Selection of equipment should be made within the confines of specifications that delineate (*a*) the objectives of each system, (*b*) the data-processing requirements, and (*c*) any special EDP requirements.
2. The selection process should accord equal opportunity and appropriate consideration to all vendors who offer equipment capable of meeting the specifications.
3. Prime factors to be considered in the selection of equipment are its capability to fulfill systems requirements and its overall costs.

Item	Incurred to Date	Projected (Additional)
1. Feasibility study/proposal preparation (salaries)	$12,000	....
2. Training (TDY, excluding salaries)	400	$1,500
3. System development (TDY, salaries)	....	30,000
4. Programming/testing (TDY, salaries)	....	36,000
5. Other travel	....	1,200
6. Service bureau or similar contracts	....	10,000
7. Site preparation (include fire protection in compliance with AR 420-94)	....	110,000
8. Initial supplies, tapes, tape cabinets, furniture, etc.	....	40,000
9. Preparation for operation under new system (file purification, etc.)	....	9,000
10. Other (explain)	....	....
Total	12,400	237,700

Figure 23-1. Estimated one-time costs format. Source: Office of the Controller General.

	First Year		Second Year				Third Year
	Third Quarter	Fourth Quarter	First Quarter	Second Quarter	Third Quarter	Fourth Quarter	First Quarter
1. Old system base	177,000	177,000	177,000	177,000	177,000	177,000	177,000
2. Total projected cost (sum of 3 and 4)	187,500	277,620	252,240	225,240	212,740	189,640	165,240
3. Old system—phase-out:							
(*a*) Personnel (ADPE/PCM Org.):							
"M" spaces	(30)	(30)	(20)	(15)	(10)	(5)	(0
"C" spaces	(20)	(20)	(15)	(10)	(5)	(2)	(0
"M" costs	27,000	27,000	18,000	13,500	9,000	4,500	0
"C" costs	24,000	24,000	16,000	12,000	6,000	2,400	0
(*b*) Rental:							
ADPE	....	....	....	....	....	....	....
PCM	18,000	12,000	8,000	6,000	6,000	2,000	0
(*c*) Supplies	6,000	6,000	2,000	1,500	1,000	500	0
(*d*) Other Personnel (coding, adm., etc.):							
"M" spaces	(60)	(60)	(30)	(20)	(15)	(10)	(0
"C" spaces	(40)	(40)	(25)	(15)	(10)	(5)	(0
"M" costs	54,000	54,000	27,000	18,000	13,500	9,000	0
"C" costs	48,000	48,000	30,000	16,000	12,000	6,000	0
(*e*) Other costs (specify)	....	....	....	....	....	....	....
Total	177,000	171,000	101,000	67,000	47,500	24,400	0
4. New system—phase-in:							
(*a*) Personnel (ADPE/PCM Org.):							
"M" spaces	(5)	(10)	(10)	(10)	(10)	(10)	(10
"C" spaces	(4)	(10)	(15)	(15)	(20)	(20)	(20
"M" costs	4,500	9,000	9,000	9,000	9,000	9,000	9,000
"C" costs	6,000	15,000	23,000	23,000	30,000	30,000	30,000
(*b*) Rental:							
ADPE	....	50,000	70,000	70,000	70,000	70,000	70,000
PCM	....	120	240	240	240	240	240
(*c*) Supplies	....	2,500	4,000	5,000	5,000	5,000	5,000
(*d*) Other Personnel (coding, adm., etc.):							
"M" spaces	(........)	(20)	(25)	(30)	(30)	(30)	(30
"C" spaces	(........)	(10)	(15)	(20)	(20)	(20)	(20
"M" costs	....	18,000	22,500	27,000	27,000	27,000	27,000
"C" costs	....	12,000	22,500	24,000	24,000	24,000	24,000
(*e*) Other costs (specify)	....	....	....	....	....	....	....
Total	10,500	106,620	151,240	158,240	165,240	165,240	165,240
5. Quarterly costs (−) or savings (+)	−10,500	−100,620	−75,240	−48,240	−35,740	−12,640	+11,760

Figure 23-2. Comparative cost-analysis format—estimated recurring direct operating costs, including net direct savings or increases. Source: Office of the Controller General; contains minor changes in format.

Computer Costs: One-Time and Recurring

To evaluate costs and potential savings requires statements of both one-time and recurring costs. Figure 23-1 shows a breakdown of typical one-time costs, which, as the name suggests, are those for establishing a new installation. Recurring costs, on the other hand, are those that will be continually incurred to keep the installation operating. Within this grouping fall the costs of annual equipment rental, operating personnel, equipment maintenance, and operating supplies. Figure 23-2 presents a comparative-cost-analysis format for presenting estimated recurring direct operating costs, including net direct savings or increases.

PURCHASE-LEASE OF EDP EQUIPMENT

Lease versus purchase determinations are an important aspect of the feasibility study. Actually, three methods of acquisition are possible: pur-

Item of Cost	Costs by Fiscal Year					
	First Year	Second Year	Third Year	Fourth Year	Fifth Year	Sixth Year
(*a*) Purchase:						
1. Purchase (cumulative)	$632,900	$632,900	$632,900	$632,900	$632,900	$632,900
2. Maintenance (cumulative)	17,292	34,584	51,876	71,160	90,444	109,728
3. Total	650,192	667,484	684,776	704,060	723,344	742,628
(*b*) Lease including maintenance (cumulative)	158,244	316,488	474,732	632,976	791,220	949,464
(*c*) Purchase exceeds lease	491,948	350,996	210,044	71,084	....	....
(*d*) Lease exceeds purchase	....	....	....	....	67,876	206,836

NOTE.—Item (*a*)1. Purchase costs should reflect planned equipment modifications. Item (*a*)2. Maintenance charges generally increase after third year.

Figure 23-3. Lease versus purchase—representative EDP system (based on 200 hours' operational use per month). Source: Office of the Controller General.

Item of Cost	Costs by Fiscal Year					
	First Year	Second Year	Third Year	Fourth Year	Fifth Year	Sixth Year
(*a*) Lease with option to purchase (option exercised at end of first year):						
1. Lease (cumulative)	$158,244	$61,164	$61,164	$61,164	$61,164	$61,164
2. Lease credit upon purchase	−97,080	......	......	......	......	......
3. Purchase (cumulative)	....	632,900	632,900	632,900	632,900	632,900
4. Maintenance for purchase (cumulative)	....	17,292	34,584	53,868	73,152	92,436
5. Lease with option to purchase cost (cumulative)	61,164	711,356	728,648	747,932	767,216	786,500
(*b*) Lease including maintenance	158,244	316,488	474,732	632,976	791,220	949,464
(*c*) Lease with option to purchase exceeds lease	....	394,868	253,916	114,956	......	......
(*d*) Lease exceeds lease with option to purchase	....	......	......	......	24,004	162,964

NOTE.—Item (*a*)3. Purchase costs should reflect planned equipment modifications. Item (*a*)4. Maintenance charges generally increase after third year.

Figure 23-4. Lease versus lease with option to purchase (exercised at end of first year)—representative EDP system (based on 200 hours' operational [one-shift] use per month). Source: Office of the Controller General.

chase, rental, or rental with option to purchase. Generally an option-to-purchase clause is obtainable in a lease contract without charge. Figures 23-3, 23-4, and 23-5 show formats for setting forth costs figures for (*a*) lease versus purchase, (*b*) lease versus lease with option to purchase—exercised at the end of the first year, and (*c*) lease versus lease with option to purchase—exercised at the end of the second year.

Item of Cost	Cost by Fiscal Year						
	First Year	Second Year	Third Year	Fourth Year	Fifth Year	Sixth Year	Seventh Year
(*a*) Lease with option to purchase (option exercised at end of second year):							
1. Lease (cumulative)	$158,244	$316,488	$122,328	$122,328	$122,328	$122,328	$122,328
2. Less credit upon purchase	....	−194,160	....	....	....	....	....
3. Purchase (cumulative)	....	....	632,900	632,900	632,900	632,900	632,900
4. Maintenance for purchase (cumulative)	....	....	17,292	36,576	55,860	75,144	96,839
Total	158,244	122,328	772,520	791,804	811,088	830,372	852,067
(*b*) Lease including maintenance (cumulative)	158,244	316,488	474,732	632,976	791,220	949,464	1,107,708
(*c*) Lease with option to purchase exceeds lease	....	....	297,788	158,828	19,868	....	....
(*d*) Lease exceeds lease with option to purchase	....	....	....	....	....	119,092	255,641

NOTE.—Item (*a*)3. Purchase costs should reflect planned equipment modifications. Item (*a*)4. Maintenance charges generally increase after third year.

Figure 23-5. Lease versus lease with option to purchase (exercised at end of second year)—representative EDP system (based on 200 hours' operational [one-shift] use). Source: Office of the Controller General.

REPORTING FINDINGS AND RECOMMENDATIONS

As part of the feasibility study, the usual practice is for the study group to submit its findings and recommendations to management. A thorough report includes the following information:

1. Recommended action:
   a. Justified need for the equipment (principal applications);
   b. Elements of return on expenditure: outflow of cash and utilization of other assets, savings (costs comparisons between the present and proposed systems), operating costs, taxes, and the time value of money;

c. Additional benefits to be derived.
2. Supporting information:
   a. Cost-breakdown charts;
   b. Determination of break-even point (see Figure 23-6).
3. Plans for developing and implementing recommendation:
   a. Installation date—proposed;
   b. Planned shifts and operations.

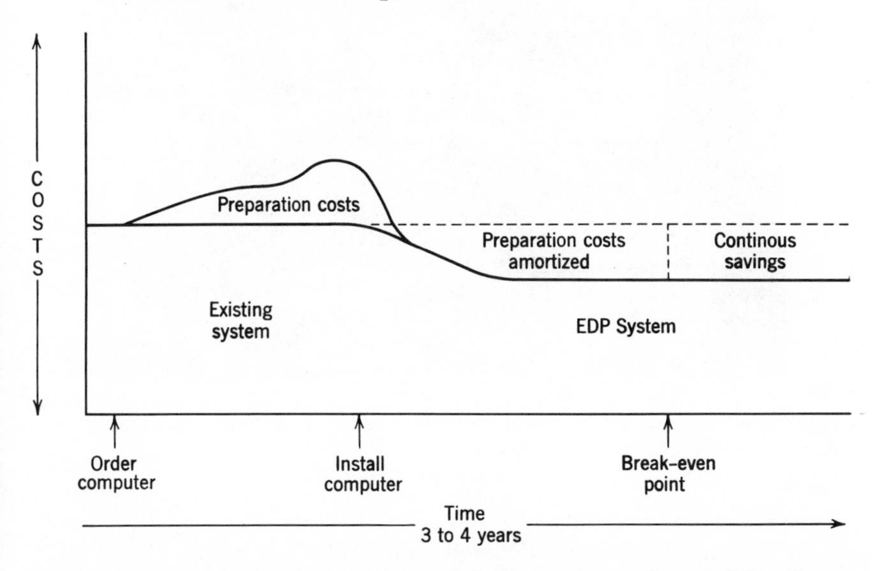

Figure 23-6. Representation of cost trends for EDP installation. (This illustration assumes that the EDP system will accomplish the same work as the existing system but at less cost. In many cases the EDP system will cost more than the previous system, but it will enable more work to be done than before or will effect other management improvements.) Source: Bureau of the Budget.

PREPARATION

Following the final approval of the equipment selection by management, procurement action begins. The interval between the placement of the order and the time of delivery allows for advanced planning, including such factors as—

1. Building the staff to handle the computer operation—through recruitment, training, and retraining.
2. Establishing the physical facilities—space allotment, construction, power, air conditioning, and furnishings. (See Figure 23-7.)

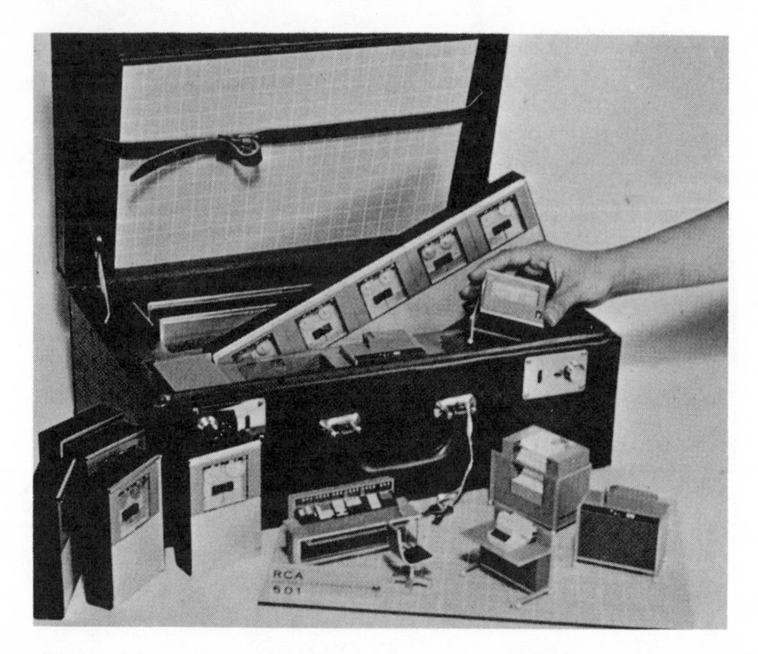

Figure 23-7. Three-dimensional scale models for laying out a computer installation. Courtesy: Industrial Models, Inc.

3. Making organizational assignments and adjustments.
4. Buying supplies.
5. Block diagramming, programming, and testing—through arrangements with the equipment manufacturer.
6. Outlining conversion procedures—running parallel operations while phasing out the old system.

READINESS REVIEWS

Several weeks prior to the scheduled delivery date of the EDP equipment an on-site readiness review should be conducted. The purpose of the readiness review is to examine the adequacy of the preparations and to decide what action is necessary in the event that conditions are not satisfactory. Sufficient time should be allowed so that, when necessary, the company and the manufacturer can reschedule delivery and adjust the installation date accordingly.

Typical considerations, including both completed and planned actions, that must be covered by a readiness review are as follows:

1. Site preparation.
   a. Facility construction or renovation.
   b. Power requirements.
   c. Elevated flooring for cabling and wiring systems.
   d. Air conditioning.
   e. Humidity control.
   f. Fire protection.
   g. Acceptance of the site by the EDP manufacturer, in writing.
   h. Scheduled completion.
2. Organization, personnel, and training.
   a. Authorized and actual strengths.
      1) At inception of EDP study.
      2) At date of readiness review.
      3) Projected on rental date.
      4) Schedule of phased release of personnel and spaces.
   b. Orientation and training of EDP operating personnel.
      1) Numbers of employees by types of training conducted.
      2) Number of existing vacancies by classification.
      3) Background and experience of personnel.
3. Systems development.
   a. Systems and procedures.
      1) Systems design.
      2) Estimated dates of completion.
   b. Programming.
      1) Number of steps required for each program.
      2) Percent of total number of steps completed for each program.
      3) Tested status of programs and machine time on no-charge basis.
      4) Estimated dates of completion.
   c. Machine timing.
      1) Initial estimates for each program run.
      2) Current estimates for each program run.
      3) Added or deleted program runs.
      4) Volume or work load.
4. Plans and schedules for purification of data.
5. File conversion and parallel operations.
   a. Plans and schedules for file conversion and machine loading.
   b. Plans and schedules for parallel operations and checkouts.

6. Need for and planned use of each machine product.
   a. Reports control management procedures for EDP equipment.
   b. Distribution and disposition of punched cards or tapes.
   c. Report frequency and justification.
7. Maintenance of records and costs.
   a. Management benefits and improvements.
   b. "Before and after" costs for personnel, machines, and supplies.
   c. Estimated-versus-actual costs for personnel, machines, and supplies.
   d. One-time costs.
   e. Status of procurement order and amendments to date.
   f. Familiarity with contract terms and provisions.
      1) Methods of machine timekeeping and computing rentals.
      2) Specifications of number of on-site service people.
   g. Start and completion dates on preinstallation steps.
8. Plans for operations in event of loss of data or equipment.
   a. Program check-point procedures.
   b. Local retention of prior cycle files.
   c. Alternative site plans.
9. Validity of procurement timetable.
   a. Delivery date.
   b. Installation date.
   c. Date on which equipment rental will begin.
   d. Insurance against "overlap rentals."
10. Equipment data.
   a. List of on-site punched-card machines and rentals.
   b. List of EDP equipment, on-site or to be delivered, and rentals.
   c. List of punched-card machines to be used in direct support.
   d. Statement covering phased release of remaining punched-card machines.[1]

[1] Adapted from a listing: Office of the Controller General.

part 8

Presentation and Implementation of Recommendations

24

Writing Effective Systems Reports

This part culminates the preceding discussion of a systems-and-procedures study. It is concerned with the concluding steps; that is, the presentation of results, the action to be taken, and the follow-up work. The time and money spent in conducting the preparatory and formative stages of a project are wasted unless the results are made known and utilized.

PRESENTATION OF THE RESULTS

Too frequently the presentation stage of a systems-and-procedures study does not receive the attention it deserves. Many studies that contain valuable information are discarded because of poor presentations. There are various reasons for this. Quite often the presentations are too long. Sometimes they are unduly technical, excessively intricate in their composition, verbose, or simply boring. Many exhibit a lack of consideration for the interest and point of view of the persons to whom they are addressed. A large portion of the effectiveness of a systems-and-procedures study depends on how accurately and skilfully the presentation is made.

This chapter discusses both subject matter included in systems-and-procedures presentations and the ways in which this material can be made to convey information and ideas effectively.

FORMS OF PRESENTATIONS

There are two basic forms in which presentations are made—the written report and the oral report. Each plays an important part in the business of conveying systems-and-procedures messages.

Written reports are categorized as (*a*) memorandum, or letter, reports and (*b*) formal, or treatise, reports. A memorandum report is a relatively short presentation of information essential for the comprehension of a situation, problem, or proposal. It often serves as the nucleus for the development of a more comprehensive report at some future date.

Memorandum reports range in composition from one-page presentations to lengthy letters with numerous attachments. Charts, diagrams, and written procedures are examples of material that is frequently appended to memoranda.

In contrast with the memorandum form of presentation, formal reports are customarily longer and more detailed. As the name implies, the material of such reports is substantially refined and polished before presentation. The format of this type of report is in accordance with rather conventional standards (discussed in Chapter 25).

WRITING SYSTEMS-AND-PROCEDURES REPORTS

The presentation of information and recommendations on systems-and-procedures studies usually requires the preparation of a written report. The consideration of elements and factors that enter into the construction of good reports is necessary.

All too often reports are as dry as dust, and needlessly so. It is often a foregone conclusion that a report must be dull. Reports should be written in an interesting style. Practically all subject matter can be presented in this way. The late Gilbert Keith Chesterton, author and literary critic, went so far as to say, "There is no such thing as an uninteresting subject; the only thing that can exist is an uninteresting person."

Writing ability is not some mysterious gift bestowed only on a few individuals. The ability to express ourselves interestingly is a human and cultural enterprise to be practiced and learned; it is a practical necessity for the performance of systems-and-procedures work.

To prepare a good report we need a command of the English language, a brisk imagination, and a genuine understanding of human nature. We also need to be willing to do some hard work. Anybody who ignores the importance of these factors does not understand the extent to which they have contributed to interesting and effective reports.

Good reports are seldom more difficult to prepare than mediocre or poor ones. With a little more effort commonplace information can be made to yield surprising results not only in style of writing but also in originality of composition.

Most reports are closely identified with the decision-making and action-taking functions of a business. They present information that enables management to know the exact state of affairs with respect to systems operations and to determine intelligent courses of action.

Any report that may result in action or influence management personnel is of sufficient importance to warrant thought and endeavor.

CONCERNING CLARITY OF EXPRESSION

The purpose of writing a report is to communicate information to a reader. Regardless of how simple or complex the information may be, it should be stated clearly. Successful communication demands clarity of expression.

Clarity is largely a matter of stating what you wish to say in the sense and spirit in which you wish it to be understood, and in such a way that your reader will comprehend both the substance and the spirit of the message. The elimination of ambiguity, or the possibility of misinterpretation, is one of the difficult problems that confront the systems analyst in writing a report. It is not sufficient to write a report that can be understood. The aim should be to write a report that cannot be misunderstood.

This discussion will omit grammatical topics such as punctuation, spelling, and the parts of speech. It is assumed that the reader possesses mastery of them. We will focus our attention on *substance,* not *form.* It is the message that is significant. Punctuation, correct spelling, and clear grammatical statements are necessary tools, but it is rhetoric—as distinguished from grammar—that is both the art and the heart of the language. Rhetoric may be defined as the skill that makes compositions expressive and effective. It deals with words, figures of speech, and phraseology. We must acquire rhetorical skill before we can use the medium of language successfully in writing.

Skill in writing can be developed. Progress in all communication, both oral and written, requires practice and determination. It is sad to see persons with valuable thoughts—some with brilliant and excellent ideas —who have not developed the skill necessary to communicate them to others. Sadder perhaps are those who think they are getting their ideas across when in reality they are not.

Keep the Reader in Mind

One of the most generally accepted principles of writing is, "Keep the reader in mind." So important a rule is this that very few publications on the art of writing do not devote space and attention to it. However, like many another principle whose merit is conceded, *keeping the reader in mind* does not always receive the consideration it deserves.

The tone, tenor, and content of a report should be geared to the attitude and knowledge of the readers to whom the report will go. A systems analyst ought to make his statements truly understandable by putting himself in the readers' place and viewing the message as they would view it. What are the things that characterize the persons who will receive the report? What are their dispositions? Their attitudes? Their habits? Their likes and dislikes? What will they be interested in learning from the report? Such questions will reveal psychological and motivating factors that influence the audience, and the understanding that results will enable the systems analyst to write effectively and convincingly.

Many systems analysts find that they can write better reports by imagining themselves to be talking to the persons involved rather than writing to them. This technique is sound. The systems analyst who uses it will find himself almost intuitively answering questions that might be asked if all the persons were sitting next to him around a conference table. We do something very similar when we prepare a speech and rehearse the presentation, imagining the audience and its reactions.

Nothing is quite so important as establishing contact with your reading audience. This is done by writing in the language of your readers. Avoid the use of words they are not likely to understand; if you must use such words, explain them. In addition guard against the use of flowery, overornate, and pretentious wording. Use words and sentences that will communicate the message with the least amount of distraction on the part of your reader. Be careful, however, not to express yourself in a condescending manner. Treat your reader as an equal.

Present more than the bare facts required in a situation. Reach out to accommodate your readers. Frequently an idea can be made clearer by just adding a point of information. In other cases an example, comparison, or contrast might be used to further the cause of intelligibility.

The Responsibility of the Systems Analyst in Writing

Systems analysts who write reports have an obligation to be intelligible. They are not writing to impress their readers but to impart information.

A report is of little use if it is not understandable to others. To write intelligibly the systems analyst must know what he wants to say and how to say it. He is expected to know what he means and to be able to express what he means in writing. He cannot rightly contend that the reader should make his own interpretations. Reports do not fall within the classification of interpretative literature. The primary aim of a report is to achieve a clear understanding of the information presented.

We must not delude ourselves by believing, when our ideas fail to get across to someone, that the fault lies in his inability to grasp them. We improve our writing not by conceiving ourselves to be perfect already but by acknowledging our deficiencies and taking affirmative action to surmount them in straightforward fashion. Negative attitudes toward readers are not helpful. What is needed is a positive approach—a willingness and eagerness to refine one's writing so it will be more meaningful.

Think and Organize for Clarity

Clarity in writing begins with clarity in thinking. Insufficient thought —not "thinking through"—is the most ordinary cause of writing failure. All attempts at clarity will be in vain unless you understand what you are writing about. Clarity, in a very true sense, begins at home. Mastery of one's own thinking is essential for attending to, and getting at, the real task of writing. We should first direct our thinking along orderly lines and then expect our writing to follow suit.

It is necessary to think clearly to avoid going off on a tangent—and more so in writing reports. If your writing is to be clear, your thoughts must be clear. Your mind must shape the language at every point along the writing path.

Because you cannot write an effective report about something you do not fully understand, a comprehensive knowledge of the subject matter is indispensable. Horace said, "Knowledge is the foundation and source of good writing." [1] He might well have added, "For there is no substitute for knowing what you want to say." The better we know the subject, the easier it will be to write naturally and intelligibly.

No quality that the systems analyst can possess is more valuable than that of striking at the heart of the matter. Effective writing calls for discerning the essential points, sorting out the irrelevant, and presenting the results directly and distinctly. For maximum effectiveness this process must be carried out step by step—sentence by sentence.

As we write, we should examine the fruits of our thinking. Is there a natural and logical unfolding of ideas? Have the essential points been duly covered? Does the writing display the attributes of unity, orderliness, and coherence?

Words, sentences, and paragraphs should be arranged so that there is

[1] Horace was a Roman writer of odes and compositions—a contemporary of Julius Caesar, Mark Anthony, and Virgil. Critics consider his poems perfect in method and style. (Quotation from *Ars Poetica*—line 309.)

no question about the information they are intended to convey. When writing, you are in the position of a storyteller, and each paragraph should help build the story along logical lines. Do not make the mistake of trying to tell the whole story at once. Good writing entails the well-reasoned and orderly development of ideas. The points of the message should be selected and presented one at a time.

Do not deliver a barrage of conglomerate ideas, facts, and opinions on readers. Our aim is to clarify, not to mystify. We should present the material so as to guide the thoughts of our readers naturally and progressively from one point to the next.

The correlation of ideas is important. Each idea should flow smoothly into the succeeding one. The information needs to be viewed broadly from beginning to end, with its elements placed in their proper perspective.

Negligent and loose thinking causes the unsound arrangement of ideas. A message is coherent when the sequence of its ideas is set forth in logical order or arrangement.

About Words

Efficiency in writing depends to a marked degree on a person's vocabulary. A stock of good words is of inestimable value, because it is the means whereby you can express yourself fluently and interestingly. More than this, it is the key to understanding.

Words have been described as the building blocks, the foundation stones, and the tools of language. Still, in another sense, words have plastic qualities that make them malleable with regard to their workability.

Many words have two kinds of meaning—denotative meaning and connotative meaning. The *denotation* of a word is the actual, or literal, meaning as given in the dictionary; the *connotation* is the suggestive or implied significance that has become associated with its matter-of-fact meaning. Think of the connotation there is, for example, in the word "sell." For some persons "sell" suggests unpleasant experiences associated with tactless persuasions put forth by presumptuous salesmen. For others it suggests an interesting challenge, a matching of wits. For still others the word "sell" suggests helpful service in the matter of making an intelligent and wise purchase.

"Sell" is but illustrative of a long line of words in the English language with connotations, or overtones of meaning. Some connotations are pleasant, others unpleasant, and between these extremes there exist still others with various shades of suggestive meanings. Following are examples of word groups, each with the same denotative meaning but with diverse connotations:

obese, fat, portly, stout, chubby
perfect, faultless, accurate, correct, unerring
confer, converse, talk, chat, gossip
contend, strive, endeavor, struggle, flounder
manage, guide, handle, contrive, manipulate

Pure mathematics is about the only language that comes reasonably close to being without overtones of meaning. Except in the minds of numerologists and soothsayers, the figure "2" means 2, no more and no less. However, mathematics, pure or applied, is no substitute for the English language, and although a report may involve the presentation of a good deal of mathematical material in the form of charts, graphs, and calculations, it is seldom that the written word assumes secondary importance. In fact, we usually think of the mathematical material as supporting the written word.

Although it is virtually impossible to write reports without using words with connotative meanings, our writing can be constructed so as to minimize unwarranted suggestiveness.

Generally the preparation of a single report will entail a combination of both factual and persuasive writing. Factual or scientific writing aims to be objective, to be as free from connotative meanings as possible, and to transmit its intelligence logically and unemotionally. On the other hand, persuasive writing seeks to convince or influence the reader in an honest and forthright manner; words with connotative meanings are employed because they lend a certain charm, force, and effectiveness to the conveyance of thoughts.

Inasmuch as the scientific study of systems and procedures is fundamentally an objective process, it may seem strange that persuasive writing should not be disapproved in reports. The use of persuasion must be strictly confined to the presentation step of a systems-and-procedures study, and the precepts governing objectivity must be vigorously upheld in performing all other phases of the work.

Often the outline or arrangement of the material in a report will provide a natural demarcation line between factual and persuasive writing; for example, factual writing is appropriate to the "findings" section of a report, whereas the "recommendations" section is suited to writing of a persuasive nature. The important thing in assuring the scientific integrity of a report is to write so as to make it clear to the reader what is factual information and what is not. We must guard against presenting facts as opinions and opinions as facts. Whatever is said must be said honestly and with an awareness of its consequences.

On Simplicity

Be continually regardful of your readers. We cannot be expected to make every idea comprehensible to everyone who can read; however, what we write should be sufficiently simple for the level of knowledge that is typical of the average member of the reading audience. We should write on the assumption that if we express ourselves in a simple enough manner for the "average" reader to understand, then we will at the same time be reaching a more judicious reader, who will appreciate the simplicity and clearness of style.

Some persons are mistaken in thinking that it is relatively easy to write simply. This is not so. As Charles Poore, long-time journalist for the *New York Times,* put it, "There is nothing quite so complicated as simplicity. Infinitudes of distractions and irrelevancies must be forced into perspective to achieve it."

It is a paradox that simplicity is a manifestation of profound thought. Eminent men of all ages and diverse fields of endeavor have written simply. Shakespeare used concrete and simple word images; he wrote for the general public of his day. It was he who said, "An honest tale speeds best, being plainly told." This is a message of particular significance for the writer of a report. If we have trouble understanding Shakespeare, it is not because he wrote in an elaborate style, but perhaps because of the difference beween Elizabethan and modern English.

Many novelists, including Charles Dickens, Robert Louis Stevenson, and James Fenimore Cooper, are known for their individualistic but simple and careful style. In the form of the short story, simplicity of expression is found in the works of Guy de Maupassant and Edgar Allan Poe. Regarding verse, Emily Dickinson succeeded in putting the deepest of thoughts on love, death, and immortality into well-chosen, simple expressions.

Similarly, outstanding men of natural science have written simply. William Harvey, discoverer of the circulation of the blood, is superb in his simple treatment of complex material on the motion of the heart and the blood. Charles Darwin, who has been called "the maker of the greatest epoch in the study of life," presented the results of his findings on the origin of species in a simple and interesting style.

To write a good report it is not necessary to have the talents of a Shakespeare, a Dickens, or a Harvey. The important point is that we can take a cue from these writers, not only with respect to their simplicity of style but also in the manner they make their writing interesting. The principles they employed are as apropos today as they were in the past.

No one expects a report to be a literary gem, but it should not bore the reader with its language, and he has every right to expect it to be clear, accurate, and illuminating.

FUNDAMENTAL CHARACTERISTICS OF GOOD REPORTS

A Good Report Is Readable. By "readable" we mean the standard definitions of the word; namely, *easy to read* and *interesting*. A well-written report has movement; it flows along naturally and smoothly. It does not ramble or falter.

We should not presume that our report will be given consideration simply because management has shown an interest in the subject matter. Many reports go into the files unread or get only a cursory look. To sustain attention we need to raise it above the commonplace by the use of imagination and skill.

Reports are prepared from material that differs considerably with respect to its intrinsic originality. At one extreme the writing of a report may entail presenting that which is mainly creative or, at the other extreme, setting forth material of a most ordinary nature. However, in practically every presentation that is written, it is necessary to take certain thoughts and perceptions and arrange them in interesting and informative statements.

Some who write in a style appropriate for the treatment of highly technical material produce reports that are sophisticated and stiff. The preparation of an effective report is contingent on the ability of the writer to employ the principles applicable to all good expository writing, which has as its attributes clearness, orderliness, correctness, and consideration.

A Good Report Is Realistic. The precept of realism, when applied to writing, means presenting things as they exist in actuality. A realist within this context is true to the facts of a situation. He is concerned with the scientific, as opposed to the visionary. It follows that the systems analyst, as a writer of the *realistic*, must see things as he finds them, not as he would deem them.

A Good Report Is Informative. Reports should be as complete and informative as possible. Completeness is a quality that varies from one report to another. Some reports are intended to convey only a portion of the facts and data, such as interim reports; others are meant to render an extensive account on a matter. In either case, they should be thorough within their limits.

A Good Report Is Concise. Too often readers are under the wrong impression that conciseness invariably means shortness and that there is no place in good writing for stressing a point by the use of repetitive wording, employment of examples, instances, and analogies. The fact of the matter is that a report may be voluminous, or very short, and possess or lack conciseness. Length is not the determinant.

True conciseness in writing consists in stating only what is of value and omitting that which is unserviceable and inefficient. Conciseness is not an absolute criterion; it is relative. What may be concise to one person may be verbose to another. For this reason instruction on the matter must be of a general nature. In good writing we need to concern ourselves with the significant and to shun irrelevancies. The answers to these questions will assist us in appraising the conciseness of what we write: Does it help rather than hinder the understanding of the thought expressed? Are there any words or phrases that are without a purpose? Is it sufficiently interesting to hold attention?

A Good Report Is Direct and Clear. A well-written report comes to the point. The effectiveness of reports depends greatly on the proper organization of material, and the logical continuity and connection of ideas. There must be no unabridged gaps and no inappropriate statements. The reader should feel a sense of order and unity.

Writers of good reports avoid using language that is hackneyed. They shun overworked expressions such as "powers that be," "it stands to reason," "by the same token," and "needs no introduction"—phrases that have been worn thin by overusage to the extent of being almost meaningless.

Nothing detracts from writing more than obscure passages and wording that lacks significance. Readers tire of a report that is replete with idle words and stale language.

Strive for some measure of distinction in report writing. Rather than being satisfied with trite terms and rubber-stamp phraseology, use originality with new and fresh wording that will be clear and meaningful to the reader.

A Good Report Is Factual and Reliable. To maintain scientific integrity a report must distinguish sharply between two kinds of material: what is known to be true and can be substantiated; and what may, could, or should be true.

Good reports are free from errors of content. One mistake of this kind may cause the reader to question the validity of the entire presentation. Errors to be avoided include contextual meanings. We must not only know what we mean but we must also say what we mean. Moreover, it is very important that any computations presented be accurate. Even a small arithmetical error can discredit a complete report.

A good report embodies accurate spelling, correct punctuation, proper capitalization, and proper grammar. Any deficiencies in these matters will reflect badly upon a report and its writer, giving the general impression that the writer has been careless. The reader may well reason, "Is it likely that a person who has been careless in these aspects has been careful in his study of the problem?"

A Good Report Is Timely. To be effective reports must be up to date. A project, including the preparation of a report, may take only a few hours or many months to accomplish. The extent of time is not the measure. Timeliness does not mean hurry, though expeditious action is often necessary in performing certain studies. What it does mean is suitability to a given occasion. In this respect a good report is ready at a time when the information is of value.

Much information becomes obsolete with the passage of time, and some facts and data lose their value quickly. As a consequence late reports can be almost useless. It is wise, therefore, to have a schedule that has been determined with care—that can be met no matter how urgent the requirements happen to be.

For many projects that require expeditious rendering of information interim reports serve the immediate needs until the issuance of a final and complete account.

One further point remains to be covered and that has to do with the *tone* of the report. Tone, which is the manner of expression, affects the way a report is received. Tone is related to the attitude of the writer and the feelings he wishes to convey. Quality of tone, of course, is contingent upon such considerations as the kind of topic, nature of the subject matter, and the personal characteristics of the reading audience. In general, however, the dominant tone of a report should be positive, constructive, and helpful; not negative, hypercritical, or sour.

REVISE AS NECESSARY

For best results a critical review of one's own writing should not be undertaken immediately after it has been written. Give your mind a chance to refresh itself and regain objectivity. If you attempt to remedy faults and make improvements directly after finishing the work, you can easily miss errors and deficiencies in what you wrote from a subjective view.

When revising, ask questions such as: Does my report say what I mean? Have I developed my thoughts fully and clearly? Have I presented the information correctly, intelligibly, and in an appropriate manner? Does my report read easily?

In revising the writer should try to assume the role of the impartial critic. He must try to forget that the language is his own, especially the phrases and sentences that seem to warrant pride of authorship. With objectivity he must reject sentences without substance, denounce that which is awkward and indistinct, prune away offensive and purposeless repetitions, cancel inept embellishments, cross out what is not in accord with the truth, recast that which is vague and misleading, and make improvements wherever improvements are necessary.

Some persons find that reading their writing aloud helps them to evaluate it. In this way they obtain a better impression of the swing and balance of the language. When they come upon a "road block" to clear and smooth reading, the word or passage that does not sound right rings clear for correction.

WAYS OF ENLIVENING A REPORT

Here are six effective ways in which to enliven a report.

Develop Only the Main Point or Points. In most assignments there will emerge some information or ideas that are more important than others. These should be developed by amplification of significant details and judicious use of explanatory material.

Emphasis is the keynote. Subordinate information and ideas should be kept in their proper relationship regardless of how interesting they may be.

Use Comparison and Contrast. Comparison and contrast are two forceful methods of securing attention and evoking action. To illustrate their use, let us suppose that the systems analyst is about to propose a new system. One of the best ways of promoting its utility is to show conditions that are likely to exist in the future if the proposal is not put into effect and then realistically relate conditions that will correspondingly exist if the proposal is adopted. This "before and after" method is most effective when it is based on a solid foundation of specific evidence.

The comparison and contrast methods may also be effectively used to bring out the essential relativity of new problems to others that have been solved; for example, certain features of a proposed system may be related to similar features of an existing system that is efficiently in operation.

We should not neglect the comparisons and contrasts that might also be drawn with respect to the operations of other companies, both those within and without the industry. An alert top management is very interested in what other companies are doing, especially competing organizations. If it can be shown that the installation of a new system will afford an advantage over a competitor—or will in some way rectify an adverse situation—it will amplify the presentation and thus heighten interest in it.

Be Appropriately Concrete and Specific. This principle applies to descriptions, examples, and incidents. The value of concrete and specific insertions is to convey distinct and positive images to the mind of the reader. When properly used they can be effectual means for instilling the presentation with interest and conviction.

Consider the contrast in the mental impressions engendered by the following sentences:

Indefinite: The procedural changes that are being recommended will result in saving the company a great deal of money.

Specific: At a conservative estimate the recommended procedural changes will effect an annual saving of $140,000.

It is not always possible to be so specific, but we should try to inject enough concrete and specific material into our presentations to make them vivid and interesting.

Employ Headings and Subheadings. Headings serve three main purposes: to draw the attention of the reader, to arouse his interest, and to enhance the readability of the presentation. They are especially useful in the case of long reports, where paragraphs alone do not provide sufficiently emphatic divisions of thought.

Headings relieve the monotony associated with masses of printed matter. When properly employed they function as a safeguard against valuable points of information going unnoticed in the text. They also enable the reader to select the portions of the text in which he has a definite interest.

Most headings are expressed in the form of a statement. However, a heading may occasionally be presented very effectively in the form of a question, a command, or an exclamation. Long reports might well contain a mixture of all four forms.

Show Proof: Facts, Data, Cause-and-Effect Relationships, Assertions of Experts, and Examples of Successful Applications. Although you may have possession of valuable information and ideas that you know are sound, it is still another matter to convince management of their worth. For this purpose the systems analyst needs to offer proof. One of the most striking ways of presenting proof is to describe a situation by showing logical relationships that lead from one event to another. A chain of reasoning is thereby constructed to direct the reader from known causes to inferred effects or from known effects back to inferred causes. The origin, history, and development of conditions and proposals are apt subjects to explain in terms of cause-and-effect relationships.

Other ways to give support to ideas and proposals is to cite testimonial evidence obtained from reliable authorities. Such testimony includes material from trade journals, statistical tables, government pamphlets, etc., as well as direct information that might be in answer to mail or telephone inquiries. Keep in mind that no material is more reliable than its source.

Another way of rendering proof is by example. Whenever possible cite concrete and practical applications as exemplified by model situations and analogous cases.

Make sufficient use of visual aids. Sketches, blueprints, charts, diagrams, forms, and other pictorial means not only help to explain matters but also contribute toward making reports concise. To quote Ivan Turgenev, in *Fathers and Sons:* "A picture may instantly present what a book could set forth only in a hundred pages." By the skillful employment of visual aids, the systems analyst can present abundant information in a report of moderate length.

25

The Composition of Reports

Written reports are broadly classified as *memorandum,* or *letter, reports,* and *formal,* or *treatise, reports.* The terms relate to the general make-up and arrangement of reports, and are not indicative of the nature or relative importance of the material they present.

THE MEMORANDUM REPORT

Brevity is a typical quality of the memorandum report, but not a universal one. A memorandum report may consist of numerous pages. Often expediency dictates the issuance of a short report, such as interim accounts on an assignment.

The memorandum report is an extremely popular form of presentation. Much of its popularity is due to the similarity it has to ordinary correspondence, particularly to the business letter. It is comparatively easy to prepare, lends itself to informality in style and composition of writing, and has a wide range of applications.

The Elements of a Memorandum Report

The elements of memorandum reports correspond to the elements of interoffice memoranda. These are as follows:

1. The heading.
   a. The address (the "to" line, possibly an "attention of" line, and sometimes a "cc:" notice. The *cc: notice* designates who is to receive *courtesy copies.*
   b. The date.
   c. Occasionally a "from" line.

   d. Occasionally a "subject" line.
   e. Occasionally a "reference" line.
   f. Occasionally a file number.
2. The body.
3. The signature(s).
4. The writer's and transcriber's initials.
5. An enclosure (attachment) reference (if required).

The layout of these elements for interoffice memoranda, and accordingly for memorandum reports, is often standardized within companies. In fact many companies have special printed forms for their interoffice memoranda that reduce most of the elements to a uniform pattern to facilitate the writing, typing, and reading of messages.

Styles of Memorandum Reports

Memorandum reports are arranged in one of two styles—the integrated style and the segregated style.

In the integrated style the body of the report begins under the heading—immediately following the introductory material. An illustration of an integrated style of memorandum report is shown in Figure 25-1. The accompanying attachments are illustrated so that the reader may visualize the overall composition of an entire report of this style.

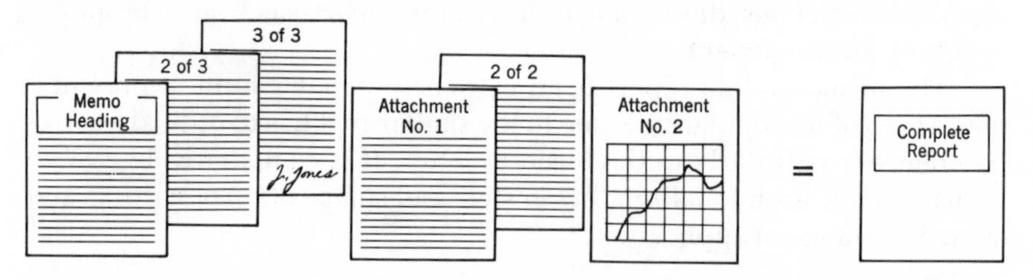

Figure 25-1. Arrangement of an integrated style of memorandum report.

The segregated style of memorandum report is illustrated in Figure 25-2. Observe that the body of the document is separate. In this style of report the cover memorandum usually contains prefatory material such as an acknowledgment of authorization for the report, a statement of the scope and limits of the work covered, a description of special techniques employed, unusual problems encountered, and significant matters to be settled. It may summarize conclusions based on information uncovered in the study, and highlight recommendations for action based on those conditions.

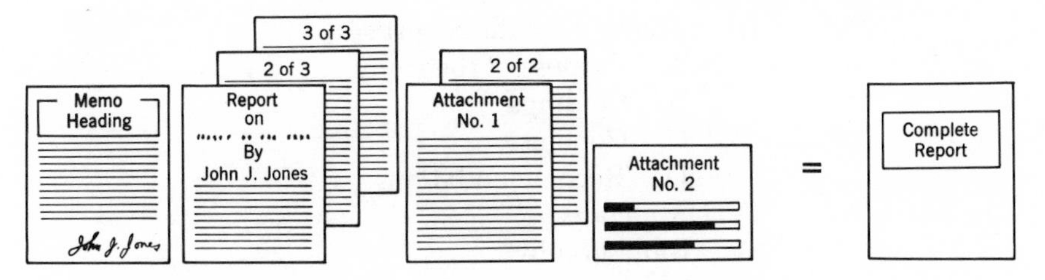

Figure 25-2. Arrangement of a segregated style of memorandum report.

THE FORMAL REPORT

The systems analyst has fewer occasions to prepare formal reports than memorandum reports. Each form has its own value in terms of time and place, however, and there are numerous instances when either can be employed with equal propriety.

In practice the form of written report that is used in a particular application depends on one or more of the following factors: the nature and composition of the material to be presented, the personality of the reading audience, the number of copies to be made, and possible directions from the authority endorsing the study.

Based on structure, there are two main types of formal reports—the *unabridged type* and the *abridged type*. The former is characterized by an elaborate, specific outline; the latter is distinguished by a relatively condensed outline that contains fewer sections but retains the essentials of the full-content presentation.

The Sections of an Unabridged Report

The following is a sample outline for preparing a formal systems-and-procedures report of the unabridged type:

I. Preliminary matter
   A. Title page
   B. Letter of authorization
   C. Letter of transmittal
   D. Preface
   E. Table of contents
   F. List of illustrations
II. Synopsis

III. The body of the report
    A. Purpose of the study
    B. Findings
    C. Conclusions
    D. Recommendations
IV. Appendix
V. Bibliography
VI. Index

Rarely is it necessary to make an outline more comprehensive than this one.

The Sections of an Abridged Report

Only occasionally does a formal report require as elaborate an outline as the one outlined above. For most reports a condensed outline will serve the purpose. Here is an example of this type of arrangement:

I. Preliminary matter
    A. Title page
    B. Letter of transmittal
    C. Table of contents and table of illustrations
II. Synopsis
III. The body of the report
    A. Purpose of the study
    B. Findings
    C. Conclusions and recommendations
IV. Appendix (when necessary)
V. Bibliography (when necessary)

This arrangement reflects the elimination and combination of certain parts of the longer outline given for an unabridged report.

BINDING THE REPORT

Lengthy memorandum reports and all formal reports should have a suitable cover or binder. First impressions are important, and this applies to the binding of a report. Although it is true that we should not judge a book by its cover, the truth is that there is an initial tendency to do just that—after all, why do publishing houses spend millions of dollars a year on elaborate bindings and attractive book jackets?

In today's world of modern packaging we expect things to be appealingly packaged, including reports. A simple but attractive cover is generally best.

Many companies keep a supply of covers on hand. These are usually satisfactory for most reports, but when the systems analyst has something special to offer in a report, he should not overlook the idea of presenting it in a different, unique cover. This technique should be reserved for reports that are truly felt to warrant special attention.

Covers of many types and styles are obtainable from stationery stores. Regardless of the source, use one that is in keeping with the character and the physical construction of the material being presented.

26

Oral Presentations

The preceding chapters have been concerned with written reports. However, in many situations the systems analyst will have occasion to present information orally to management personnel, either individually or in a group.

The composition of the listening audience—the level of management to be reached—will depend in each case on the nature and importance of the subject matter, the number of departments affected, and company policy.

In the case of reports of major importance it is usual to present them to the top executives of the company before relating them to any other group of management personnel. Reports of lesser importance may be initially presented to groups of subordinate management personnel, after which—depending on the extent to which authority has been delegated—they may not need to be related to top management.

In oral reports the acceptance of systems-and-procedures proposals is often contingent on the skill with which the presentation is made. It is important, then, that the systems analyst develop proficiency in this valuable means of communication.

ESSENTIALS OF EFFECTIVE ORAL PRESENTATIONS

Just as there are well-founded principles for making written presentations effective, there are also well-founded principles for making oral presentations effective. Some of the more important ones are discussed below.

Stance

Assume a comfortable standing position. Avoid rigidity. Keep the feet apart, but not so far apart as to appear awkward. Let the weight of your body rest mostly on the balls of your feet and keep your weight slightly forward. Do not rock back and forth on your toes and heels. Most of us have witnessed this to-and-fro motion in some speakers and know how annoying it can be. Hold your chin up and your shoulders back, with the ease and bearing of one in control of himself and the situation.

This does not infer that you should remain in the same standing position or confine yourself to the same area. On the contrary, it is generally a good idea to move about a bit, to take a few side steps, or even to walk slowly from one place to the next as long as the audience can see and hear you. Such movements appear natural as they give the audience the impression that one is not glued to the floor. When the movement is for a particular purpose, such as walking over to a chart to point something out, it looks all the more natural.

Many persons are at a loss as to what they should do with their hands when making an oral presentation. Here are a few suggestions. Do not clasp your hands in front of you, because this presents a clumsy appearance. Neither is it good practice to grip them behind you, as this has a tendency to pull your shoulders back too far and to place excessive weight onto your heels. Furthermore, do not keep them in your pockets. It is not in bad taste, however, to slip one of your hands into your pocket now and then—but do not keep it there, and always have one hand showing.

When your hands are not being usefully employed let them fall naturally at your sides. Other things you might do with your hands should be of a positive nature. If there is a speaker's stand, you may loosely hold the edges of its top surface. Do not, however, grapple with any notes that you are referencing; turn the pages as they need to be turned, quickly and without fanfare.

One of the most purposeful uses to which hands can be put during an oral presentation is in the making of gestures. In this respect hands are used to (*a*) indicate directions such as by pointing "over there," (*b*) call attention to something (e.g., by using a stick-pointer or forefinger to designate a location on a flow diagram), (*c*) suggest size and shape (as "it stands this high [from the floor]," "it reaches from here over to that wall," "it is shaped like this," and similar manifestations), and (*d*) emphasize certain points to express the approval or disapproval of ideas.

It should be pointed out that gesturing need not be restricted to hand motions. Facial expressions and movements of the head and shoulders can also be powerful conveyors of meaning.

Voice

It is often true that we fail to make sufficient use of our speaking capabilities. Recognizing this basic fact, the purpose here is to point out specific qualities and principles of delivery that are necessary for the best results in oral presentations.

A good speaking voice is characterized by clear diction, pleasantness, and flexibility. Few persons will listen attentively for any length of time to a voice that is indistinct, harsh, or monotonous. Once the audience becomes weary of a speaker's disagreeable voice, the information he is trying to convey may well fall on deaf ears, and even propositions and ideas of the first caliber will go unnoticed.

Remember that what you say to the audience must be heard and comprehended if it is to be of value. Effective projection of the voice contributes to both of these requirements. Your voice should be loud enough for the person who is furthest away to hear you without strain. Vary the rate, volume, and pitch of your voice to avoid monotony and give emphasis to the strategically significant parts of your presentation. If you make a slip of the tongue when presenting your material, do not dwell on the mistake; instead, correct it immediately and continue as though no mistake had been made. The chances are that you will be the only one who feels badly about it unless you make the feeling contagious by dwelling on the matter.

GENERAL INSTRUCTIONS FOR MAKING ORAL PRESENTATIONS VIVID

The following steps will do much to promote the successful presentation of oral reports:

1. Arouse the interest of the audience in the opening statements of the presentation. Know exactly what is going to be said and how it is going to be said. The ideal beginning is one that not only gains the interest of the listeners but also leads that interest into the body of the report.

2. Have thoroughly in mind the progression of the ideas that are going to be presented. The presentation must possess movement, the onward motion of ideas, or else the interest of the listeners will soon wane.

3. Be cautious about making apologies. Most of them are uncalled for; they simply annoy the audience rather than serve any constructive purpose. This refers to needless apologies such as, "I'm sorry, but this is the best that can be done on short notice," "Pardon me while I put my notes in order," and "This is a poor illustration, but it will serve the purpose."

Apologies of this sort are negative and may even give the impression that the presentation is unworthy of much attention. Of course, there are times when apologies are in order, such as when one is detained and people are kept waiting. In these cases make the apology sincere and brief, and then get quickly to the matter at hand.

4. Inject occasional figures of speech, striking phraseology, and tactful humor into the presentation to keep it interesting and alive. We all have a relatively low saturation point for absorbing cut-and-dried information. The discerning systems analyst recognizes this and provides the variety of voice, movement, and content that is necessary to maintain attention.

5. Use visual aids whenever practical. The proverb, "One picture is worth a thousand words," applies also to oral presentations. Visual aids such as sketches, charts, and diagrams may be shown by means of a film projector, displayed from an easel, or presented on the old standby, the chalkboard.

Do not overlook the possibility of giving the listeners copies of forms, charts, and other visual aids that will be discussed. Also, copies of the entire presentation may be distributed to the audience.

6. Exhibit and explain actual devices when such demonstrations are suitable to the occasion. Most of us like to be shown how things work. We generally find this to be interesting and informative; for example, an explanation of the operations and capabilities of a new piece of equipment for use in connection with a proposed system will be more effective if arrangements can be made to demonstrate its performance. Companies that sell equipment ordinarily make both their products and the services of their representatives available.

7. Let qualified people assist in the presentation when such help is desirable. A presentation concerning a large and complex system may be broken down so that a number of other persons can each give a part of it; for example, arrangements might be made for an associate systems analyst to give one part; a sales representative of an equipment manufacturer to give another part; and certain department managers, whose operations will be effected, to give still other parts.

8. Close the presentation swiftly, pleasingly, and forcefully. Summarize the main points that were related; highlight any specific action that is considered necessary or desirable; and, as applicable, point out areas in need of further study. In the case of a proposal be sure to sum up and emphasize the reasons for its adoption.

Do not permit the ending to lag. Be prepared to put forth the spurt of energy necessary to make the delivery of the conclusion brisk and buoyant.

In many presentation situations, after the systems analyst has completed his talk, he is expected to answer questions or objections raised by the audience. Often while the presentation is in process it is common for listeners to interject queries. The systems analyst should regard such activity on the part of the audience as an expression of interest and as an opportunity to clarify the points that are of special concern to his listeners.

27

Gaining the Acceptance of New Proposals

Under favorable circumstances a sound proposal will go a long way toward gaining its own acceptance by virtue of the fact that it is sound. However, the acceptance of proposals, even the best of them, must often be acquired under less than favorable circumstances. Gaining acceptance depends on the earnestness, perseverance, and personally engaging attributes of the systems analyst.

Winning the acceptance of a proposal is normally a cumulative activity. It begins from the moment the systems analyst first meets the persons involved; from that point it will be to some extent contingent on the establishment and continuance of cordial, friendly relationships. The conscientious systems analyst is aware of this and builds acceptance for himself and his ideas.

OVERCOMING OBJECTIONS TO PROPOSALS

To make his dealings with people more intelligent and effective the systems analyst needs to have an understanding of their attitudes and motivations.

People object to proposals for one or a combination of the following reasons: (*a*) a natural inertia toward that which is new, (*b*) personal prejudices or motives, and (*c*) considerations that are believed to be valid. There are, fortunately, certain approaches to the handling of these objections. For the purpose of furnishing instruction of the kind that will be found useful, we will discuss the matter in some detail.

Natural Inertia

The inertia that proposals have to overcome is of greater importance than is commonly supposed. To begin with, it takes work to concentrate on a proposal, to evaluate its worth, and to put it into effect. Moreover, some persons are unduly conservative; they are content to "let well enough alone," to retain and embrace the status quo, and to stagnate without any disturbing thoughts of improvement or change.

By presenting new aspects of the situation, it is frequently possible to stimulate interest. Most proposals can be considered from different angles. It may be that the presentation of some other angle of a proposition will reflect it in a more favorable light and thus kindle interest where previously there was nothing but apathy. Furthermore, this method permits the systems analyst to use repetition for emphasis without appearing to do so.

Another method of surmounting the barrier of inertia is to lay stress on the need for action. Possible approaches along this line include citing dramatic illustrations, evincing striking comparisons and contrasts, and relating forceful proof in the form of facts and figures.

Still another means is to appeal to personal motives and drives. Desires for recognition, achievement, to excel one's own record or that of a competitor—these and similar motives form the basis for possible appeals that may be used to engender enthusiasm.

The grand leveler of inertia is the competitive system. Those who compete are pressured to adopt new ideas, new systems and procedures, and new equipment on penalty of commercial annihilation. In some cases pointing out this well-known fact may be of sufficient influence to dispel any tendency toward apathy.

Personal Prejudices

Not infrequently objections to proposals are based on sheer prejudice. Straw barriers to a proposition are raised typically by those who wish to discredit it and conceal such underlying reasons as envy of the systems analyst for thinking of something new, fear that he will receive recognition or that they will not, and fear that the adoption of the proposal will in some way impair their status or position.

The systems analyst needs to develop skill in recognizing and handling prejudices of this type. First, he must know how to sense a prejudiced situation. No one is likely to tell him that "I don't like the proposal because it is your idea" or "I'm against the suggestion because it will make me appear bad in the eyes of management." Nevertheless, although there is no infallible way of ascertaining a prejudicial attitude that is unexpressed, there are cogent signs that may be indicative of its presence. Look for the absence of inertia and at the very nature of the objections. People who assume a prejudicial position with regard to proposals are prone to make vehement objections on trifling grounds. When such is the case it may reasonably be suspected that there is some deeper reason for the attitude.

Secondly, it is important that the systems analyst become proficient in diagnosing the exact nature of personal prejudices. It is generally much easier to deal with a prejudice if you know its true character. Although this can be difficult to determine and at times impossible, frequently a person's expressions and deeds will reveal his real objections to a proposal. A few choice questions presented by the systems analyst will help in this direction.

Third, he must know how to deal with a prejudicial position. His concern, of course, is not so much with the objector's attitude as with the practical fact that when a person assumes a prejudicial stand, it presents a difficult obstacle to surmount.

Listed below are certain fundamental principles to keep in mind with regard to gaining acceptance of proposals that are in conflict with personal prejudices; each principle does not apply in every case, but all have a wide range of application.

1. Deal with a prejudice in a tolerant and understanding manner. Let the person know that you are aware of his position. Make a sincere effort to see his viewpoint.
2. Refuse to get angry however annoying or ridiculous the prejudice may be. Anger will only put the person on the defensive and make him all the more unyielding.
3. After having made it known that you are clearly aware of the prejudice, treat it with diplomatic indifference. This will cause the prejudice to lose influential status in the mind of its holder.
4. Tactfully express the motive that underlies the prejudice and then point out that the proposal is not incompatible with it. Give emphasis to those aspects of the situation that will conciliate the proposition and the person's motive.
5. Stress higher motives than the one that is the real cause of the objection and show that the proposal is in keeping with these higher motives.

Considerations That Are Believed To Be Valid

Proposals are frequently objected to on the basis of considerations that are believed to be valid. In such cases persons with objections usually bring them forth in a straightforward manner, along with the reasons for their beliefs. The objections may or may not be valid. With those that are not, it is recommended that the systems analyst handle them as follows:

1. Restate the objection so that the objector will know that it is correctly understood. When practicable proffer additional information in line with the objection to show him that you appreciate his point of view.
2. Direct attention to the reasoning on which the objection is based.
3. Point out—tactfully and discreetly—the fallacy or invalidity of this reasoning. Bring forward confirmatory evidence to make the point; that is, facts, figures, statistics, documents, specific illustrations, and the like.
4. Dispose of the objection to the satisfaction of the objector. Reasonable people will accede a contention when shown valid evidence that it is wrong.
5. Present the correct conclusion. Make it clear that the proposal is sufficient to produce the results claimed.

Handling Valid Objections. When faced with a valid objection to a proposal, one of two courses may be followed. First, the systems analyst might adjust the proposal to meet the objection. In such a case he should readily admit that the objection is sound and then proceed to show how the proposal can be adapted to conform to the needs of the situation. Second, if the objection is of a type that permits no adjustment, he might show that it is relatively unimportant when weighed against the overall advantages of the proposal. In this case he may well proceed in accordance with the following directions:

1. Restate the objection.
2. Accept its validity, but show it to be immaterial or of little importance.
3. Sum up the benefits of the proposal.
4. Emphasize that the advantages outweigh the disadvantages.

HUMAN RELATIONS

A point to be repeated is that the cultivation and maintenance of good personal relationships is important in conducting systems-and-procedures assignments. At the presentation stage the opposite situation—poor personal relationships—can be a great source of resistance to the acceptance of proposals. The systems analyst should remember that nearly everyone is more willing to embrace ideas presented to him by a person he likes than by one he dislikes.

Innumerable sermons have been delivered on the theme, "And as ye would that men should do to you, do ye also to them likewise." The philosophy underlying this theme is applicable to the whole subject of human relations. To develop and sustain good relationships with people the systems analyst needs to see things from the other person's point of view and to interpret the findings in terms of how he would like to be treated if he were in the other person's position. Accordingly, he will work with and through people in a spirit of cooperation and mutual aid.

Actions of the following kind can go a long way toward good personal relationships:

Seek advice. Most people will consider a sincere request for advice as a genuine compliment, which it is.

Offer assistance. This is an excellent way in which to build up good will.

Be friendly. Friendliness toward people and a good-natured manner will win respect and confidence.

Encourage people. Now and then we all have need of encouragement, and the person that gives it is usually very much appreciated.

Consult with the proper persons. Keep them informed of developments that affect their interest. Depending on the nature of the assignment, remember that the man at the bench may be among the "proper" persons to be advised.

Render appropriate recognition. It is a fact known to all of us that one of the most important things in the life of every person is his feeling of usefulness and self-esteem. Be sure that individuals who make a worthy contribution to the improvement of a system receive appropriate recognition. Establish a reputation for giving credit where credit is due, and you will gain the respect and trust of others.

28

Installation and Follow-up

After the proposal has been approved the next step is to carry it out. The particular action that is necessary will depend, of course, on the precise nature of the proposition. To make this step clear, we will consider it in connection with the specific process of installing a new system.

INSTALLATION OF NEW SYSTEM

Practically every segment of systems-and-procedures work involves planning. If a new system is not to falter or fail, particularly during the initial period of operations, plans must be worked out for its smooth installation.

Everyone knows by his own experience how plans tend to succeed or fail in proportion to the thoroughness of the preparation given them. There is little need to stress the necessity of a course of action beforehand to install a new system efficiently and economically. With an orderly plan more will be accomplished within a definite period, with no unnecessary delays, and certain occurrences that cause trouble will be eliminated. A typical schedule for installation is shown in Figure 28-1.

It is essential for the systems analyst to plan the installation in such a way that the work on one phase will fall in with that of another phase. Thus when the process of installation reaches a particular point, the necessary equipment, supplies, and manpower are ready to be put into service. Hence in planning the installation it is advisable to make a schedule showing the time requirements for each phase of the work. New equipment and supplies should be ordered in accordance with this schedule, and the systems analyst should provide whatever follow-up is necessary to assure that they are delivered and available for use in advance of the time when they are actually needed.

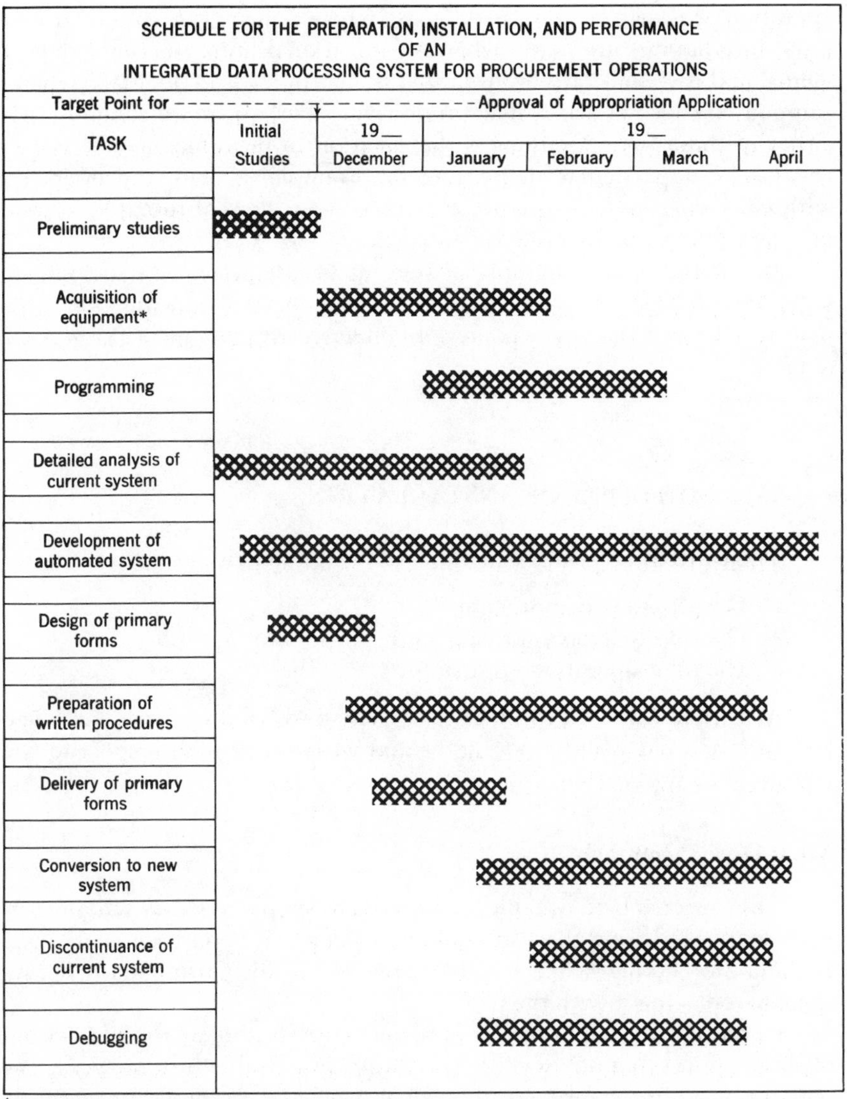

SCHEDULE FOR THE PREPARATION, INSTALLATION, AND PERFORMANCE
OF AN
INTEGRATED DATA PROCESSING SYSTEM FOR PROCUREMENT OPERATIONS

Target Point for ---------- Approval of Appropriation Application

TASK	Initial Studies	19__ December	19__ January	February	March	April
Preliminary studies						
Acquisition of equipment*						
Programming						
Detailed analysis of current system						
Development of automated system						
Design of primary forms						
Preparation of written procedures						
Delivery of primary forms						
Conversion to new system						
Discontinuance of current system						
Debugging						

*Computypers, keypunch machines, etc.

Figure 28-1. Progress chart.

THE NEW AND THE OLD

The new system may be intended to replace one that is currently in operation or it may be designed for use where none exists. Initial, or first-time, installations are made when organizational units (sections, departments, and divisions) are brought into being; they also take place when a company decides to add a new product line. Perhaps more common than either of these two situations is the decision to introduce a new system because of a particular deficiency; for example, a company operating without a cost system, quality-control system, market-research system, etc., may determine that one is necessary.

By far the greater number of systems installations are those that involve the problem of conversion; that is, the effective introduction of the new system and the correspondingly effective retirement of the existing one.

THE METHODS OF INSTALLATION

There are three primary methods of installing new systems:

1. The all-at-once approach;
2. The step-by-step approach; and
3. The pilot-operation approach.

At various times all three methods may be advantageously employed, because each has peculiar strengths and weaknesses with respect to specific areas of application.

All-at-Once Approach

This approach is suitable for relatively simple systems wherein the work load is light and there are not many people or departments involved. It embodies taking what is now being done, dispensing with it, and quickly replacing it with the new.

Certainly, the all-at-once approach is the fastest method of accomplishing an installation, but it is not always practicable or wise. Some systems are much too complex and involved for that, and should be instituted in stages so that the installation is not adversely affected by too swift a change.

Step-by-Step Method

For projects that are large in scope, of major importance, and involve intricate processing, the step-by-step approach is by far the most logical method of installation. Progress under this method is slower than under the all-at-once approach, but it is generally on the safer side. Unforeseen difficulties may be ironed out of the system while they are of relatively minor significance and before they have had a chance to accumulate.

The actual step-by-step installation is usually carried out through the progressive introduction of subsystems or other sets of closely related procedures that are reasonably independent. The nature of the operations will largely determine the order and timing of the steps. Where the character of the system does not make necessary some different order of elements, the layout of the physical plant may suggest the order of installation.

Many new systems are introduced along departmental lines; that is, a part of the system is put into operation in one department, then another part in the next, and so on, until the entire installation is accomplished. In most cases the new system supplants the old one in what amounts to a series of transitions over an extended period of time.

One guide to action is as follows: when using the step-by-step approach, the systems analyst must be careful not to spread it out over a prolonged period of time. If the schedule is unduly long and drawn out, dramatic results will be lost, interest will wane, and financial benefits will be difficult to pin-point.

Pilot-Operation Approach

Occasionally, it is possible to introduce a system on a small but representative scale; that is, to set up a miniature version of an entire system. Few situations lend themselves to this method, but where applicable it may serve two very worthwhile purposes; namely, to discover and remove any flaws in the system before it is completely instituted, and to offer proof of its capabilities to anyone who might hold reservations about them.

Instituting inspection procedures during the experimental stages of a product's development is a good example of a pilot operation. Then, when the product goes into production, all that is necessary is to expand the application of the procedures.

PARALLEL OPERATIONS

When the problem is one of conversion the new system may be installed and run concurrently with the old one until such time as the former is free from operational difficulties. Operating the two systems in parallel during the critical period of installation acts as a deterrent against risks that might be present if an immediate abandonment of the existing system was made. Because of its dualism, it is an expensive means of installation and should thus be reserved for use in situations that truly warrant this kind of hedging. To keep added expense down, it is sometimes possible to retain only parts of the old system while the new one is being phased in.

Payroll-system installations involving profound changes are typical of the kind of situations that warrant the simultaneous operation of the old and the new. Since the federal and state governments require that every business keep accurate records of each employee's earnings and the amounts withheld from earnings, it is imperative that necessary safeguards be taken to insure proper conformance.

TIMING THE CHANGEOVER

The timing of the switch from one system to another can be of considerable importance. It is occasionally possible to select a changeover period in which operations are temporarily suspended or at a low ebb, so that adverse influences furnish least resistance. Examples of such times would be holiday periods, summer-vacation periods when many companies all but close their doors for 2 or more weeks, and seasonal periods in which dips in activities normally occur.

PARTICIPANTS IN THE INSTALLATION PROCESS

In the beginning the systems analyst is largely responsible for leadership in getting the installation under way and in satisfactory working order. Gradually this leadership is relinquished and transmitted to the line organization. Here lies one of the most subtle challenges that the systems analyst faces. He must steer a wary course between directing every facet of the work, and slackening the rein so that people learn by experience, even at the risk of making mistakes.

Three types of people are ordinarily involved in an installation. These are the departmental supervisor, the systems analyst, and the workers. Each plays an important part.

The Supervisor's Role

Line supervisors are normally concerned with routine, stabilized, everyday operations. They usually lack the time or familiarity with the newly approved system to be made solely responsible for its installation. The systems analyst must recognize that whereas he is extremely familiar with the procedures involved, the supervisor may have only a superficial or broad-outline knowledge of the work.

Here, as in so many other phases of systems work, cooperation is the key word. The installation should be viewed as a joint effort, calling for teamwork on the part of the systems analyst, the department supervisor, and the operating personnel. The systems analyst should render every bit of assistance he can.

In the last analysis, however, it is the supervisor or foreman who will have the continuing responsibility and authority for making the system work. The people who will carry out the tasks report directly to him, receive their instructions from him, and largely owe their allegiance and loyalty to him.

The Systems Analyst's Role

Usually department supervisors are conscientious about their work and their responsibilities. They want to do a good job, possess an abiding willingness to learn, and will accept all reasonable help.

The systems analyst should be prepared to offer whatever support he can, including instruction to acquaint the supervisors and the workers with the new procedures. Much will be accomplished if he approaches the job in the spirit of being a coach who kindles interest, teaches, assists, corrects, and encourages.

During the course of the installation the systems analyst will want to continue an active surveillance over the procedural activities, their results and progress, and to confer regularly with the persons in charge.

The Worker's Role

Basic to the success of any system is a consideration of the workers—the clerks, operators, and others whose job it is to carry out the program in detail. Although these persons are not charged with responsibility for the installation, each is influential in a moderate but important way for its ultimate success or failure. If they fall short, the system suffers. This is generally caused not by a lack of willingness or enthusiasm, but by a lack of understanding; of their not knowing precisely what to do. On behalf of the systems analyst it should be evident that he has a role to play in stimulating supervisors and workers, in letting them know the relationship of their individual tasks to the entire system, and in making certain that they have a real grasp of their own jobs.

ORIENTATION AND TRAINING OF PERSONNEL

The installation schedule should take the orientation and training of personnel into account and provide the time necessary to indoctrinate the employees in any unfamiliar activities. Remember that no system can be more effective than the know-how of its participants.

The importance of properly instructing the personnel responsible for the operations of a new system is too often neglected in the enthusiasm and haste to get the work under way. Educational considerations are set aside in view of saving time. This is a shortsighted view—in the long run it commonly leads to operational setbacks and delays, requiring more time for corrective measures than was ever initially saved.

If a new system installation is to achieve its fullest effectiveness, it must be manned by people who know what they are doing. Certain basic principles and techniques of employee training that help to bring about this performance are discussed below.

Training Principles

No two training situations are exactly alike. Each case is different. Both persons and occupational tasks differ greatly in their characteristics. Individuals differ in their abilities to learn, their attitudes, their interests, and their physical competence. Specific jobs, of course, have an endless range of variation. However, there are universal principles of training that apply to practically every case.

Training by doing is widely recognized as the most effective form of instruction. Learning in this instance takes place through meaningful activity. The employee is shown and told how the tasks are performed and then proceeds to follow directions. In this way understanding comes from real experience.

Motivation supplies the desire of persons to learn. If the employee is to learn expeditiously, it is absolutely essential that he be adequately motivated. This implies that the employee ought to be made to feel that a knowledge of the tasks is important, that the work has its own dignity, and that his contribution to the successful operation of the system will be appreciated.

To make the learning process interesting and meaningful use comparisons with things already within the experience and knowledge of the employee. Tie the known in with the unknown. The emphasis here is in showing the employee, wherever possible, how the knowledge of one task assists in the learning of another.

It is advisable to begin with simple explanations and proceed to complex ones. This approach directs the course of development of interest in a natural and progressive manner.

Since individuals are different, each learns at his own pace. Recognition of this is of special significance when dealing with the employee who is a slow but capable learner. Be patient with him and you will be rewarded for your efforts. Many employees, though slow at learning new tasks, show increasingly faster improvement in time and ultimately turn out to be highly competent in their work.

In some training situations it is possible for the employee to develop accuracy and speed at the same time. On occasions where one must be

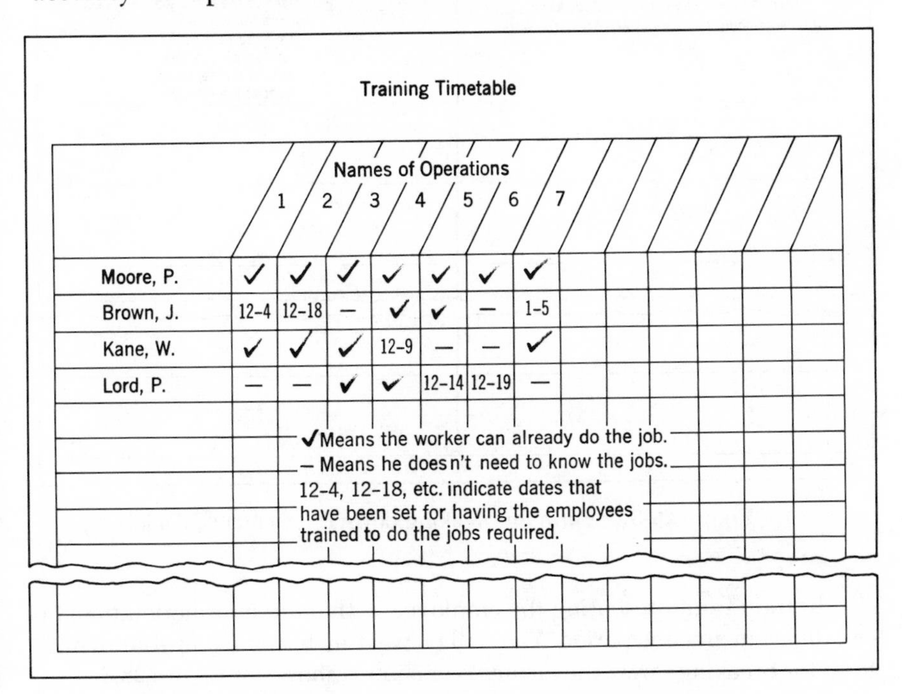

Training Timetable

	Names of Operations 1	2	3	4	5	6	7
Moore, P.	✓	✓	✓	✓	✓	✓	✓
Brown, J.	12-4	12-18	—	✓	✓	—	1-5
Kane, W.	✓	✓	✓	12-9	—	—	✓
Lord, P.	—	—	✓	✓	12-14	12-19	—

Figure 28-2. Example of a training timetable.

temporarily sacrificed, it is usually best to defer the requirement for speed. Make haste slowly; it is of fundamental importance that effort be made to understand the work rather than to gain immediate speed without adequate preparation. No job performance that is not accurate should be considered satisfactory.

An example of a training timetable is given in Figure 28-2.

The information imparted in the course of an instruction is of no value unless it is remembered. The human mind has its limitations; it can grasp only a few ideas at a time. Therefore the length of the individual periods of instruction must be governed largely by the employee's aptitude and ability. As far as possible they should be short enough not to produce fatigue or monotony but long enough to sustain interest and continuity in the subject matter.

JOB BREAKDOWN SHEET FOR TRAINING MAN ON NEW JOB

Part: ______________________ Operation: ______________________

IMPORTANT STEPS IN THE OPERATION Step: A logical seqment at the operation when something happens to ADVANCE the work.	KEY POINTS Key point: anything in a step that might—Make or break the job. Injure the worker. Make the work easier to do; i.e. "knack", "trick", special timing, bit of special information

Figure 28-3. Form for breaking down the elements of a job.

Refrain from presenting the employee with more information than he can digest in any one period. You will be wasting his time and your own. A form for breaking down the elements of a job is shown in Figure 28-3.

A Practical Method of Training

During World War II the Training Within Industry Service of the War Manpower Commission developed a basic training program that proved to be extremely practical and successful. Key points of the program, as given below, provide an excellent means for training employees in the operations of a new system.

HOW TO INSTRUCT

Get Ready To Train [1]

1. *Determine your objectives.* Decide specifically what you want to accomplish.
2. *Determine what training is needed;* what added knowledge and skill, and what improvement in attitude.
3. *Decide what is the best method of teaching and the best order of presentation.*
4. *Know the subject yourself.* Prepare carefully for each training session. Practice until you can do easily any demonstration you plan to give.
5. *Have everything ready.* Have tools, blueprints, chalk, erasers, etc., ready before you start.

First Training Step

Prepare the Learner

1. *Put the learner at ease* with a friendly, informal, encouraging comment.
2. *Make the learner want to know* by showing him what the knowledge will do for him.
3. *Use praise rather than criticism as an incentive.* There will always be something you can praise.

Second Training Step

Instruct by Telling and Showing

1. *Start with the known; lead into the unknown.* This captures attention and gives the learner confidence.
2. *Teach the simple first; lead up to the complicated.* This prevents discouragement and provides a challenge.

[1] For a thorough discussion of the program see *Job Instructor Training* (1942) and *Job Instruction* (1944), publications of the Training Within Industry Service, War Manpower Commission, Washington, D. C.

3. *Keep your explanation to the point.* Telling of unrelated incidents distracts attention.

4. *Give reasons for each step.* Knowing why increases the probability of remembering.

5. *Demonstrate by doing correctly and exactly what the learner will later be asked to do.* Go slowly; be sure that the learner sees and understands each thing you do.

6. *Encourage questions.* This is the best way to be sure the learner understands.

Third Training Step

Try out Learner's Knowledge and Skill, and Correct Deficiencies

1. *Give the learner an opportunity to demonstrate the operation as if he were teaching you.* This greatly strengthens the original impression. Also it enables you to discover and correct misunderstandings.

2. *Keep the learner informed as to his progress.* He does better work when he knows how he is getting along.

3. *Give less and less supervision.* Put the learner on his own as soon as you are sure he can continue without your help.

Fourth Training Step

Follow-up

1. *Check from time to time to see how well the information is retained and used.* Supply further on-the-job coaching when needed.

Remember: *Training has been futile unless the learner uses what he has learned.*

Where several employees are to be instructed in identical or similar operations it is often advisable to supplement individual training in technique with the so-called "classroom method" of teaching. Some practices and procedures lend themselves nicely to being taught to a group, and indeed are more easily and economically explained in this manner.

In holding group-instruction sessions it is good practice to keep them free from the traditional scholastic atmosphere. Make them informal. Try to avoid any suggestion of teacher-student relationship. In this connection it is preferable to use the words "instruct" rather than "teach," "group" rather than "class," and "worker" or "learner" rather than "student."

Do not overlook the possibility of employing visual aids in a training situation. As suits the occasion, operational manuals, instruction sheets, written procedures, diagrams, and the like may be utilized to impart information and facilitate understanding.

Vendor Training Programs

Of special importance are the training schools and educational centers operated by machine and equipment vendors. They function to satisfy the instruction and guidance needs of customers and potential customers. When a new system entails automated procedures it is a good idea to seriously consider sending planned operators to this type of instruction.

FOLLOW-UP ON THE ASSIGNMENT

Good judgment will suggest the need for follow-up activity on an assignment. This concluding step may serve a number of worthwhile purposes. In the first place it permits the systems analyst to determine whether the action being taken conforms to that which was recommended and approved. Are any procedural steps being omitted? Is the work being performed in the manner and order in which it should? The aim here is to uncover and rectify any misunderstandings that may have arisen on the part of the people who are carrying out the work.

Second, inasmuch as faults are sometimes found in what are thought to be the best of plans, the follow-up step functions to bring these to light so that they may be corrected. Are there points in the system where adjustments or modifications will afford refinements? Is it desirable to alter a form when it comes up for reordering? Applying the finishing touches indicated by these and similar questions is often the master stroke that crowns the success of an assignment.

As part of the follow-up step the systems analyst should look for places where the operators and other people utilizing the system have made improvements that were not contemplated at the time the installation was planned. Each improvement should be appraised in terms of its contribution to the overall system. If the improvement truly benefits the entire system—in whole or in part—then the systems analyst will make whatever adjustments are necessary, including the updating of written instructions.

By reviewing actual performance it is possible to determine the extent to which compliance has met with original predictions. Are the costs savings in line with those anticipated? What are the reactions of the personnel to the new system? Do the supervisors show satisfaction? Are the employees content?

Finally, the follow-up step may consist of reporting on the status of the assignment; that is, the issuance of a report to management.

WHEN TO FOLLOW UP

In the case of a new system or major change, it should receive frequent and constant checking in the early period of its operation, the surveillance being tapered off as it becomes increasingly clear that recurring performance of the procedures will be satisfactory. By contrast, a set of procedures that have undergone only a minor change may require but one check at an early date to see if they are functioning as intended.

Some systems analysts make it a practice to follow up on a project whenever they are in the area or whenever they come in contact with personnel with a knowledge of the situation. Others review progress constantly and regularly at selected times.

TICKLER FILES TO INSURE FOLLOW-UP

To exercise control over the follow-up activities on a project many systems analysts find it of value to maintain a tickler file. Files of this kind are commonly set up in the form of a collection of cards, each containing the following information: identification (title and project number), notations with respect to any points in need of special checking or evaluation, and the date or dates on which to take follow-up action. It is customary to file the cards under the follow-up dates so that they can be easily referred to at the specified times.

Postface

This book has provided an excursion into a common body of knowledge to be possessed by those who practice in the systems-and-procedures field. The full professional knowledge that the systems analyst must have, however, neither begins nor ends with the reading of one subject-matter book. A basic understanding of the practical areas of business is necessary, including marketing, finance, production, and business management. Beyond this, there is the ever-present need to focus attention on professional development; to acquire, and to continue to acquire, administrative breadth and knowledge.

Professionals in the field recognize the need for continual self-improvement. They know that they must be prepared to grow with changing conditions and ideas. Toward this end they enroll in special on-campus courses, read trade magazines and journals, attend conferences and seminars, and discuss experiences with colleagues at meetings held by professional societies. The inference for the beginning practitioner is clear: *Go forth and do likewise.*

Index